Bat Calls of Britain and Europe

Bat Calls of Britain and Europe

A GUIDE TO SPECIES IDENTIFICATION

Edited by Jon Russ

PELAGIC PUBLISHING

Published by Pelagic Publishing
PO Box 874
Exeter
EX3 9BR
UK

www.pelagicpublishing.com

Bat Calls of Britain and Europe: A Guide to Species Identification

ISBN 978-1-78427-225-8   *Hardback*
ISBN 978-1-78427-226-5   *ePub*
ISBN 978-1-78427-227-2   *PDF*

Front cover: Daubenton's bat *Myotis daubentonii* © Jens Rydell
Rear cover: Brown long-eared bat *Plecotus auritus* © René Janssen

Typeset by BBR Design, Sheffield

Printed and bound in Wales by Gomer Press Ltd.

Contents

About the Editor

Jon Russ first became interested in bats in 1994 while completing research on pipistrelle social calls as part of a degree in zoology at the University of Aberdeen. This led to a PhD at Queen's University Belfast investigating the community composition, habitat associations and echolocation calls of Northern Ireland's bats. Since then he has been involved in a wide variety of bat-related projects which have taken him from the freezing rain of northeast Scotland and the fine soft nights of Ireland to the humid rainforests of Madagascar, Thailand and Myanmar. Jon is the Director of Ridgeway Ecology Ltd, a specialist bat consultancy, and for several years he worked for the Bat Conservation Trust coordinating the iBats project in the UK and eastern Europe. He has written a large number of articles in scientific journals, and his other publications include the widely used book *British Bat Calls: A Guide to Species Identification* published by Pelagic Publishing. After more than 25 years of involvement in bat research and conservation, he continues to be fascinated by these remarkable mammals.

Preface and Acknowledgements

Following the surprising success of *British Bat Calls*, published in 2012, Nigel Massen of Pelagic Publishing kindly waited a few years before tentatively suggesting I collate a European version of the book. My immediate reaction was very positive – it would be a simple matter to 'crowbar in' the other species and I could probably have the whole thing wrapped up in six months. However, it soon became clear that I was being a little bit naive, and that incorporating 22 additional species (plus four that were added to the European list during the writing of this book) was well outside the scope of my experience – and available time. After shelving the idea for around a year it occurred to me that it would be better to identify people who record and come into contact with those species for which I have limited or no knowledge and ask them to write the chapters instead. Taking up the role of editor as well as author, I began the task of finding volunteers, researchers and enthusiasts from around Europe who were willing to give up their time to assist with the project. It did not take as long as anticipated, and thanks to the excellent network of bat workers throughout Europe, I soon had a list of contributors – and from then on the book began to take shape. Although at times it felt as if I was manoeuvring a large oil tanker into a small harbour, I think it has been well worth the effort by all involved. I hope it will be useful to volunteers and professionals alike.

This book would not have been possible without the efforts of all the authors who have given their valuable time and expertise. I have been overwhelmed by the generosity of my co-authors and cannot thank them enough. They are listed in the separate chapters and species sections, but all of them deserve a mention here. I am extremely grateful to Arjan Boonman for Chapter 3 (*Echolocation*), Grace Smarsh for Chapter 4 (*An Introduction to Acoustic Communication in Bats*), Philip Briggs, Arjan Boonman, Martijn Boonman, Jeremy Froidevaux and Kate Barlow for Chapter 5 (*Equipment*), Kate Barlow and Philip Briggs for Chapter 6 (*Call Analysis*) and Yves Bas, Charlotte Roemer, Arjan Boonman, Alex Lefevre and Marc Van De Sijpe for Chapter 7 (*A Basic Echolocation Guide to Species*). The species accounts in Chapter 8 were written by Francisco Amorim, Leonardo Ancillotto, Maggie Andrews, Peter Andrews, Arjan Boonman, Erika Dahlberg, Johan Eklöf, Péter Estók, Gaetano Fichera, Joanna Furmankiewicz, Panagiotis Georgiakakis, Clara Gonzalez Hernandez, Julia Hafner, Daniela Hamidović, Amelia Hodnett, Pedro Horta, Artemis Kafkaletou-Diez, Andreas Kiefer, Erik Korsten, Alex Lefevre, Mauro Mucedda, Stephanie Murphy, Jorge M. Palmeirim, Eleni Papadatou, Ricardo Pérez-Rodríguez, Ermanno Pidinchedda, Ana Rainho, Helena Raposeira, Orly Razgour, Hugo Rebelo, Dina Rnjak, Danilo Russo, Jens Rydell, Horst Schauer-Weisshahn, Claude Steck, Sérgio Teixeira, Marc Van De Sijpe, Carola van den Tempel and myself.

I would like to thank (again) Marc Van De Sijpe and Alex Lefevre. Not only have they contributed to more than their fair share of the species chapters, as well as helping to write the basic echolocation identification guide, they have also unhesitatingly and generously provided me with hundreds of echolocation and social calls, which have vastly improved the book. They were always available to help when I was struggling with a particular species, and if they didn't have calls themselves they would find someone who did. Their knowledge of European bat species has been a rich seam to mine, and the book would have been much poorer without their input.

Arjan Boonman assisted enormously from the very beginning with his great technical knowledge – and he put me in touch with Grace Smarsh, who vastly improved the acoustic communication chapter from my section on this topic in the *British Bat Calls* book.

Many people were kind enough to provide echolocation and social calls: Daniel Fernández Alonso, Francisco Amorin, Leonardo Ancillotto, Maggie Andrews, Paulo Barros, Yves Bas, Yannick Beucher, Kirsten Bohn, Arjan Boonman, Erika Dahlberg, Jonathan Demaret, Christian Diez, Simon Dutilleul, Bengt Edqvist, Péter Estók, Rich Flight, Joanna Furmankiewicz, Panagiotis Georgiakakis, Julia Hafner, Daniela Hamidović, Amelia Hodnett, Sally-Ann Hurry, Iain Hysom, David King, Erik Korsten, Karl Kugelschafter, Karri Kuitunen, David Lee, Alex Lefevre, Harry Lehto, Risto Lindstedt, Jochen Lueg, Kari Miettinen, Mauro Mucedda, Stephanie Murphy, Ian Nixon, Eleni Papadatou, *Plecotus* (Estudos Ambientais, Unip), Sébastien Puechmaille, Ana Rainho, Helena Raposeira, Phil Riddett, Ricardo Pérez-Rodríguez, Danilo Russo, Jens Rydell, Horst Schauer-Weisshahn, Tricia Scott, Grace Smarsh, Graeme Smart, Michael Smotherman, Claude Steck, Congnan Sun, Sérgio Teixeira, Marc Van De Sijpe, Carola van den Tempel, Anton Vlaschenko, Liat Wicks, Tina Wiffen and Bernadette Wimmer.

Several people provided photographs: Leonardo Ancillotto, Martyn Cooke, Klaus Echle, Péter Estók, Panagiotis Georgiakakis, René Janssen, José Jesus, Boris Krstinić, Harry J. Lehto, Mauro Mucedda, Dragan Fixa Pelić, Ana Rainho, Angel Ruiz Elizalde, Jens Rydell, James Shipman, Sérgio Teixeira and Daniel Whitby. René Janssen deserves a special mention for the considerable number of stunning bat images he generously donated to this project, including the superb photograph of a brown long-eared bat *Plecotus auritus* on the back cover.

The front-cover photograph of a Daubenton's bat *Myotis daubentonii* was taken by Jens Rydell – who, in 1995, taught me the basics of bat echolocation by scratching sonograms in the mud with a stick in Seaton Park in Aberdeen, and who introduced me to the wonderful world of pipistrelle social calls. It was a shock to us all when Jens suddenly passed away in April 2021 during the publication of this book. He has been a big influence on bat workers throughout the world for decades and leaves behind an indelible legacy of knowledge, photos and memories. He will be sorely missed by all his friends and colleagues.

I am grateful to Tom McOwat for producing the beautiful illustrations of bat wing shapes, ear shapes and habitats in Table 3.3.

I am hugely indebted to Nigel Massen of Pelagic Publishing, who has supported this project throughout its four-year development. I would also like to thank Hugh Brazier, copy-editor, for his thoughtful attention to detail and ability to tease out meaning from a mess of technical jargon written in a variety of styles.

As always, I would particularly like to thank Paul Racey, who not only inspired my interest in bats but also enabled me to pursue a career that has been so rewarding. Without his enthusiasm and support in the early days, I wouldn't be in the fortunate position I find myself in now.

I also wish to thank the following people, who assisted in various ways during the long drawn-out process of creating this book: Sébastien Puechmaille, without whose help the *Myotis crypticus* chapter would have been sound-free; Kati Suominen, for information about *Myotis brandtii* in Finland; Vicent Sancho, Toni Alcocer and Jasja Dekker, who assisted with emergence times for *Myotis emarginatus*; Szilárd Bücs, for giving me the opportunity to record a wide variety of bat species during a brief visit to Romania; Yves Bas and Charlotte Roemer, for helping enormously with the species identification section; Panagiotis Georgiakakis, for his help with several species; and the following people who provided assistance, encouragement and comment along the way: Andy Allsop, Martyn Cooke, Chris Corben, Christian Diez, Hazel Gregory, Dave Russ, Steve Russ, Tricia Scott, Jackie Underhill and Alison Warren.

The software programs Batsound v4.4 (Pettersson Elektronik AB, Uppsala, Sweden) and Avisoft SASLab v5.2 were used to construct the sonograms, oscillograms and power spectra displayed in this book.

I would like to thank Eimear for all her love and support during the writing of this book, which is dedicated to our two wonderful daughters, Ellen and Anna, and to my dear friend and fellow wildlife enthusiast, Darren Bradley.

Brown long-eared bat *Plecotus auritus* © René Janssen

Please help!

A book of this nature is unlikely to be perfect, and there are bound be omissions and errors – but hopefully only a small number. The editor welcomes all comments, and would be grateful for any recordings of bat vocalisations or any other information that could be used to update future editions. Of particular interest would be information to separate similar species, social calls (including their function) and geographical variation in both sympatic and allopatric populations (email batcalls@ridgewayecology.co.uk).

The Sound Library

To accompany the text, the sound files used to create the sonograms presented in Chapter 8 are provided in a downloadable Sound Library. These are available via the following link:

www.pelagicpublishing.com/pages/batcalls-sound-library

Calls have been resampled from their original form to mono files with a sample rate of 384 kHz and a bit size of 16 bits. They are provided in *.wav format.

Each filename includes the figure number, author, species, country it was recorded in and a small amount of detail about the recording.

1 Introduction

Jon Russ

In 1793, Lazzaro Spallanzani, an Italian biologist, physiologist and Catholic priest, demonstrated that bats were able to avoid obstacles without the aid of vision. He stretched thin wires with small bells attached across a completely darkened room and observed that bats were able to fully navigate between them without causing the bells to ring. Blinding the bats also did not impair their ability to manoeuvre around them. Meanwhile, a Swiss zoologist, Charles Jurine, revealed that blocking one of the ears of a bat spoiled its ability to navigate, a finding that Spallanzani then pursued. A series of experiments which involved blocking the ears or gluing the muzzle closed led him to conclude that while bats did not have much use for their eyes, any interference with their ears that adversely affected hearing was disastrous, resulting in them colliding with objects they could usually avoid and being unable to forage for prey. He concluded that 'The ear of the bat serves more efficiently [than the eye] for seeing, or at least for measuring distance.' At the time, Spallanzani's findings were met by his fellow scientists with ridicule and scepticism, as bats were believed incapable of producing any sound and therefore such results defied logic.

Nearly 150 years after Spallanzani's work, Donald R. Griffin, while an undergraduate at Harvard University in the 1930s, took an interest in the 'bat problem'. New advances in technology allowed him to use a 'sonic receiver', designed and built by Harvard physics professor George Washington Pierce. This device captured high-frequency sounds that were beyond the range of human hearing and reduced the pitch to an audible level. For the first time, it became apparent that bats emit short, loud, ultrasonic clicking sounds. Along with a fellow student, Robert Galambos, who was an expert in auditory physiology, Griffin designed a set of further experiments which showed that bats were avoiding obstacles by hearing the echoes of their ultrasonic cries. Further experimentation revealed that bats were able to adjust the structure of their calls for prey search and capture and collision avoidance. Griffin named this acoustic orienting behaviour 'echolocation'.

A bat's echolocation system is highly sophisticated. By emitting short high-frequency pulses of sound from their mouths or noses, bats can use the information contained within the echoes returned from a solid object to construct a 'sound picture' of their environment. Not only are they able to identify the size, position and speed of objects within three-dimensional space, they are also able to differentiate between forms and surface textures. However, as there is no single signal form that is optimal for all purposes, bats have evolved a large number of signal types. This diversity of echolocation signals is likely to reflect adaptations to the wide range of ecological niches occupied by different bat species. For example, in Europe, the noctule *Nyctalus noctula*, which largely forages high over parkland, pasture and woodland in an uncluttered environment, tends to produce extremely loud low-frequency calls of relatively long duration, narrow bandwidth and low repetition rate. Conversely, Bechstein's bat *Myotis bechsteinii*, which often forages very close to or within woodland vegetation in a very cluttered environment, usually produces relatively quiet, very broadband calls of short duration with a high repetition rate. Thus, the calls of different bat species are shaped by the habitats in which they usually forage, and the resulting different call types can often be used to separate species in the field. However, echolocation call shape is not fixed for a species and shows a certain degree of plasticity depending on the habitat within which an individual is currently located. In addition, although habitat is a significant factor determining the 'shape' of bat echolocation calls, they may also vary with sex, age and body size, geographic location and presence of conspecifics. Finally, species that occupy similar niches may use similar echolocation call types, and there is often significant overlap in calls between species. An understanding of these different levels of variation both within and between individuals and species is essential to the successful use of echolocation calls for bat species identification.

Social calls produced by bats are often more structurally complex than echolocation calls used for orientation. Social calls are used to communicate with other bats (including other species) and for many species consist of a wide variety of trills and harmonics, comparable in many respects to bird song. Social calls may have several functions. Some are used to defend patches of insects against other bats or to sustain territorial boundaries. Others function in attracting a mate or, in the case of distress calls, to initiate a mobbing response. Perhaps the most astounding are the isolation calls emitted by young bats, which allow their mothers to identify them. At Bracken Cave in Texas, for example, millions of Mexican free-tailed bats *Tadarida brasiliensis* cluster in a large maternity colony. After the mothers have given birth the walls of the cave are covered with young bats packed tightly together. Each of these young bats has an individual call that is in some way different from that of all the other young bats. These variations enable a returning mother to distinguish her offspring from all the others.

Since Griffin's discovery, several techniques have been developed to allow us to listen to the ultrasonic vocalisations of bats. These range from relatively cheap 'heterodyne' detectors, which convert a narrow range of frequencies into an audible signal in the field, to 'real-time full-spectrum' recording, which has become possible through the development of high-speed analogue-to-digital converters built into or connected to computers or solid-state recorders. These high-tech devices utilise a sufficiently high sample rate to enable the ultrasound to be captured digitally and allow later processing and analysis of recordings. Since around 2010, bat enthusiasts and researchers have been taking advantage of the explosion in the availability of smartphones. These devices can be used as recording devices when connected to a bat detector that converts the ultrasound into the audible range, and with the development of small inexpensive USB ultrasonic microphones 'real-time' ultrasound recording is now possible. Smartphone apps can even incorporate classification algorithms that assign calls to species, providing instant identification (with limitations) of bats in the field in a readily available, cost-effective hand-held device.

Donald Griffin referred to his discovery of echolocation as 'opening a magic well' from which scientists have been extracting knowledge ever since. Echolocation provides a window into the lives of bats, giving us access to a previously unknown world. It has been used, for example, to help us identify individuals to species; locate roost sites; find commuting routes and foraging areas; study foraging behaviour; establish species distributions; and monitor annual variations in bat populations. In addition, the study of the social calls of bats has allowed us to investigate the vocabulary of bat communication. Not only can these calls be used to identify species of bat and individuals, but some calls can also be used to assess male territoriality and female selection of mates, as well as providing a measure of male reproductive success, while others can give us an insight into interactions between females and their young, food competition at foraging sites and levels of distress. However, although a great deal has already been learned about the vocalisations of bats, much remains to be discovered.

The importance of sound to bats cannot be underestimated. They rely upon sound to locate food, to find their way around in the dark, and to seek out and communicate with other bats. By using ultrasonic detectors to eavesdrop on them we can investigate their behaviour in the field without disturbing and endangering these remarkable mammals. In this book, we provide a guide to listening to, recording and analysing the echolocation and social calls of bat species found in Europe to identify these calls to species. Although it is not always possible to reliably identify all bat species from their echolocation calls, we have tried to give as much information as possible on how to identify bats from their calls using different types of bat detectors.

2 The Basics of Sound: Properties, Acquiring, Representing and Describing

Jon Russ

2.1 Properties of sound

Sound is a form of energy which travels through a medium such as a solid, liquid or gas. It is produced when the medium is disturbed in some way by a moving surface such as a loudspeaker cone. As the cone moves forward, the air immediately in front is compressed, causing a slight increase in air pressure. It then moves backwards, past its rest position, and causes a reduction in the air pressure (rarefaction). The process continues so that a wave of alternating high and low pressure radiates away from the speaker cone at the speed of sound in air ($340 \, \text{m·s}^{-1}$) (Figure 2.1a). This process can also be thought of as a wave travelling through the air (Figure 2.1b).

The speed the waves travel depends on the medium, and in air it largely depends on air temperature, yielding $337 \, \text{m·s}^{-1}$ at $10 \, °\text{C}$ to $350 \, \text{m·s}^{-1}$ at $30 \, °\text{C}$. The wavelength (λ) is the length of one cycle of the wave (e.g. from one high-pressure peak to the next high-pressure peak) and the amplitude is the height of the wave, which is related to the amount of energy the wave contains. If the speed of sound in a medium is fast, the distance a wave propagates away from the speaker in one second is greater than that covered by a wave of the same frequency in a slow medium, hence its wavelength in a fast medium is longer than in a slow medium.

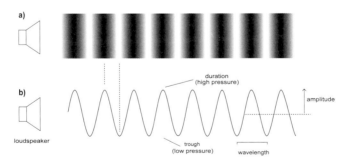

Figure 2.1 (a) Sound travelling through air produced by the vibration of a loudspeaker. The darker bands represent areas of high pressure and the light bands represent areas of low pressure. (b) The same sound represented by a wave.

2.1.1 Amplitude

Amplitude is a measure of the intensity, loudness, power, strength or volume level of a signal. This is most commonly expressed in terms of the sound pressure level (SPL), which is measured in units of decibels (dB). The decibel is a logarithmic unit ($\log_{10}$) used in several scientific disciplines. In acoustics, the decibel is most often used to compare sound pressure, in air, with a reference pressure of 20 micropascals (µPa) (Figure 2.2).

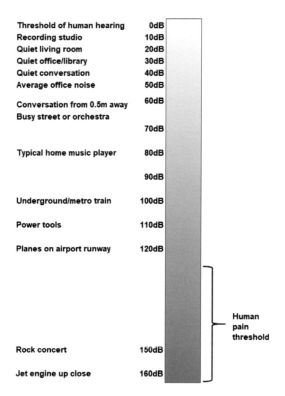

Figure 2.2   Decibel range chart.

A source, such as a bat or a loudspeaker, has a power, which is a measure of the amount of energy produced per second (joules/second = watts). We can measure the sound when it is propagating through the air as intensity (joules/second/m²). If we sum the intensity over a time window (e.g. over an entire bat pulse) this is the total energy (joules). Intensity contains both particle velocity (speed of air particles due to local pressure variations) and pressure (newton/m² = pascal). Most microphones only measure variations in pressure, which is then converted into variations in voltage.

The amplitude of the wave is related to the amount of energy contained within the wave (Figure 2.3). In other words, the energy of the wave is proportional to the amplitude (A) squared (i.e. A^2). In terms of the human voice it is the difference between a loud (high-amplitude) and a quiet (low-amplitude) voice. To produce a wave with higher amplitude, the cone of the loudspeaker moves further away from the rest point in both directions (and therefore requires more energy to move the cone).

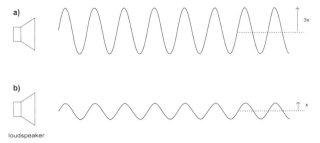

Figure 2.3 Assuming that the duration of the waves is the same in both cases, wave (a) has amplitude 3 times that of (b) and in effect has 9 times the energy.

2.1.2 Frequency

If, instead of altering the distance moved by the cone, we increase the rate at which it moves back and forth, in effect decreasing the wavelength, the frequency of the wave will increase; in other words, the number of waves (or areas of high and low pressure) that are produced per unit of time will increase (Figure 2.4).

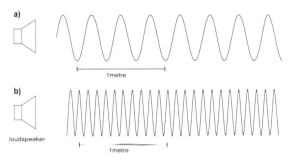

Figure 2.4 Two waves of different frequencies. The wavelength of (a) is 0.33 m (1/3 m) and the wavelength of (b) is 0.01 m (1/10 m). Therefore (a) has a frequency of 1,030 Hz (340/0.33) and (b) has a frequency of 3,400 Hz (340/0.1).

Two waves of the same duration and amplitude but with different frequencies will contain the same amount of pressure (Figure 2.5). However, since pressure is rebuilt more frequently per unit time for high frequencies than for low frequencies, locally, particles attain higher velocities and more work is done on the medium, and therefore losses (e.g. heat) tend to be higher at high frequencies compared to low frequencies.

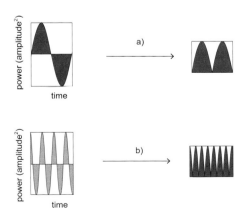

Figure 2.5 The shaded areas of each wave represent the pressure contained within each wave: (a) contains the same amount of pressure as (b).

2.1.3 Attenuation

In reality, as we move away from the loudspeaker the amplitude of a wave becomes smaller as the energy dissipates (Figure 2.6) – a process referred to as attenuation. Increasing the amplitude of a sound means that the sound (pressure differences) can be detected at greater distances. It is the difference between a quiet and a loud voice. This is why we need to raise the volume of our voices to enable someone to hear us at the end of a long room. The two main mechanisms behind the loss of sound energy are spherical spreading and absorption.

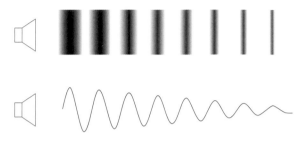

Figure 2.6 Attenuation. The greater the distance from the loudspeaker, the lower the amplitude (and therefore the energy) of the wave.

The drop in sound intensity as the sound spreads out from its source is due to spherical spreading. If we imagine sound propagating from a source as a sphere that expands as it moves away from that sound source, we can see that as the sphere expands, its area must increase. Therefore, the number of molecules over which the fixed amount of power must spread increases. This means that the amount of energy transferred to any single molecule decreases as the sphere expands. The amount of power per unit area decreases with the square of the distance from the sound source. Attenuation is generally proportional to the square of the sound frequency.

Losses are also due to absorption. Sound compresses and decompresses air molecules in repetitive cycles. Air, like any other medium, is viscous, which means that if you try to compress it, its molecules will move faster, and its temperature will rise (in accordance with the ideal gas law). The problem is that some of the fast molecules may escape to cooler regions and then no longer contribute to the consecutive decompression. Even more importantly, air molecules do not bounce back with the identical force with which they were compressed (shear/viscous losses). Both types of 'pressure leak' lead to linearly increasing losses (in dB) with the square of sound frequency and are summed together. In addition to this main effect, molecules of oxygen and nitrogen of air may also be set into a short vibration due to the sound frequency used (relaxational processes). These vibrations cause modifications to the 'linear losses' described above. In the range of 20–100 kHz, it is mainly oxygen molecules that slightly modify the relationship with attenuation. Water vapour has a strong effect on how oxygen molecules vibrate, increasing sound attenuation. Attenuation is greatly lessened when the humidity drops below 15%, which can happen in desert climates on several days of the year.

2.1.4 Doppler shift

Suppose you are facing a loudspeaker producing a sound wave of 10 hertz (Hz; i.e. 10 waves per second). The wave 'peaks' will be reaching you at a rate of 10 per second. Now, imagine you start moving towards the loudspeaker at speed. The rate at which the waves reach you will increase. Although the loudspeaker is still producing waves of 10 Hz, they will be reaching you at a greater number of waves per second, so that from your point of view the frequency appears to be higher. The opposite is true if you start to move away from the loudspeaker. The waves will be reaching you at a slower rate, so the frequency will appear to be lower. This is similar to the effect produced by a siren on an ambulance as it drives past, except it is the ambulance that is moving whereas you are stationary. As it moves towards you the frequency you hear is higher because the waves appear 'squashed together'. When it is level with you, you hear the true frequency produced by the siren. Then, as it drives away, the waves appear further apart and the perceived frequency drops. For bats, an echolocating common pipistrelle using a pure 46 kHz signal and flying at an average speed of 3.9 m·s⁻¹ would result in a Doppler shift of 0.54 kHz (towards observer 46.54 kHz, away from observer 45.46 kHz), whereas a Leisler's bat using a pure 24 kHz frequency and flying at 7.9 m·s⁻¹ would result in a Doppler shift of 0.56 kHz (towards observer 24.56 kHz, away from observer 23.34 kHz). The effect depends on the speed of the bat and the frequency it is emitting, and it can be calculated by the following equations:

$$fd = [(c + v)/c]f \text{ for a bat flying towards an observer}$$
$$fd = [(c - v)/c]f \text{ for a bat flying away from an observer}$$

where fd is Doppler frequency change (Hz), c is the speed of sound in air (m·s⁻¹), v is the speed of the bat (m·s⁻¹) and f is the echolocating frequency (Hz).

Examples of this variation for common pipistrelles and Leisler's bats are presented in Figure 2.7.

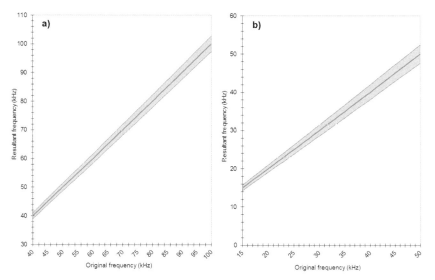

Figure 2.7   Variation in recorded frequency compared to emitted (original) frequency of (a) a common pipistrelle and (b) a Leisler's bat flying at an average speed of 4.4 m·s⁻¹ and 5.8 m·s⁻¹ respectively. The central line represents a stationary bat, the upper line represents a bat flying towards the observer at its average speed, and the lower line represents a bat flying away from the observer at its average speed.

2.2 Signal acquisition

To analyse any type of sound the signal must be converted from an analogue signal (a continuous time-varying signal) to a digital one (which represents the sound in the form of discrete amplitude values at evenly spaced points in time). This digital signal is then available for manipulation and analysis using a computer program. However, for this conversion, there are two important parameters when digitising sound that can affect the 'recorded' signal. These are the sampling rate and the sampling size.

2.2.1 Sampling rate

The sampling rate is the number of times a signal is 'sampled' (or a data point is recorded) over a period of time. Figure 2.8 illustrates the data points sampled from a wave. The sampling rate must be high enough so that an accurate picture of the input is recorded. The following figures illustrate how an inadequate sampling rate affects the representation of the original signal. Figure 2.9 shows the same signal sampled at a rate lower than that in Figure 2.8, but still sufficiently high to give an accurate representation of the original. If we now look at Figure 2.10 (which contains an overlay of the original sound), we can see that if the sampling rate is too low, the actual wave is inadequately sampled because the points could be fitted to a different wave with another wavelength and hence frequency.

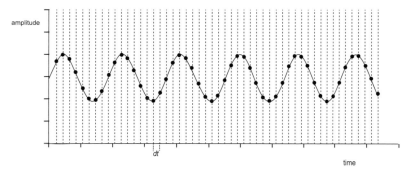

Figure 2.8   Sampling to create a digital representation of a pure tone signal. Each dot represents a single sample taken at evenly spaced time intervals, *dt* (vertical lines).

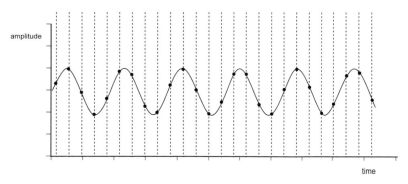

Figure 2.9   The same original signal as in Figure 2.8, but with a lower sampling rate. Data points still provide an accurate representation of the original signal.

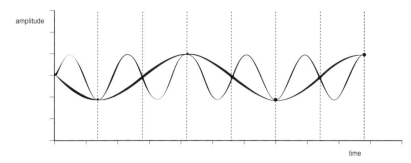

Figure 2.10   The same original signal as in Figure 2.8, but with a further reduction in sampling rate, showing 'aliasing' resulting from an inadequate sampling rate. The thin line represents the original signal and the thick line represents the 'perceived' signal due to aliasing.

This sampling error is known as aliasing. To avoid this problem, we use the simple rule that the sampling rate must be more than twice that of the highest frequency in the original signal. Thus when recording lesser horseshoe bats, which have a maximum frequency of around 110 kHz, the time expansion on the bat detector divides this by a factor of 10 (depending on the type of detector), resulting in a signal with a maximum frequency of 11 kHz (11,000 Hz). To obtain an accurate signal we must use a sampling frequency of twice this, i.e. 22,000 Hz. In fact, many sound cards in computers are fixed at specific sampling rates (i.e. 11,025 Hz, 22,050 Hz and 44,100 Hz). In this case, we would use the 22,050 Hz or perhaps the 44,100 Hz sampling rate to be on the safe side. Most sound cards have built-in anti-aliasing filters (i.e. a filter that cuts off at half the sampling frequency). The quality of these varies, but usually they are good enough to avoid severe effects from aliasing.

2.2.2 Sampling size

The sampling size is the actual number of amplitude points to which the original signal can be fitted, and it also depends on the type of sound card. We usually measure this size in terms of 'bits'. If a sound card is 8-bit then it can measure 2^8 or 256 discrete amplitude points, whereas a 16-bit card can measure 2^{16} or 65,536 points. However, the higher the sampling size, the more memory is needed to record the signal. Figure 2.11 illustrates an error caused by a low sampling size. There are only four discrete amplitude levels that can be measured, so the amplitude of the original wave has to be 'forced' into one of these levels.

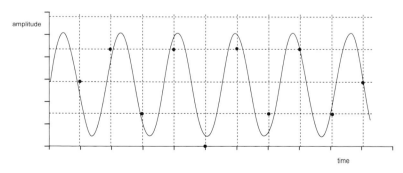

Figure 2.11   Digitising error produced by a sampling size that is too low. A 2-bit sampling size only represents four discrete amplitude levels to which the original signal has to be fitted.

2.3 Representing and describing sound

2.3.1 Converting the digital data into a sonogram

Any acoustic signal can be represented in one of two forms – a frequency-domain graph, which shows how much of a signal lies within each given frequency band over a range of frequencies, and a time-domain graph, which shows how a signal changes over time. In Figure 2.12a a pure tone of 20 kHz is represented in the frequency domain, and in Figure 2.12b the same signal is represented in the time domain.

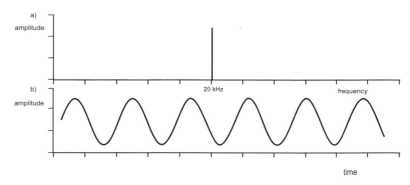

Figure 2.12   The same signal as (a) frequency representation and (b) amplitude representation.

To combine these diagrams and represent acoustic signals as a representation of frequency, time and amplitude (or sonogram, also known as a spectrogram) we use a mathematical method called the Fast Fourier Transform (FFT) to calculate the frequency-domain representation from the time-domain one. In basic terms, this is a mathematical formula that converts the data into frequency data. If we look again at the sample points in Figure 2.8, which are based on time and amplitude measurements, these are the actual data points used to calculate the frequency values. If we have a pure tone of a single frequency such as this one, we can just use a single FFT analysis to calculate the frequency of the wave. Since the frequency content of bat sounds varies over time, however, we need to calculate several FFTs to see this variation.

2.3.1.1 Short-time Fourier analysis: 'windows'

All bat sounds comprise more than one frequency, so we use a short-time Fourier Transform (STFT). This is in effect a series of 'windows' across the waveform within which the frequency (and amplitude) data are calculated. Putting these windows together builds up a complete representation of the signal. In Figure 2.13 we can see one of these windows. The length of the window contains a certain number of data points (or samples), which is usually a power of 2 ($2^2 = 4$, $2^6 = 64$, 128, 256, 512, 1,024, 2,048, 4,096, etc.).

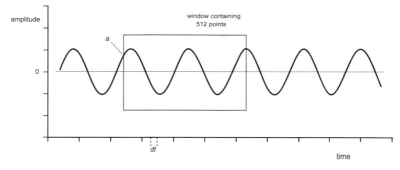

Figure 2.13   A sample 'window' containing a certain number of data points. Putting these windows together builds up a complete representation of the signal.

2.3.1.2 Windows functions

To calculate an accurate FFT for each window, the amplitude of the first and last data point needs to be at zero. If this is not the case, as indicated by (a) in Figure 2.13, the result is a series of sidelobe frequencies shown as a broadening of the original frequency (Figure 2.14). These occur as a result of the waveform being truncated, giving instantaneous pressure 'jumps'. These jumps imply extremely fast changes in pressure resulting in high frequencies, which are only present for a single and infinitely short time.

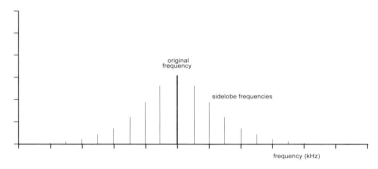

Figure 2.14   'Sidelobe' frequencies.

To solve this problem we use a window function. This is a waveform that is multiplied by the original waveform within each FFT window. As you can see in Figure 2.15, the resultant wave (within the Fourier window) tapers to zero at each end of the window, which significantly reduces the sidelobes (Figure 2.16). There are many different window functions (e.g. Hanning, Hamming, Blackman, Bartlett) and they all give slightly different results. The important point is that they all produce a resultant window waveform that tapers to zero. The exception is the rectangular window function, which would in effect give the same result as having no window function at all.

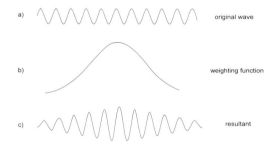

Figure 2.15   The result of multiplying the original wave within a 'window' by a 'window function'. The resultant wave tapers off to zero at either end within the window.

Figure 2.16   A wave without any sidelobes present.

So one window gives a representation of the wave in terms of time and frequency, which is plotted on the axis as in the section in Figure 2.17a. Note the shading, which represents the amplitude of the wave. When all the separate FFT windows are placed together they give us a representation of the overall sound in terms of frequency, time and, to a certain extent, amplitude (Figure 2.17b). To 'smooth' the sonogram we usually use some kind of overlap between these separate sections, or windows, which produces a more 'refined' diagram (Figure 2.17c).

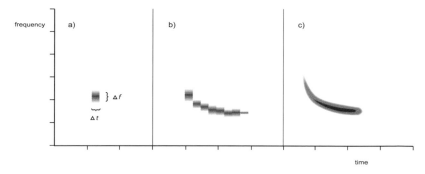

Figure 2.17 The addition of FFT windows to produce a sonogram, where (a) shows the time and frequency representation of a single window, (b) shows the time and frequency representation of several windows together and (c) shows the results of 'smoothing' the sonogram by overlapping windows.

2.3.1.3 The trade-off between time resolution and frequency resolution

Ideally, we would like to have both a very good time resolution and a very good frequency resolution, but these two are related. To understand this, it is useful to view the sonogram in two ways.

The first is to view the image as a series of frequency 'bins'. You can think of it as dividing the call into a series of successive short time intervals or 'frames'. Each frame gives information about the spectrum of the signal at one moment in time. To display the whole spectrum, all these slices or frames are plotted side by side with frequency running vertically and amplitude represented by shading or different colours (Figure 2.18). The accuracy of the frequency data within this frame is related to the number of sample points (amplitude points) within the sampling window (e.g. 1,024, 2,048, etc.). This is known as the 'spectral slice' model.

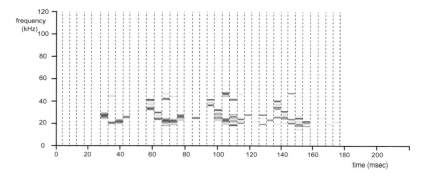

Figure 2.18 Spectral slice model of analysis.

An alternative is to think of the spectrum as a bank of 'bandpass filters' plotted one on top of the other that filter out all the frequencies except one small range (Figure 2.19).

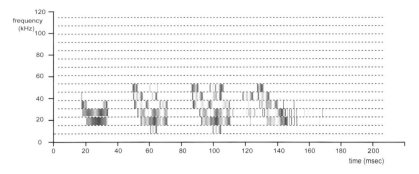

Figure 2.19   Bandpass filter model of analysis.

Putting these two views together, it is apparent that frame length and filter bandwidth are inversely proportional to each other. Unfortunately, therefore, our ideal sonogram, with fine resolution in both time and frequency, is impossible to achieve. Although a short frame length (e.g. 128 points) yields a sonogram with finer time resolution, it also results in wide bandwidth filters and therefore poor frequency resolution.

2.3.2 Oscillograms, power spectra and sonograms

For analysing bat calls we generally use three main graphical representations of a sound wave: an oscillogram, which displays sound pressure (amplitude) against time (Figure 2.20a), a power spectrum, which displays sound pressure (amplitude) against frequency (Figure 2.20b), and a sonogram (also called a spectrogram), which displays frequency against time with sound pressure (amplitude) being represented by colour intensity (Figure 2.20c).

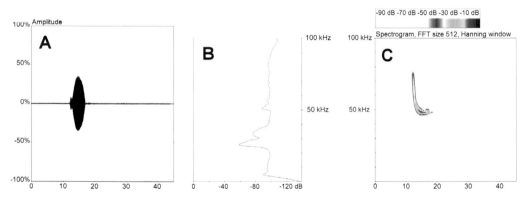

Figure 2.20   A single echolocation pulse represented by (a) an oscillogram, (b) a power spectrum and (c) a sonogram.

2.3.3 Call shape

As discussed in section 2.1, sound waves can vary in their amplitude and frequency; they can also vary in duration. It is useful to view these three types of variation separately, in both oscillographic and sonographic forms. For example, discrete sound pulses may be of very high amplitude, or of low amplitude, or they may vary in their amplitude with time (Figure 2.21). Similarly, they may vary in duration (Figure 2.22) or frequency (Figure 2.23).

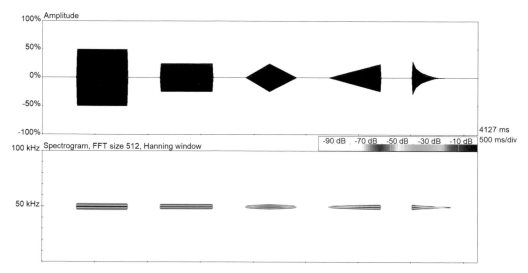

Figure 2.21   Signal of similar duration and frequency but varying in amplitude, displayed as an oscillogram (above) and a sonogram (below) in which changes in amplitude are colour-coded.

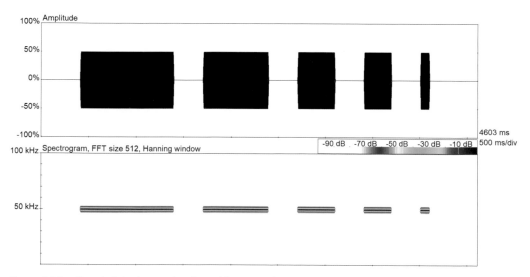

Figure 2.22   Signal of similar amplitude and frequency but varying in duration.

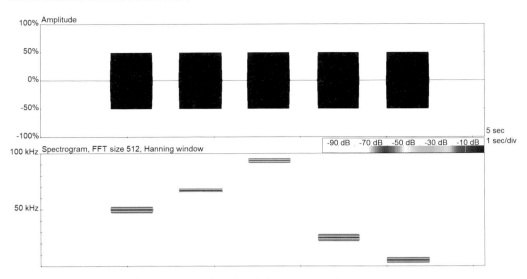

Figure 2.23   Signal of similar duration and amplitude but varying in frequency.

However, bats combine this variation within their echolocation pulse to create different 'call shapes' (Figure 2.24).

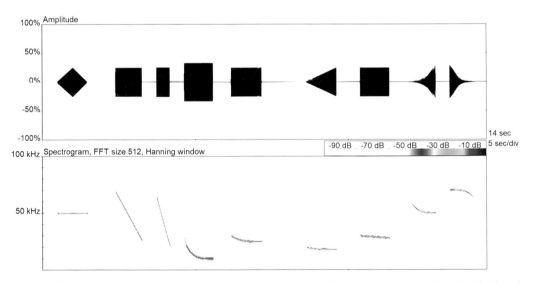

Figure 2.24   Example of discrete sound pulses varying in their amplitude, frequency and duration, displayed as an oscillogram (above) and a sonogram (below).

These call shapes can be described in terms of the degree of frequency modulation (FM), constant frequency (CF) and quasi-constant frequency (qCF) components they contain (Figure 2.25). Quasi-constant frequency is defined as less than 5 kHz bandwidth.

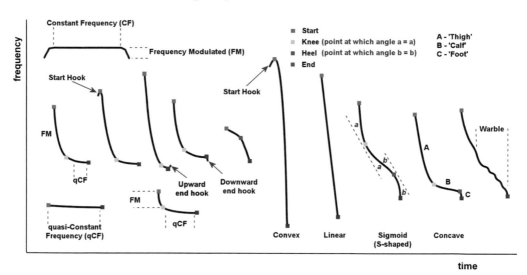

Figure 2.25   A stylised view of features of echolocation pulses that make up 'call shape'.

3 Echolocation

Arjan Boonman and Jon Russ

3.1 Why did echolocation evolve mainly in bats?

Echolocation has been identified in bats, toothed whales, 24 species of swiftlet, the oilbird, tenrecs, rats, pygmy dormouse and possibly shrews (Thomas and Jalili 2004, Brinkløv *et al.* 2013, Panyutina *et al.* 2017). Three species of fruit bat possess a crude echolocation ability (Boonman *et al.* 2014). Since crude echolocation is relatively simple to learn, even by individual humans (Thaler *et al.* 2017), it is assumed that this form of echolocation has evolved and de-evolved many times in many species of fruit bat, and possibly in other mammals.

Ranging

Bats echolocate by producing and projecting ultrasonic sounds from their mouths or noses and then detecting the echoes that return from any solid object within range. Bats produce these pulses in rapid succession to receive a regularly updated picture of their environment. In other words, a single call provides the bat with a single snapshot of its environment whereas a series of calls provides a series of snapshots, in much the same way as a strobe light provides us with a series of staggered images. As the strobe rate (or pulse rate) increases, the separate images begin to be perceived as a continuous image.

As sound travels at a constant speed in air, bats can measure how far away an object is by determining the difference between the time at which the call was emitted and the time at which the echo returns. If an object is far away, the sound waves will take longer to return to the bat than for a nearer object. For example, in the figure, (a) will take 1 m/340 m·s^{-1} = 0.0029 seconds, whereas (b) will take 2.5 m/340 m·s^{-1} = 0.0074 seconds.

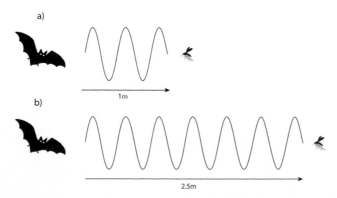

a)

1m

b)

2.5m

Ranging: bats measure distance by using the time it takes for an echo to return.

The echolocation of cave-dwelling birds is much more salient and purposeful than the rudimentary wing-claps of fruit bats or the tooth-clatter of terrestrial mammals, which makes it more likely to have been more consistently present during the evolution of cave-dwelling birds. Although it has been shown that echolocating swiftlets are capable of avoiding rods with a diameter of 6.3 mm (Griffin and Thompson 1982), their exact capabilities remain unknown. The main handicap of birds, however, is that they are unable to hear ultrasound, which could only have been achieved by a radical evolutionary redesign of their hearing system (Necker 2000, Manley 2012). Consequently, birds are still restricted to using audio-echolocation instead of ultrasonic echolocation. This is limited by poor or no focusing of the sound beam, poor reflectivity off small objects and poor information on the structure of targets, and avian echolocation can therefore be classified as 'crude' as well.

Visual information is far more information-dense than acoustic information, and nightjars, owls and flying squirrels get by just fine without using echolocation. This raises the question of why bats evolved sonar in the first place, especially seeing as many bat species are tree dwellers or cave-entrance dwellers and not deep-cave dwellers.

Measurements and simulation of visual and echolocation acuity have shown that echolocation can provide insect-detection and tracking abilities superior to what is offered by vision (Fenton *et al.* 1998, Altringham and Fenton 2003, Boonman *et al.* 2013). Nightjars and frogmouths are nocturnal insect-eating birds, but studies have revealed that their hunting is restricted to twilight or moonlit nights and to a diet consisting mainly of medium-sized to large insects (Jetz *et al.* 2003, Taylor and Jackson 2003), restrictions that do not apply to bats. Unrestricted access to flying nocturnal insects opened up a niche to bats from the equator northwards to the Arctic circle and southwards to 55°S in a wide variety of habitats and altitudes. The biomass of flying nocturnal insects is considerable (Jackson and Fisher 1986, Gray 1993, King and Wrubleski 1998, Lynch *et al.* 2002). The oldest known fossils of echolocating bats also reveal that all of them were insect eaters (Simmons and Geisler 1998). A range of other diets, including fruits, only emerged later, among bats that already had an advanced insect-echolocation system.

None of these facts precludes the possibility that dark caves played a role in the evolution of echolocation. All 25 (of the world's 10,000) bird species that echolocate are cave dwellers. Echolocation in birds, however, is found in only two genera, of which other members with no echolocation do not differ markedly from the echolocators. None of the insect-eating echolocating birds use echolocation to detect prey. Clearly, (cave) echolocation in birds did not cause a speciation boom as it did in bats. This is why it is presumed that, given the ease with which mammals (contrary to birds) can emit and detect ultrasound (an ability they had possessed for millions of years before bats evolved), the speciation boom in bats is related to their superior ability to detect and track insect prey by ultrasound as compared to vision in twilight and darkness (Boonman *et al.* 2013). Genetic studies have shown that ultrasonic echolocation (as opposed to crude echolocation) has evolved independently in bats at least twice (Jones and Teeling 2006).

3.2 Extant echolocation types

The echolocation discussed in the remainder of this book is 'advanced echolocation', meaning echolocation that initially evolved to detect insects and which we will simply refer to as 'echolocation'. One exception to the 'insect detection first' evolution of echolocation is found in the fruit bats of the genus *Rousettus*, which have precise echolocation using well-controlled 60-microsecond clicks, but whose echolocation probably never evolved to detect insects. All other airborne echolocation is tonal echolocation: not clicks but signals lasting from 300 microseconds (μs) to as long as 100 milliseconds (ms) and therefore consisting of many periods whose rate changes only gradually over time. A useful grouping of echolocation types of bats in the world comprises (1) CF-Doppler (second harmonic) bats, (2) uni-harmonic FM bats and uni-harmonic FM-qCF bats, and (3) multiple-harmonic FM bats and second-harmonic qCF bats. Examples of species that use these echolocation types are shown in Table 3.1.

It immediately becomes apparent that these echolocation types have not evolved evenly over the planet, or not even evenly relative to the equator. While South America can pride itself on its many multiple-harmonic FM bats, Southeast Asian jungles boast a huge diversity of CF-Doppler bats. The European bat fauna nearly exclusively exhibits FM uni-harmonic and FM-qCF bats, with *Plecotus* species exhibiting features of multiple FM echolocation and *Rhinolophus* species being CF-Doppler bats.

Table 3.1 provides some interesting insights into the types of echolocation that have evolved. Of all 1,066 echolocating bats (excluding all fruit bats), 25% use mainly the second harmonic to echolocate, and since this feature is present in at least six different groups of bats this type of echolocation, which relies on a specialised formant filtering, therefore probably evolved independently six times. Echolocation using broadband FM chirps evolved strongly in *Myotis*, *Kerivoula* and *Murina* in the Vespertilionidae, which may be due to their common ancestry. However, Furipteridae in South America also evolved a chirp echolocation system independently. Multiple-harmonic broadband FM pulses appear to have evolved at least four times independently in very different groups of bats. This type of echolocation is very dominant in South American rainforests but is rare in Old World rainforests, where clutter-adapted bats mainly rely on FM chirps or Doppler echolocation. In extreme clutter, we find the small high-frequency Hipposideridae occupying this niche in the Old World.

Half-open to open space echolocation types make up at least 40% of all echolocators. However, since all *Myotis* species are grouped as FM (and not all *Myotis* are equally clutter-adapted), the FM group is somewhat inflated, as is the multiple-harmonic FM group, where some Phyllostomidae are more adapted for half-open space. In short, open space and cluttered space seem to have spurred similar diversification. Jones and Teeling (2006) independently arrived at a grouping similar to that presented in Table 3.1, and also reached the same conclusions about the independent evolution of many different types of echolocation systems.

The highest constant frequencies can be found in Hipposideridae, followed by Rhinolophidae, and among the qCF bats *Miniopterus pusillus* has qCF components up to 75 kHz (*Rhynchonycteris naso*, main 2nd harmonic at 100 kHz), but the main bulk of qCF components are within the range 23–45 kHz, with *Otomops martiensseni* sweeping down to as low as 9 kHz. The highest echolocation frequency bats produce with the larynx is 250 kHz (*Kerivoula hardwickii*) as the start of a rapid FM down-sweep. The highest FM sweep rates can be found in *Murina* bats. The great majority of functional bat sonar pulses appear to be FM down-sweeps and not upsweeps.

In Europe and northern Asia, the multiple-harmonic echolocation system is much rarer than it is at more southern latitudes (Table 3.2), and the same applies in North America. The genus *Myotis* contains several broad-bandwidth species that are lacking in Southeast Asia, where *Kerivoula* appears to be filling the broad-bandwidth niche. The highest frequency emitted in Europe is the start frequency of *Myotis emarginatus* at 175 kHz, and the highest constant frequency is 114 kHz of *Rhinolophus hipposideros* (disregarding hipposiderids). The highest qCF is 56–58 kHz, emitted by *Miniopterus schreibersii* and *Pipistrellus pygmaeus*, and the lowest qCF is emitted by *Tadarida teniotis*, sweeping down to 12 kHz.

Table 3.1 Bat sonar types.

		Old World / New World / Both	No. of species	Common features*
CF-Doppler		Rhinolophinae	87	High frequencies, noseleaf, no tragus, use of 2nd harmonic
		Hipposiderinae	88	
		Pteronotus parnelli	1	
Uni-harmonic	FM	Furipteridae	2	Limited vocal tract filtering, oral emission, below 1.5 ms harmonics appear
		Vespertilionidae	196	
	FM-qCF	Molossidae	102	Loud, harmonics appear in short pulses
		Noctilionidae	2	
		Vespertilionidae	236	
	FM	Phyllostomidae	192	Noseleaf, weak echolocation, passive listening, small home ranges
		Nycteridae	16	
		Megadermatidae	5	
		Natalidae	10	
		Mystacinidae	2	
Multiple harmonic	qCF (2nd harmonic)	Emballonuridae	52	Loud, limited ability to emphasize higher harmonics during buzz
		Mormoopidae	10	
		Rhinopomatidae	4	
		Thyropteridae	3	
		Craseonycteridae	1	
		Myzopodidae	2	

*Not found in all groups.
CF, constant frequency; qCF, quasi-constant frequency; FM, frequency modulation.

Table 3.2   Biosonar types found in European bats.

CF-Doppler		*Rhinolophus* – all **Asellia tridens* **Hipposideros tephrus*
Uni-harmonic	FM	*Myotis*
	FM-qCF	*Tadarida teniotis* *Eptesicus* spp. *Nyctalus* spp. *Pipistrellus* spp. *Vespertilio murinus* *Miniopterus schreibersii* *Hypsugo savii*
Multiple harmonic	FM	*Plecotus* spp. *Barbastella barbastellus*
	qCF (2nd harmonic)	**Rhinopoma cystops*
Non-tonal clicks		*Rousettus aegyptiacus*

*Possible rare vagrant

3.3 Echolocation types of European bats and their function

While we are not sure what parameters the rudimentary echolocators such as wing-clapping fruit bats or tenrecs (or even swiftlets) are measuring, all advanced echolocators measure delay in the time it takes sound to travel from the bat's head to a target and back. This time delay depends on the distance of the target, which is precisely what a bat wants to measure. Temperature can influence the speed of sound (e.g. 331 m·s^{-1} at 0 °C, 337 m·s^{-1} at 10 °C, 343.5 m·s^{-1} at 20 °C, 350 m·s^{-1} at 30 °C) and therefore affects the delay perceived by bats, for which they can presumably recalibrate to some extent. In the real world, targets reflect many more than one echo, and when a bat wants to land at its roost entrance, the emission of one pulse will generate a long series of overlapping echoes ('echo-jumble'). Researchers disagree on the nature of the 'images' bats can extract from such echo-jumbles, but what is certain is that bats more frequently hear echo-jumbles than single echoes.

CF-Doppler bats (Rhinolophidae/Hipposideridae) also measure delay to targets with the small final FM component of their pulses. The constant-frequency component of their long pulses is used to detect insect motion. Horseshoe bat calls contain constant-frequency components of very long duration and they have a filter in their ears tuned precisely to that particular frequency. The frequency of the call varies between species. When the echolocating horseshoe bat flies towards the echo returning from a prey or obstacle, the frequency it will hear will become increasingly higher in pitch. This is a result of the Doppler shift. However, there is a danger that if the frequency of the returning echo is too high, the bat will not be able to hear it (as its ears are tuned to a specific frequency) so it will counteract this problem by continually changing the frequency of its calls. A non-moving object, such as a tree, would produce no change in the original call and the echo would return as a pure tone. Fast-flying insects or the wingbeats of insects cause systematic frequency modulations in the otherwise noisy echo, which immediately draws the attention of the bat. This system enables Doppler bats to hunt in very dense ('jumbled' or 'cluttered') environments. The CF detection system even works well in flight, because the bat can lower its emitted frequency with increasing flight speed. In this way the Doppler shift these bats encounter is compensated for, and echoes still fall in the narrow frequency band to which they are highly sensitive.

Bats that hunt high in the sky in search of insects would not profit from a CF-Doppler system, because up in the sky there are no cluttering objects and the high frequencies needed for the Doppler system would limit their detection range (see below). This is why some bat species evolved qCF pulses with much lower frequencies. The reason for using long pulses is explained in the next section. The reason why most aerial hawking bats use quasi-CF (qCF) and not true CF pulses is that adding some bandwidth increases the probability of detecting insects when more than one insect is reflecting back to the bat (Boonman *et al.* 2019).

If bats want to hunt at lower altitudes, for example in half-open environments, very low frequencies and long pulses would make this impossible (see below), so they tend to use slightly higher frequencies and shorter pulses containing more bandwidth. There are two ways to generate a large bandwidth. One way is to sweep over many frequencies in one pulse (FM chirp). The other way is to utter a short pulse that only sweeps down little, but to use resonance cavities in the head to selectively amplify higher harmonics. All these harmonics together then make up a large bandwidth. This system is called multiple-harmonic FM, and in Europe it is employed by *Plecotus*.

This all makes sense, but the question remains why CF-Doppler and broad-bandwidth bats meet each other in similarly dense habitat. If CF bats are superior in dense environments, why don't they outcompete the broad-bandwidth bats there? Just by looking at what information is theoretically available to *Rhinolophus* with their limited bandwidth, we can conclude that they probably are not as adept at assessing the shape and structure of objects (e.g. vegetation) as *Myotis nattereri*, for example, with its broad bandwidth. A species of *Rhinolophus*, with its fixed frequency, will also have a fixed optimal detection range (see below). *Myotis nattereri* theoretically can detect smaller targets and generally detects targets of a broader range of sizes from a variable range of detection distances (see below). *Rhinolophus*, however, might more quickly detect a fluttering moth just in front of vegetation, which would just sound like an echo-jumble to the *Myotis*. The *Myotis* will only detect this moth from the correct angle when it is itself close to the vegetation, but its imaging may allow it to glean that moth from the vegetation after it has landed, whereas to the *Rhinolophus* it would then be 'invisible'. It is therefore not surprising that small Hipposideridae hunting close to vegetation (as do some *Myotis* and Kerivoula) also use high frequencies, a broad bandwidth and short pulses, making the informational content of their signals more similar to broad-bandwidth bats. These adaptations may explain why it is so much harder to catch Hipposiderae than Rhinolophidae in mist nets. The bottom line, however, is that in the same habitat, bats with different sonar types perceive the prey-fauna differently and therefore hunt differently. The next section provides more information on the physics behind different pulse parameters.

3.4 The function of pulse duration

Bats cannot hear much during the emission of a very loud pulse, since the middle-ear muscles contract to dampen the bat's hearing to protect it from being damaged by the loud emission. Therefore, as the start of the sound is travelling away from the bat, the bat is deaf to any reflections that may be created by an insect in its path (Figure 3.1). Only when the entire pulse has been emitted is the hearing system ready again to receive any echoes. A bat's pulse duration, therefore, sets a 'blind zone' in front of the bat. The blind zone of a pipistrelle using 4 ms pulses is a radius of 70 cm from the mouth, whereas a 20 ms pulse of a noctule leads to a 3.5 m blind zone.

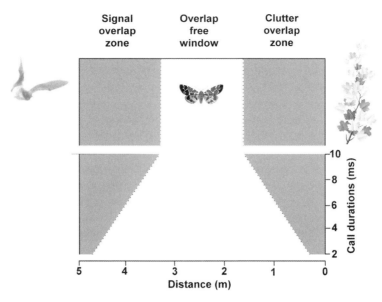

Figure 3.1   A bat operating with the pulse durations as depicted on the y-axis would face a pulse-duration-dependent blind zone in front of it, and also in front of clutter. The white zone in between is the zone in which bats can detect prey without any problems (based on Denzinger and Schnitzler 2013).

A similar blind zone exists when the target of interest (e.g. an insect) is flying close to vegetation behind it (Figure 3.1). If the insect is 18 cm in front of a hedge, after a millisecond of detecting its weak echo, the enormously loud echo of the hedge will drown the echo of the insect completely. To detect the insect without difficulty, the bat must either shorten its pulse duration to 1 ms or, if the pulse is still 4 ms, wait until the insect is 70 cm away from the background.

Long pulse durations, therefore, create wide blind zones and a restricted detection zone, but long pulses mean that more energy has been emitted, which results in a better chance detecting weak echoes. In qCF bats, long-duration pulses (> 10 ms) are usually of low frequency. Since the propagation losses of low frequencies are very low, reflections from any tree or hedge are loud and will drown any insect echo. For this reason, low-frequency long-duration bats tend to hunt in fully open space, high above the ground, where vegetation echoes become less of an issue. qCF bats that hunt closer to obstacles tend to have somewhat higher frequencies and shorter pulse durations, enabling them to hunt insects successfully. Still, when hunting flying insects, these bats will attempt to direct their sonar beam away from vegetation into relatively open areas. For horseshoe bats the situation is very different, as they listen to the echoes during each emission, using very long pulses (40 ms), but for all non-CF-Doppler bats, the principle holds.

For pulses that have a broad bandwidth (FM pulses), the blind zone in front of vegetation may be less of a problem than for a qCF bat. In FM pulses the signal duration plays an important role in the discrimination performance of the bat (see below).

3.5 Why do bats have specific echolocation frequencies?

People often ask, 'Can I identify a bat species by its echolocation frequency?' The short answer is that indeed many bats use qCF or CF components that are specific to the species, and even in bats that use only broadband sweeps the frequencies contained in this sweep are thought to be typical of a species. The reason why a bat would use a signal of restricted bandwidth (qCF or even CF) is that this improves the signal-to-noise ratio, making it easier for the bat to detect even the faintest echo from an insect. But this still does not explain why this frequency in species A should typically be, say, 20 kHz while being 40 kHz in species B.

We claimed earlier on that the prime reason for echolocation to evolve was the ability to detect insect prey. Following this logic, we can calculate the echo strength of model insects (e.g. spheres and discs), take into account atmospheric attenuation, and combine all available equations to calculate maximal detection distances of objects of a certain size, depending on the bat's frequency. Such calculations reveal that the frequency matters more for medium to large targets (diameter > 4 mm). Related to this, the exact frequency the bat uses begins to make a difference below 30 kHz. So when a bat is interested in finding moths over large distances using low frequencies, its detection distance for this moth will critically depend on frequency. The larger the moth, the lower the frequency should be, to optimise detection. However, the bat may want to focus not only on prey of a certain size, so there is still no reason to assume that a low-frequency bat should use exactly one fixed frequency.

For bats using frequencies above 30–40 kHz and eating smaller prey (a huge beetle may not even fit into its mouth) the exact frequency used only has a marginal effect on the detection range of these insects, unless the bat really whispers – but, as we know, most bats echolocate extremely loudly (> 120 dB SPL). Given these insights, must we reject the idea that echolocation frequency is related to prey detection? No. Bats interested in detecting large prey over long distances will use frequencies even below 20 kHz, depending on the size of the target insects. Bats that cannot eat such large prey anyway will stick to higher frequencies, where the exact frequency used matters less but gives louder echoes from smaller prey, as long as this prey is to be detected at short distances (< 2 m).

However, if the exact frequency a bat uses is not related to prey size, then why would a pipistrelle use 43 kHz with a variation of maybe 1–2 kHz at most? A serotine qCF really is restricted to 27–28 kHz and not all over the place, so if not by diet, how can we explain the frequency specificity we find in qCF bats? The most likely explanation is some degree of tuning in the hearing system to the qCF frequency the species evolved to use. In *Eptesicus fuscus*, increased frequency tuning was found from the cochlea, cochlear nucleus up to neurons of the inferior colliculus (Poon *et al.* 1990, Pinheiro *et al.* 1991, Haplea *et al.* 1993, Macias *et al.* 2006). The qCF of bat species such as pipistrelles is more tuned than that of the Molossidae, so the degree of neural (and possibly cochlear) tuning of the hearing system may depend on the species of bat.

Bats also evolved a mouth gape (aperture), creating a frequency-dependent beamwidth (Figures 3.2 and 3.3). Just like the frequency adaptation to detect insects, this will constrain the range of frequencies a bat can use while still being able to effectively beam the echolocation. Yet again, the frequency restrictions will be to within a few a dozen kilohertz rather than a few kilohertz. Bats can produce very loud sounds at a large range of frequencies, so the resonance properties of the bat's voice box are not related to the bat's restricted frequency tuning. In short: head size and prey detection govern the choice of a preferred qCF frequency. Species tend to adhere to this frequency more precisely than expected, because to a large degree their hearing system is fixed.

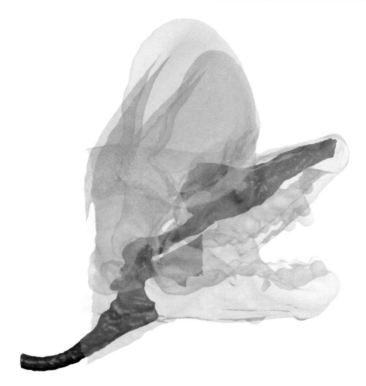

Figure 3.2   Result of a CT scan of *Vespertilio murinus*. Transparent grey: exterior/surface. Red: air pathway from lungs to nostrils and mouth. This is an example of a non-specialised vocal apparatus that has no resonance chambers. The interior of the nasal chambers (red bulge above mouth opening) is an extremely finely mazed network of turbinate tissue and is therefore used to smell and not to echolocate. Air is pumped out of the lungs through the tracheal pipe and the tracheal chambers, and is compressed in the larynx, where the vocal folds produce an extremely powerful tone (130 dB SPL) that is emitted out of the mouth. The beam shape can be regulated by the bat varying its gape width, or of course by altering the frequency of its emission. Reconstruction of head and air pipe originate from the same individual, and their relative sizes and placement are true to life.

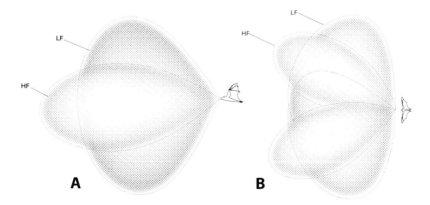

Figure 3.3   High-frequency (HF) and low-frequency (LF) lobes: (a) side view; (b) overhead view. From Kuc (1994)

We have now explained the spread and specificity of qCF frequencies from their lowest frequency of 12 kHz to the highest of 55 kHz (in Europe). However, we have not discussed why CF-Doppler bats use frequencies up to 180 kHz (in Europe up to 112 kHz, lesser horseshoe bat). CF-Doppler bats are dependent on detecting Doppler shifts. A pulse with a frequency of 80 kHz can detect velocity differences four times smaller than a pulse of 20 kHz with an identical duration. High frequencies and long durations are best suited for detecting tiny frequency shifts, which is most likely why horseshoe bats use exactly such pulses. Horseshoe bats do not avoid overlap with returning echoes, but actively use them in compensating for Doppler shift (Schuller 1977), which is another reason why they routinely use such long pulse durations (> 30 ms). The high frequencies horseshoe bats use give them a good ability to detect small insects, such as mosquitos, but it worsens their ability to detect insects at long range. The higher the CF frequency, the weaker background echoes (e.g. from vegetation) become, and the more suited this echolocation becomes for dense environments, since background echoes become less strong. Some species of small hipposiderid bats use frequencies as high as 180 kHz and fly even within very dense vegetation. Since CF-Doppler bats have restricted themselves to close-range foraging, the exact frequency a species uses matters little (see above). Secondly, when using ultra-high ultrasonic frequencies, a shift of 15 kHz barely changes the wavelength, contrary to a shift from 12 to 27 kHz, which more than halves the wavelength (the actual driver of all physical effects). These two factors may explain why in CF-Doppler bats we see a unique phenomenon, with individuals of a given species having highly specific fixed frequencies, but with strong variations of this fixed frequency over the species' geographical range (A. Lin *et al.* 2015) and even between colonies (Chen *et al.* 2016).

Most studies on geographical variation in the calls of any vespertilionid qCF bat species have been inconclusive or have had a negative result (O'Farrell *et al.* 2000, Murray *et al.* 2001; Volker Runkel, personal communication). Only *Vespadelus* species in Australia were reported to have geographical frequency variation, and *Pipistrellus pipistrellus* in Europe can exhibit a strong individual variation that is not yet understood. In the great majority of qCF bats, however, no indications of qCF frequency variations were ever detected, even across island populations and vast geographical ranges.

In short, horseshoe bats use high frequencies because those are good for doing Doppler tasks. Just like qCF bats, individual CF bats have a 'neurally fixed frequency', which is even more precise in CF bats than in qCF bats. In any given species, however, this fixed frequency may differ regionally. Most likely this is because repercussions in terms of mouth size and beamwidth don't exist for CF-Doppler bats (they use nasal emission), and regional frequency differences are also negligible in terms of insect detection (see above).

3.6 The function of FM pulses

In contrast to dolphins, which achieve a broad bandwidth by using ultra-short clicks, FM bats gradually lower frequency in a controlled manner. This system of frequency modulation (FM) or chirp sonar enables bats to resolve (hear separately) targets whose echoes overlap strongly with each other.

The potential significance of using a controlled frequency is that even long (high-energy) pulses can be used to measure distance with high accuracy. This means that even over several metres, the distance to objects can be measured reliably, which may fail with a weak click. Bandwidth can also serve to characterise the structure of objects. The smallest resolvable distance between reflectors is inversely proportional to bandwidth. Bats using a lot of bandwidth are therefore best at resolving structures, and therefore at 'seeing' patterns. Two echoes can still be resolved, even when they are more closely spaced than the resolution limit, because of the unique spectral/temporal notches they create. However, with more than two echoes this task becomes ambiguous. Methods have been invented to still accomplish a separation of more than two sub-resolution echoes in an iterative method that includes the calculation of cleaned spectra (Matsuo *et al.* 2004), but it is not known if bats can implement such a method neurally.

Besides the issue of the exact resolution limits bats have, it is often argued that bats can navigate quite well without having to assign every single echo to a target in space, as they could simply associate the sound quality of overlapping echoes with their external space (i.e. use a specific 'signature'; van der Elst *et al.* 2016). However, even if bats used a series of overlapping echoes in a 'dumb' correlative 'internal library' way to learn something about an object (e.g. could there be a crack or crevice to crawl in to?), the more bandwidth the bat has at its disposal, the more feasible this task would become.

Bats are not machines, and their hearing systems have integration times in the form of neural spikes. Unlike theoretical receivers, bats can improve the resolution of echoes not only by increasing bandwidth but also with faster sweep rates (frequency change). The perfect image, from a bat's perspective, can therefore be created with an FM pulse that sweeps down extremely fast over a very wide range of very high frequencies.

The world champion FM bats live in Southeast Asia. *Kerivoula* bats boast an incredible bandwidth, with *K. hardwickii* sweeping down from 250 to 80 kHz, which makes it also the

Harmonics

Harmonics occur when there is an increase in the air pressure used to produce sound. All harmonics are related to the frequency of the 'fundamental' or the 'base call'. For example, if the fundamental is 20 kHz, then the harmonics will be at 40 kHz, 60 kHz, etc. The main purpose of producing harmonics is that they increase the overall bandwidth of the call and hence increase the detecting resolution. Notice that the harmonic has a slightly higher sweep rate. If the call is linear frequency-modulated, as in *Myotis* bats, the harmonics are much steeper than in calls that are linear period modulated, as in *Pipistrellus* bats.

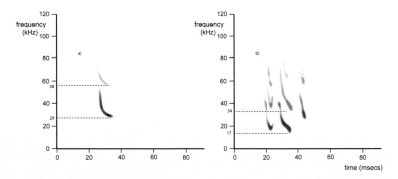

Harmonics: (a) echolocation call; (b) social call.

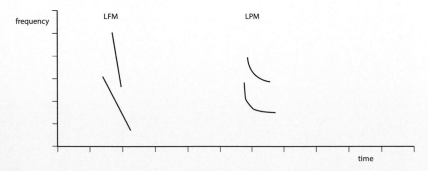

Harmonics: the difference between linear period modulated (LPM) sounds and linear frequency-modulated (LFM) sounds.

bat with the highest frequency on earth. *Murina* bats have a narrower bandwidth, but they have faster sweep rates (130–140 kHz/ms). In Europe, *Myotis emarginatus* (170 to 45 kHz) and *M. nattereri* (155 to 20 kHz) display the broadest bandwidths.

3.7 Multiple-harmonic FM

Some bat groups (Phyllostomidae, Nycteridae, Megadermatidae and, in Europe, *Plecotus*) use a series of harmonics to increase sonar bandwidth. Most of the species which do this are nasal emitters, their emissions are quieter than those of most other bat species, many of them have large ears, and they tend to hunt in relatively small areas often with dense vegetation.

Instead of sweeping down over a large range of frequencies, multiple-harmonic FM bats sweep down during very short pulse durations over a restricted range of frequencies but enforce a range of harmonics using their nasal chambers. The intensity of each harmonic – at least in some of species that have been investigated – can be regulated at will. From a signal processing point of view, the signals of these bats are different from a normal FM sweep, which has one frequency at any point in time, as these calls have several (harmonically related) frequencies at the same time. As these signals tend to be of short duration (<2 ms) and of limited intensity, they appear most suited for the examination of nearby targets. Their nasal emission is creates a narrow beamwidth and their short durations (still with full bandwidth) enable these bats to scan targets at short range. Since the sweep rate is relatively low and jumps irregularly inside the entire pulse, it is not entirely clear how these bats separate packets of echoes that bounce off targets. Many bats in this group are proficient at scanning targets accurately to get nectar from flowers, but also as predators of scorpions, big bugs or even mice and birds.

3.8 Flexibility of echolocation

So far, we have highlighted different echolocation systems among groups of bats. The variation in echolocation pulses a single individual bat can produce, however, is enormous, particularly among the non-CF-Doppler bats. As an example, let's examine the range of *Nyctalus noctula*, a high-flying low-frequency qCF bat (Figure 3.4). In open environments, its calls are long and of very low to moderately low frequency, sometimes starting with FM and ending with qCF pulses. These pulses are therefore optimised to detect medium to large insects (or possibly insect swarms) over long distances. The repetition rate of these pulses is correlated with the bat's wingbeat (Suthers *et al.* 1972, Wong and Waters

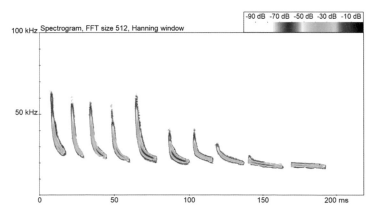

Figure 3.4   Variation in the echolocation calls of *Nyctalus noctula* moving from a closed environment (left) to an open environment (right).

Between-species and within-species variation

As discussed in Chapter 2 (section 2.3.3), the calls of bat species vary in their amplitude, duration and frequency. This variation is related not only to their insect prey but also to their foraging strategy and the habitat in which they commonly forage. Generally, bat species foraging primarily in a cluttered environment usually put more emphasis on the FM components of their calls, while those that forage primarily in an open environment tend to put more emphasis on the qCF components of their calls. For example, consider a species that commonly forages within woodland (a cluttered environment) such as a Natterer's bat. The bat's priority is to collect detailed information about its environment and distinguish fast-moving insects within the confined space. Thus, this species produces an extremely broadband call that provides a very detailed picture of the environment. Calls are of very short duration to minimise overlap between the emitted pulse and the returning echo. In addition, as the bat is foraging in clutter, the time it takes for the echo to return from an object to the bat is relatively short, and therefore the repetition rate is very high. Conversely, the noctule commonly forages high in the open (an uncluttered environment). As the echolocation calls must travel a long way, this species produces low-frequency echolocation calls of very narrow bandwidth, almost constant frequency. The repetition rate is very low as it takes a long time for the echo to return from the nearest object. Of course, some bats which forage in an edge (both open and closed) situation, such as pipistrelle species, use a combination of FM and qCF components. It is this variation that in many respects sets the species apart from each other.

Thus, the calls of different bat species are shaped by the habitats in which they usually forage. However, echolocation call shape is not fixed and can vary within a species (and an individual). For example, a common pipistrelle typically produces an echolocation call with an FM-qCF structure. However, if the bat moves into a more cluttered environment its echolocation calls become more broadband, the duration and inter-pulse interval shorten, and the frequency containing maximum energy shifts slightly upwards. Conversely, as the bat moves to a more uncluttered environment, the calls become very narrowband, the duration and inter-pulse interval lengthen, and the frequency containing maximum energy decreases.

2001) as this saves the bat energy to emit each call (Speakman and Racey 1991). In very open environments, the bat – for unknown reasons – may skip pulse emission every other wingbeat. When the bat has detected a possible prey, it will all of a sudden switch to FM pulses of progressively shorter durations as it approaches the insect. The repetition rate at which pulses are emitted also increases markedly, starting with four pulses per second before detection to two pulses every wingbeat (16–20 pulses per second), going up to 80 pulses per second and well over 100 pulses per second in a final buzz. In the final buzz, pulses lose much of their bandwidth, become progressively shorter, and are accompanied by a strong second harmonic just before the bat captures the target (Jones 1995). During the approach to the target, it appears that the bat attempts to prevent receiving the echo while still emitting the pulse that caused it, by reducing pulse duration progressively with decreasing distance.

Each of the pulses from search to capture seems to originate from a 'prefabricated' set designed to suit all potential distances to targets around. For example, a pulse of 8 ms has a specific frequency range and shape that is more or less standardised, no matter whether it was directed at a prey target or obstacles. Even though the bat may be interested here to see more detail, the bandwidth of its first harmonic is limited from 70 to 25 kHz

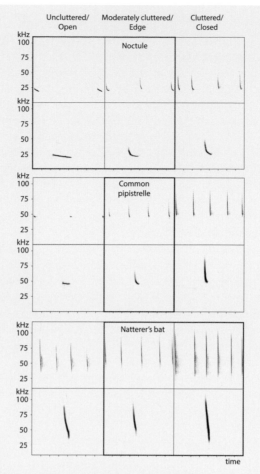

Examples of within-species and between-species variation in the echolocation calls of bats in different environments. Calls commonly emitted by bats flying in the habitat in which they are usually found are surrounded by a black border.

(Jones 1995) and will always have a curved shape at 8 ms. When the noctule returns to its roost and flies closer to obstacles we will see the same shorter (5–10 ms) pulses with broad bandwidth as we saw during the approach of the bat to the target. When the bat is circling its tree-roost and finally lands, we see pulses very similar to those we record during the bat's feeding buzz. Despite the huge variation in pulse duration, bandwidth and pulse design that the bat exhibits, it seems to be limited (or wants to be limited) in the specific parameter combinations it uses at any one time. The use of 'templates' and parameter limits in the echolocation of each species forms the basis of species identification of echolocating bats.

3.9 A viable strategy for identifying sonar calls of bats

Despite nearly 10-fold changes in bandwidth and pulse duration within seconds, many bat species are still identifiable, as they appear to use specific pulse templates that suit their needs in each situation. A general trend we can observe in Vespertilionidae and Molossidae is that low-frequency open-space foragers do not have the perfect pulse templates to assess extremely dense environments, while species that are adapted to dense environments lack

Feeding buzz

A good example of plasticity in call shape within an individual is seen in the 'feeding buzz'. This occurs when a bat catches an insect. Consider a bat foraging in an edge environment producing FM-qCF calls similar to the pipistrelle calls in part (a) of the figure. When it detects insect prey from a returning echo, it moves towards that area. At first, the echoes take a relatively long time to return to the bat. However, as the bat closes in, the distance between the bat and its prey is shorter and therefore it takes less time for the echo to return. So the bat needs to produce echolocation pulses at a faster rate to receive useful information. Equally important, as the bat gets closer to its prey, it does not need long-distance ranging to locate prey, as the insect has already been found, so the qCF signals are no longer necessary. It can therefore 'free up' some of the energy put into the long-distance call and gradually turn it into an FM call, which will give it more detailed information about the prey. At first, these sweeps are long in frequency range or bandwidth, as at point 1 in part (b) of the figure, since the bat does not yet have an accurate picture of the size or properties of the insect. However, as it closes in even more, the FM sweeps become noticeably shorter as the bat 'tunes in' to the size of the prey, maximising the information returned. In some species, there is a noticeable decrease in the duration of these pulses at this stage, which prevents overlap between the emitted pulse and the returning echo. This leads to point 2 (terminal buzz I). Finally, leading up to capture, the last few pulses are extremely rapid as the bat is very close and the overall frequency range always drops, as shown at point 3 (terminal buzz II). We are not sure why there is a final drop in the bandwidth at the terminal stage, but it may be related to a limitation of the vocal cords due to the high repetition rate. Immediately after insect capture, there is a gap while the bat consumes the insect, and then the calls return to those of normal search mode as it begins to hunt for new prey.

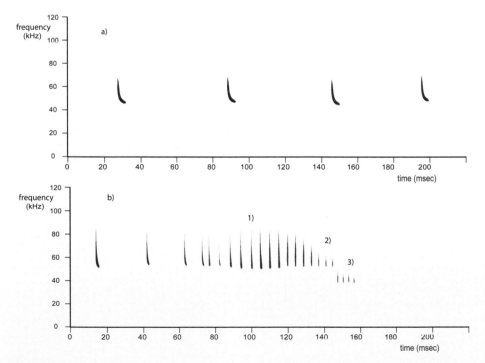

Feeding buzz of the common pipistrelle: (a) search calls of the bat leading up to the feeding buzz; (b) calls as the bat captures an insect.

templates in their repertoire suitable for long-range detection of targets in open space. Even the shortest call of a noctule (buzz) or the pulse with the fastest sweep rate (final approach) is still crude for a dense-space specialist such as *Myotis nattereri*. We can transfer this same idea to species that are ecologically closer in terms of their adaptation to dense versus open environments. For example, we can compare the echolocation pulses of *Eptesicus serotinus* with those of *Vespertilio murinus*. What is fairly open to the serotine is still perceived as cluttered by *Vespertilio*. Since, in the field, we may not always know at what altitude the bat is flying, it may not always be possible to assess the environment parameter. Still, from our recordings, the degree of bandwidth the bat uses in its calls will tell us something about how the bat is perceiving its surroundings; for example, extremely narrowband qCF pulses mean that the bat is in a very open environment.

We can correlate bandwidth with pulse duration in the two species (Figure 3.5). Because the species are ecologically close, we see a strong overlap in the plots, but *Eptesicus* has already reached minimal bandwidth at 10–12 ms, whereas *Vespertilio* goes flat out only beyond 12 ms. We see that serotines rarely use this minimal bandwidth and usually still have some FM component in their calls, characteristic of a species that is a bit more adapted for half-open rather than fully open environments (a recurring difference between species). We see an overlap in qCF frequency between the two species, but when *Vespertilio* uses very long pulse durations, its qCF frequency can be as low as 22 kHz, way below *Eptesicus*. For shorter pulse durations we see that *Vespertilio* does not start any higher than 58 kHz, a full 10 kHz below *Eptesicus*. These observations teach us to use sonar parameters in conjunction with each other to identify a particular bat species. Identical start frequencies, end frequencies or pulse durations alone are bound to be found in pulse templates used by a species that is ecologically close, but that very same template may not match any template used by our species.

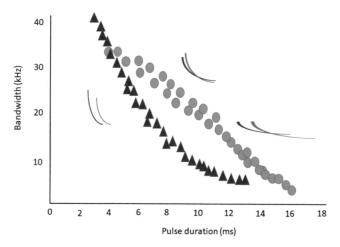

Figure 3.5   Theoretical comparison between *Vespertilio murinus* (blue ellipses) and *Eptesicus serotinus* (red triangles) (see Jensen and Miller 1999, Schaub and Schnitzler 2007). The first harmonic of *E. serotinus* has a higher starting frequency (thus higher bandwidth) than *V. murinus*, and it also reaches its minimal bandwidth at a lower pulse duration than *V. murinus*. *V. murinus* has a slightly lower terminal frequency, which is only present at very long pulse durations. *E. serotinus* only rarely uses pulses longer than 14 ms. In reality, the clouds of points will overlap frequently. The plot is to indicate the rough trend that applies to many species comparisons (more versus less clutter-adapted bats).

Table 3.3   Relationship between habitat, wing shape, emergence time, echolocation and tragus shape.

Habitat	Uncluttered/open	Moderately cluttered/edge	Cluttered/closed

Wing shape			

Emergence time			

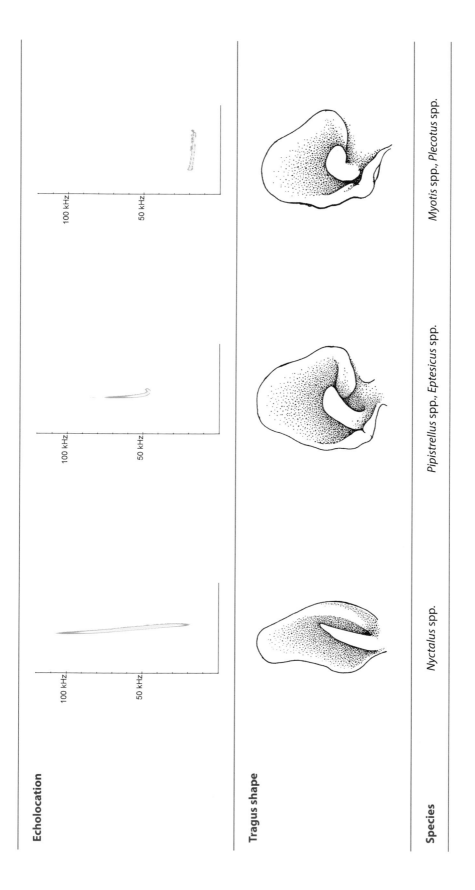

Echolocation

100 kHz

50 kHz

100 kHz

50 kHz

100 kHz

50 kHz

Tragus shape

Species

Nyctalus spp.

Pipistrellus spp., *Eptesicus* spp.

Myotis spp., *Plecotus* spp.

The relationship between habitat, emergence, echolocation, wing shape and tragus shape

The nature of a bat species' echolocation calls is related to its habitat, the shape of its wings and the time of emergence. In addition, echolocation call structure can often be predicted by the shape of the tragus, which is involved in vertical localisation of sound. Bats that have long narrow wings, such as noctule and Leisler's bat, are fast fliers and spend most of their time in open environments. Therefore, the energy put into their echolocation calls is concentrated around a very narrow band of frequencies and the overall frequency is low so that the calls can travel long distances. Because of the flight capabilities of the bat, which may enable it to avoid predators more easily than slower bats, it emerges earlier. On the other hand, the more manoeuvrable bats with broad short wings, such as brown long-eared bats, fly much more slowly. Their echolocation calls are usually FM sweeps, optimised for flying in clutter, and they are usually the last species to emerge. A summary of these relationships is presented in Table 3.3. For many European species, it is possible to predict a bat species' echolocation, habitat preferences, wing shape, tragus shape and the time of emergence based on knowledge of any one of these characteristics.

Apart from the systematic differences described above, there are also more random differences between echolocation calls of species (Chapter 2, Figure 2.25). Examples include the bent appearance of the FM component in *Miniopterus* (sometimes with starting hook), the position of the 'heel' in *Myotis* species or the sweep rate of the end hook. Pulses can have many features that are not directly functional echolocation-wise but that are nonetheless useful for identification. These features will be discussed in detail in the section for each species in Chapter 8.

4 An Introduction to Acoustic Communication in Bats

Grace Smarsh

Bats are well known for the echolocation behaviour that many species use to navigate the environment and target their prey. Increasingly, however, attention has also turned to investigating and describing acoustic communication repertoires: signals produced by a sender and detected by one or more receivers (Bradbury and Vehrencamp 2011). Because of the nocturnal nature of many species, acoustic signals are an efficient signalling modality to communicate with other individuals, as compared to vision or olfaction. There are over 1,400 species of bat, and they are long-lived animals with a great diversity of social and mating systems (Fenton and Simmons 2015). Additionally, bats often display fidelity to social groups and are increasingly impressing researchers with their social learning abilities (e.g. Page and Ryan 2006, Ramakers *et al.* 2016, Omer *et al.* 2018). Bats thus exhibit rich, diverse vocal-social repertoires ranging from isolation calls produced by pups (Bohn *et al.* 2007) to contact calls to recruit individuals to hard-to-find roosts in forests (Chaverri and Gillam 2016), and even songs used to attract and defend mates (Behr and von Helversen 2004, Behr *et al.* 2006). Furthermore, the communication abilities of bats can be complex, including vocal learning at young and adult stages (Boughman 1998, Knörnschild 2014, Prat *et al.* 2015, 2017, Vernes 2017), perception of call rhythm (Janssen and Schmidt 2009) and production of syntactically arranged sequences of syllables (Ma *et al.* 2006, Jahelková *et al.* 2008, Bohn *et al.* 2013).

Daubenton's bat *Myotis daubentonii* © René Janssen

While there have been numerous studies on the sophisticated use of echolocation within and across species since the 1940s and 1950s (Griffin 1944, 1958), we still know comparatively little about the variability and function of different communication vocalisations in bats. Much of this knowledge gap stems from technical constraints and a historical focus on sensory ecology in bats (Smotherman *et al.* 2016). We hope that this brief introduction to acoustic communication will give readers a better sense of how bats fit into natural soundscapes, and provide pointers towards potential avenues of exploration in this field.

4.1 Technological constraints and observer bias

Communication in bats has been noted for centuries, with early comments on bat 'squeaking' in Homer's *Odyssey* (eighth century BCE) and Isidore of Seville's book *Etymologies* (seventh century CE) (Lewis and Llewellyn-Jones 2018). However, bat calls and songs have not gained as much attention historically as bird song, as bats lack the conspicuous songs, brightly coloured plumage and attractive behavioural displays that birds possess. Bats have been challenging to observe and track because of their small size and nocturnal nature, and have been difficult to record acoustically because of their ultrasonic repertoires, requiring recorders with larger sample rates and data storage capacity (Smotherman *et al.* 2016). Many bat communication calls are partially within human hearing range (less than 20 kHz), as Homer and Isidore documented. However, the audible part of the social call may sound like squeaking, buzzing, honking, squawking, chirping or even strumming. A passer-by may not realise that they are hearing a bat and may mistake the call for a frog or insect. Alternatively, the sound may not be detectable to humans at all.

As technological advances allow us to better track bats and collect longer periods of vocal signals, researchers will continue to uncover the intricacies of bat acoustic signalling behaviours. The ability to acoustically record for long periods is a key advance, because communication vocalisations are produced in specific social contexts and may be seasonal (Catchpole and Slater 2008, Bradbury and Vehrencamp 2011, Smotherman *et al.* 2016). Recording throughout the day and across different seasons is therefore important for the collection of full repertoires. Once discovered, these communication signals can improve acoustic libraries. Acoustic monitoring is a powerful but limited method employed in many ecological studies (Jones *et al.* 2013). Communication-specific acoustic signals can greatly aid in targeting bat species of interest, differentiating species with similar echolocation pulses, and determining composition (sex, age, number of individuals) of a group of bats at a foraging site or day roost (e.g. Gjerde 2004). Combining large acoustic datasets with more targeted studies aimed at discerning the behaviour associated with communication calls can greatly improve wildlife management strategies through improved monitoring of soundscapes with a better understanding of underlying bat behavioural ecology.

4.2 Understanding and describing bat acoustic repertoires

Bat repertoires (the number and type of communication signals a species uses) can be described in two main ways, distinguishing vocalisations based on signal temporal and acoustic structure, and by the function of the vocalisations. Describing a type of acoustic vocalisation is best achieved using a combination of approaches. While it is often quite easy to tell one vocalisation from another by eye on a spectrogram, sometimes vocalisations can look similar and may exhibit fine variability within individuals that can complicate classification of the sound. To statistically define a call, the classic approach involves measuring a basic suite of temporal and acoustic parameters of the signal (such as maximum frequency, minimum frequency, frequency containing maximum energy, and duration) and more complex parameters that describe spectrotemporal curvature of the signal. These parameters can then be used in a clustering analysis to get an estimated number of call types. MANOVA and discriminant function analyses further solidify statistical separation of call types. These methods are still relevant, but more complicated, when the signals are multisyllabic, in which analyses include inter-syllable interval and the number of syllables, and examination of syllable types. Additionally, it may be difficult to separate multisyllabic call types in *.wav files if the starts and ends of the calls are not clear, in which case a statistical bout analysis may be needed to estimate the temporal cut-off parameter between vocalisations (e.g. Bohn *et al.* 2008).

Improved statistical approaches and automated analysis can greatly influence our ability to decipher bat communication repertoires, through more rigorous vocal detection and acoustic processing scripts. Automated scripts in MATLAB and R can quickly allow us to target communication signals, and extract relevant parameters for further analysis. Increasingly these scripts are available online. Unsupervised and supervised machine learning is a powerful way of finding patterns in large acoustic datasets, using previously labelled parameters of the analysed signals (supervised machine learning), or unprocessed datasets (unsupervised machine learning). For very noisy vocalisations, frequently observed in pteropodid bats, human speech processing parameters such as cepstral coefficients can be used to statistically quantify call types and the information they may carry (e.g. emitter identity, behavioural context), as expertly demonstrated in Egyptian fruit bats *Rousettus aegyptiacus* by Prat *et al.* (2016).

Understanding the behavioural context and function of the call type is important in describing a communication repertoire. Two statistically distinct call types, for example, could both be used in territorial contests but one may be used as a warning signal and the other when the fight escalates. Seba's short-tailed bats *Carollia perspicillata*, for example, produce down-sweeps, warbles and trills during the escalation stages of boxing matches (Fernandez *et al.* 2014), whereas great Himalayan leaf-nosed bats *Hipposideros armiger* use one frequency-modulated call type with fine-scale acoustic modifications (Sun *et al.* 2018). Who produces a signal (age and sex of the individual), who is nearby and supposedly receiving the signal, spatial proximity of individuals, and any associated behaviours all provide contextual information for describing the call type. Vocal information can be linked with a well-conducted ethogram (a catalogue or table of all the different kinds of behaviour or activity). These ethograms are best created in the wild, where the full suite of natural behaviours can be observed, but many behaviours can still be observed in captivity, as long as the bat colony mimics the natural composition (proportion of males, females, and juveniles) in the wild.

Finally, the function of the call(s) is confirmed through experiments, testing the use of the signal through acoustic playback and analysis of individual response. While most of the research on the function of bat repertoires is conducted in captivity or in the roost, owing to constraints in observing individual bats' responses, improved tracking, recording and visualisation technology or a combination of clever approaches can allow us to continue to explore the use of commonly observed but little-understood communication repertoires in the field (Greif and Yovel 2019).

4.3 Common types of communication calls

Across families, there are several types of commonly observed communication signals, described by emitter and receiver identity, spatial context, and associated behaviours. Despite variation in morphology across bat groups and evolutionary distance, there are often similarities in the structure of these communication signals emitted, and they are used in similar ways. Bats often produce far more vocalisations than can be easily categorised, but below are some commonly recognised call types.

4.3.1 Pre-emergence calls

Much of our knowledge of bat communication stems from day roosts, where acoustic behaviour is often easily recorded. Bats may roost in groups for several reasons, including limited roost availability and physiological regulation (Kerth 2008), creating social activity centres where various social-vocal behaviours can be observed. In the evening, prior to emergence, as bats become more active, social calls can often be heard (Figure 4.1). Individuals begin moving about and jostling, often resulting in calls as bat interact with their neighbours and compete for space. For much of the year, a variety of agonistic and affiliative calls can be recorded (Altringham and Fenton 2003). Several studies have attempted to describe bat repertoires in the roost, including *Megaderma lyra*, *Murina leucogaster*, *Pteronotus parnellii*, *Rhinolophus ferrumequinum*, *R. clivosus*, and *Tadarida brasiliensis* (Leippert 1994, Leippert *et al.* 2000, Ma *et al.* 2006, Bohn *et al.* 2008, Clement and Kanwal 2012, H.-J. Lin *et al.* 2015, Peterson *et al.* 2019). These studies attempt to define vocalisations as related to specific behaviours, such as landing near another individual, rubbing faces in close contact with a conspecific, or boxing for a chosen spot (Peterson *et al.* 2019). However, it can be difficult to assign distinctive sounds to a discrete observed behaviour in some very vocal species (e.g. *R. ferrumequinum*).

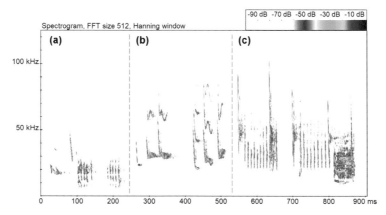

Figure 4.1 Examples of pre-emergence calls: (a) *Pipistrellus nathusii* (Jon Russ), (b) *Eptesicus serotinus* (Sally-Ann Hurry), (c) *Myotis nattereri* (Marc Van De Sijpe and Alex Lefevre).

Geoffroy's bat, *Myotis emarginatus* © Klaus Echle

4.3.2 Contact calls

Contact calls allow individuals out of sight from one other to exchange information on individual, sex or group identity, as well as information regarding the location of the signaller and a nearby resource (Kondo and Watanabe 2009). Contact calls can be recorded in foraging areas and often within or near a roost. Contact calling may help individuals target roosts. Pallid bats *Antrozous pallidus*, for example, produce calls in flight when they return to their rock-crevice roosts after foraging (Vaughan and O'Shea 1976). Playbacks of these low-frequency, multisyllabic calls attract investigative flights and response calls (Arnold and Wilkinson 2011). Additionally, individuals are more responsive to calls of their groupmates, suggesting that these calls allow individuals of this highly roost-switching species to maintain group cohesion (Arnold and Wilkinson 2011). Similar behaviour has been investigated in multiple species of leaf-roosting bats, including Spix's disc-winged bat *Thyroptera tricolor* (Chaverri *et al.* 2010, Gillam *et al.* 2013). Bats produce individualised 'inquiry' calls on the wing near a roost, and enter after receiving 'response' calls (Chaverri *et al.* 2010, Gillam and Chaverri 2012). This species uses highly ephemeral, but sparse roosts, and thus contact calling may be an energy-saving strategy in targeting roost locations (Chaverri *et al.* 2010). Contact calling has been observed to facilitate roost convergence in Bechstein's bats *Myotis bechsteinii*, common noctules *Nyctalus noctula*, and Natterer's bats *M. nattereri* (Schöner *et al.* 2010, Furmankiewicz *et al.* 2011).

A remarkable use of contact calling between roost mates is in vampire bats, where it facilitates cooperative behaviour. Vampire bats engage in food-sharing behaviour, regurgitating blood to unsuccessful foragers. Because blood is nutrient-poor, missing a blood meal would have rapid dire costs (Carter *et al.* 2008, 2012, 2017, Carter and Wilkinson 2013). Contact calls of vampire bats vary from one individual to another. Common vampire bats *Desmodus rotundus* have been shown to respond more often to the calls of recent food donors, regardless of relatedness (Carter and Wilkinson 2016). Thus, reciprocity is generally more important in this specialised cooperative behaviour then how closely individuals are related.

Bechstein's bat *Myotis bechsteinii* © René Janssen

4.3.3 Distress calls

Many researchers note that bats captured in nets and held in the hand emit noisy calls. These are known as distress calls, and they have been observed to attract individuals in multiple species, suggesting that mobbing a predator may be an important function of these calls. Distress calls can be quite similar in spectrotemporal structure, as easily seen in the pipistrelle species of Britain (Figure 4.2). They are multisyllabic, consisting of short-duration, steep frequency-modulated sweeps. Accordingly, responses to the calls are not species-specific, unlike responses to courtship calls. *Pipistrellus pygmaeus*, *P. nathusii* and *P. pipistrellus* respond to the distress calls of each other (Russ *et al.* 2004). Similarly, recordings collected from hand-held individuals of 11 species from four different families (Vespertilionidae, Rhinolophidae, Hipposideridae and Miniopteridae) revealed convergence in distress-call structure: noisy, broadband, long-duration calls (Huang *et al.* 2018).

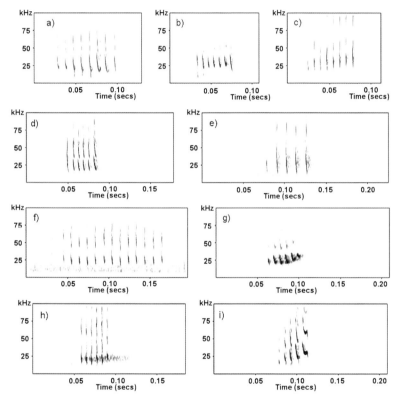

Figure 4.2   Examples of structurally similar distress calls of European bats: (a) *Myotis bechsteinii*, (b) *M. brandtii*, (c) *M. daubentonii*, (d) *M. mystacinus*, (e) *M. nattereri*, (f) *Plecotus auritus*, (g) *Pipistrellus nathusii*, (h) *P. pipistrellus*, (i) *P. pygmaeus* (Jon Russ).

Testing the mobbing hypothesis of distress calls is difficult. Actual observations of bats mobbing a predator are rare and have been opportunistic, such as *Phyllostomus hastatus* mobbing a large spectacled owl *Pulsatrix perspicillata* near their roost (Knörnschild and Tschapka 2012), *Taphozous nudiventris* attacking a barn owl *Tyto alba* (Lučan and Šálek 2013), and several *Nyctalus noctula* harassing a young peregrine falcon *Falco peregrinus* during falconry training (Sedláček and Kolomaznik 2015). A playback experiment of velvety free-tailed bat *Molossus molossus* calls found that individuals flew past, but against the expectations under the mobbing hypothesis, individuals seldom approached the playback source (Carter *et al.* 2015). This suggests that responses to distress calls may be an information-gathering behaviour about potential predators rather than mobbing. However, there may be additional factors needed for bats to have heightened motivation to engage in mobbing, such as visual confirmation of a predator, or a certain location of the distress calls (Eckenweber and Knörnschild 2016).

Noctule *Nyctalus noctula* in the hand © Harry J. Lehto

4.3.4 Mating signals: courtship and mate defence

Bats are a fascinating group in which to study courtship behaviours. They have diverse mating systems in this large family with a variety of multimodal displays, including body movements, vocalisations, pheromones from scent glands and visual ornaments (e.g. fur patches) (McCracken and Bradbury 2000, Altringham and Fenton 2003). Males, the dominant vocalisers in the courtship context, can be heard producing advertisement calls (e.g. *Myotis myotis* (Zahn and Dippel 1997), *Nyctalus leisleri* (von Helversen and von Helversen 1994)), song-like calls (e.g. *Pipistrellus pipistrellus*; Barlow and Jones 1997a, 1997b) (Figure 4.3) and even complex songs (e.g. *P. nathusii, Vespertilio murinus, Tadarida teniotis, Nyctalus noctula*; Weid 1994, Pfalzer and Kusch 2003, Zagmajster 2003, Jahelková *et al.* 2008, Balmori 2017b) (Figures 4.4 and 4.5).

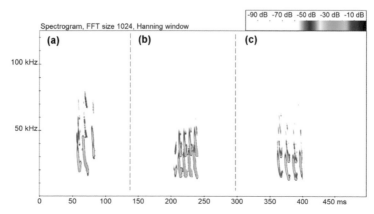

Figure 4.3   Mating calls of (a) *Pipistrellus pygmaeus* (Jon Russ), (b) *P. pipistrellus* (Jon Russ), (c) *P. kuhlii* (Danilo Russo).

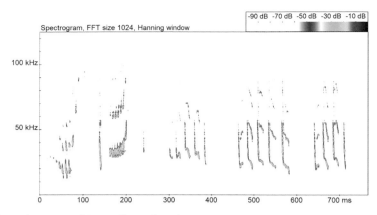

Figure 4.4   Courtship songs of *Pipistrellus nathusii* (Jon Russ).

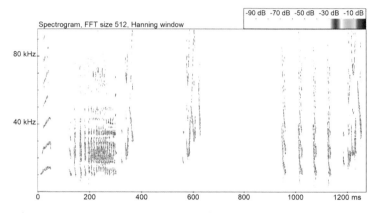

Figure 4.5   Part of the courtship song of *Nyctalus noctula* (Grace Smarsh).

Songs are typically considered to be vocalisations that are multisyllabic and arranged in a temporal pattern, produced in bouts in a courtship or territorial context (Catchpole and Slater 2008). European bats have a variety of polygynous mating systems that are seasonal, either involving multi-female/multi-male 'swarms', or dispersal of males to additional roosting locations to display to choosy females in autumn. These displays fall on a continuum between resource-defence polygyny and resource-independent lekking (Gerell and Lundberg 1985, Gerell-Lundberg and Gerell 1994, Weid 1994, McCracken and Bradbury 2000, Zagmajster 2003, Jahelková and Horáček 2011, Toth and Parsons 2013). Thus, depending on the species, males may be found spaced within the same roosts attracting females, in separate roosts, or on the wing. Many *Pipistrellus* species, for example, have a resource-defence polygynous mating system that involves song-flight behaviour. Males sing on the wing to attract females to a suitable day roost that they defend from other males. Those that sing the most attract more females (Gerell and Lundberg 1985, Lundberg and Gerell 1986, Gerell-Lundberg and Gerell 1994). Similarly, male *Vespertilio murinus* produce bouts of audible song flight near urban roosts from autumn into winter, and can be observed chasing others away (Rydell and Baagøe 1994, Ahlén and Baagøe 1999, Zagmajster 2003).

Bats may use one courtship call type or song type to advertise their location, identity, quality and motivation to mate. Others have multiple signals. For the greater sac-winged bat *Saccopteryx bilineata*, males attract and defend females in a harem, but there are frequently satellite males nearby in the roost (Fulmer and Knörnschild 2012). In this species, dominant males sing a male-directed territorial song and a female-directed courtship song that consists of different phrase types (Behr and von Helversen 2004, Behr *et al.* 2009). For the lekking lesser short-tailed bat *Mysticina tuberculata*, males sing long strings of variable song. Males have two different singing strategies for attracting females to their singing roost: smaller males occupy a roost alone but sing more, versus multiple males taking turns singing in the same roost. The latter strategy may attract more females at the cost of shared paternity (Toth *et al.* 2015). For *Tadarida brasiliensis*, males sing songs spontaneously (undirected) with high variation in phrase order, but sing targeted songs (directed) with syntactically arranged phrases when another individual passes by the roost (Bohn *et al.* 2013).

4.3.5 Mother–pup communication

In maternity or mixed-sex colonies of bats of several families (Emballonuridae, Molossidae, Noctilionidae, Phyllostomidae, Pteropodidae, Rhinolophidae, Vespertilionidae), recordings of frequency-modulated calls rich with harmonics have been recorded from young pups when separated from their mothers, termed 'isolation calls' (Gould 1971, Gelfand and McCracken 1986, Scherrer and Wilkinson 1993, Parijs and Corkeron 2002, Bohn *et al.* 2007, Liu *et al.* 2007, Knörnschild and von Helversen 2008, Engler *et al.* 2017) (Figure 4.6). These vocalisations are a type of contact call, and behavioural studies have shown that they help a mother to target her offspring upon return to the roost from foraging, and that these calls are individualistic (Thomson *et al.* 1985, Gelfand and McCracken 1986, Jones *et al.* 1991, Rasmuson and Barclay 1992, Scherrer and Wilkinson 1993, Fanis and Jones 1995, Parijs and Corkeron 2002, Knörnschild *et al.* 2007, 2013). Mothers may also produce 'maternal directive calls' to the pups, and the pair may engage in back-and-forth antiphonal calling to facilitate reunion (Brown *et al.* 1983, Esser and Schmidt 1989, Balcombe 1990, Bohn *et al.* 2007, Knörnschild and von Helversen 2008, Knörnschild *et al.* 2013). Experiments examining pup perception of mother signals are limited, but have demonstrated that pups are responsive to mother directive calls and may recognise their mothers based on their communication calls, echolocation, and/or odour cues (Turner *et al.* 1972, Balcombe 1990, Balcombe and McCracken 1992, Jin *et al.* 2015). Mothers may use odour recognition in combination with isolation calls to target individual pups (Turner *et al.* 1972, Thomson *et al.* 1985, Gustin and McCracken 1987, Balcombe 1990, Fanis and Jones 1996).

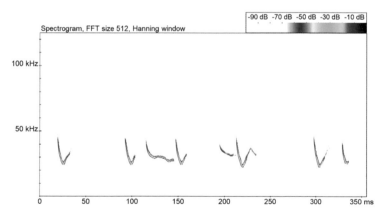

Figure 4.6   *Pipistrellus pipistrellus* infant isolation calls (Jon Russ).

Isolation calls can be precursors of adult communication calls (Knörnschild *et al.* 2006, Monroy *et al.* 2011, Prat *et al.* 2015) or echolocation (Gould 1971, Fanis and Jones 1995, Mayberry and Faure 2015, Mehdizadeh *et al.* 2018). Pups may produce intermediate calls during development, and even 'babble' as greater vocal-motor control develops (Jones *et al.* 1991, Knörnschild *et al.* 2006, Liu *et al.* 2007, Monroy *et al.* 2011, Mayberry and Faure 2015). Excitingly, vocal learning has been demonstrated in several bat species and remains an

Brown long-eared bat *Plecotus auritus* colony © René Janssen

intriguing area for more investigation to determine how widespread this complex behavioural trait is and which features pups may learn from adult exposure (Knörnschild 2014, Vernes 2017).

4.3.6 Cooperative foraging calls

Studies of how bats use communication signals to mediate social interactions during foraging are limited. Foraging guild and resource type are suspected to influence both the foraging strategy and the social-vocal signals we observe outside of the roost. Bats that forage for unpredictable food sources may benefit from foraging cooperatively as a group to target prey items, and thus may use signals to maintain cohesion of individuals during searches (Egert-Berg *et al.* 2018). Molossid bats may eavesdrop on the echolocation pulses of others while navigating the environment to more effectively target clouds of mobile insects (Dechmann *et al.* 2009, 2010). While this remains a promising area of research, the use of active communication rather than eavesdropping on echolocation to maintain foraging groups has been established in bats as well, as demonstrated by earlier work in greater spear-nosed bats *Phyllostomus hastatus* (Boughman 1997, Boughman and Wilkinson 1998). Female greater spear-nosed bats have stable social groups and use group-specific screech calls to maintain group cohesion while foraging. While vocal learning is usually associated with young animals, Boughman (1997, 1998) demonstrated that adults learn the screech calls of the group. Why some species use communication calls as opposed to echolocation cues to coordinate foraging likely involves an intersection of echolocation pulse type, prey preferences, morphological constraints and social selection.

4.3.7 Territorial and patch-defence signals

Conversely, territoriality can arise when a species consumes predictable prey items with a more even distribution (Egert-Berg *et al.* 2018). Megadermatid bats are probably the best examples of bats that vocally defend territories, either on the wing (*Lavia frons*; Vaughan and Vaughan 1986) or from various perches in the territory (*Cardioderma cor*; Vaughan 1976, McWilliam 1987, Smarsh and Smotherman 2015, 2017). Home-range studies have suggested territoriality in other species as well (e.g. *Macroglossus minimus*; Winkelmann *et al.* 2003), but further behavioural studies are needed to determine the social and vocal mechanisms of territory maintenance. At a local level, bats may engage in resource defence when competition levels are high enough at a food patch. Barlow and Jones (1997a) demonstrated the use of vocal defence of prey in *P. pipistrellus* through observation and playback experiments in the field. At low insect levels, individuals used more social calls, and when social calls were played, bat activity lowered (Barlow and Jones 1997b). Using a microphone array in the field, Götze *et al.* (2020) tracked flight paths of pipistrelles in foraging patches. These social calls were emitted by individuals before chasing and evicting an intruding bat, demonstrating their territorial function. Wright *et al.* (2013, 2014) found that flight social calls in *Eptesicus fuscus* are used in prey defence. When in competition for the same prey item, a competitor will produce a 'frequency-modulated bout', resulting in a change in flight behaviour of the opponent and successful prey capture for the emitter.

Studies in captivity and the field have provided great insight into the types of social interactions and the relevant vocal behaviours in which bats engage. More research is needed to understand how vocal communication signals, patchily observed and documented, tie in to foraging strategies and other behaviours outside the roost. With this information, we will be in a better position to understand the evolution of call use in bats across guilds and families, and the links between environment and behaviour, including responses to habitat modification.

4.4 Selective pressures and constraints on communication signals

Multiple selective pressures can drive and constrain variability in acoustic signals, affecting the spectral and temporal patterns of vocalisations we observe in our recording files. Echolocation shape is under strong natural selection to effectively perform specific sensory tasks to target prey items. Pulse frequency and temporal structure are thus evolutionarily constrained by prey type and size, foraging style and habitat (e.g. foraging guild) (Schnitzler and Kalko 2001, Siemers and Schnitzler 2004, Denzinger and Schnitzler 2013). Because the function of communication, however, is to transmit signals to other individuals for a range of purposes such as mating and competition for resources, such signals are subject to social and sexual selection as well as natural selection, giving rise to broad acoustic repertoires of varying spectral shapes and temporal patterns (Bradbury and Vehrencamp 2011). Multiple hypotheses seek to generalise spectrotemporal patterns of calls and repertoire sizes across species, families, taxa, and social or spatial context, and while these hypotheses can be predictive, there are always exceptions in highly speciose and diverse groups such as bats. A good review summarising selective constraints and drivers of acoustic signals can be found in Wilkins *et al.* (2013).

Communication signals are largely species-specific, ensuring the right receiver acquires the signal information. Species identity can often be inferred more readily from communication signals than from echolocation, and they can therefore be useful for detecting cryptic species in an area (Russo and Papadatou 2014). However, among related species, acoustic signals can have patterns based on genus or family (phylogenetic constraints). Pipistrelles, for example, are known to generally exhibit high similarity in echolocation and some similarity in social calls (see Figures 4.3 and 4.4) (Barlow and Jones 1997a, 1997b, Russo and Papadatou 2014). Despite similarities in their social vocalisations, playback experiments demonstrate the ability of species to discriminate each other, and thus the role of stabilising selection to maintain species-specific differences crucial for communication (Russo *et al.* 2009). Voigt-Heucke *et al.* (2016), for example, conducted call playbacks during the autumn mating period to multispecies swarms comprising *Pipistrellus pipistrellus*, *P. pygmaeus*, and *P. nathusii*. *P. pipistrellus* increased their social call rate in response to conspecific calls but not heterospecific social calls. Similarly, Russ and Racey (2007) showed that greater numbers of *P. nathusii* echolocation pulses were recorded during playback of male *P. nathusii* advertisement calls than during playback of congeners' advertisement calls or control sound.

These phylogenetic patterns can carry across large geographical regions, supporting the use of ancestral vocalisations by a common ancestor millions of years ago (Smotherman *et al.* 2016). *Tadarida brasiliensis*, for example, produces complex multiphrasic courtship songs and is found in the southern USA and Central and South America. Smotherman *et al.* (2016) observed that the little free-tailed bat in East Africa, *T. pumila*, also sings complex songs from baobab trees, as well as the European free-tailed bat *T. teniotis*, primarily found within rock crevices or buildings in Europe. Both *T. teniotis* and *T. brasiliensis* sing songs with multiple phrases consisting of trills, buzzes and downward harmonic syllables, while *T. pumila's* multiphrasic songs include syllables that are slightly more divergent in shape (Smotherman *et al.* 2016) (Figure 4.7). The extent of genus or family-level similarities in bats has yet to be quantified. This can be addressed with larger sets of recordings of different species, and a greater understanding of other factors influencing vocalisation variability, including the morphological differences across species (e.g. sound source (larynx), sound filter (e.g. nose/mouth shape) and neurological substrates), variation in mating systems and social organisation, and habitat effects (Wilkins *et al.* 2013).

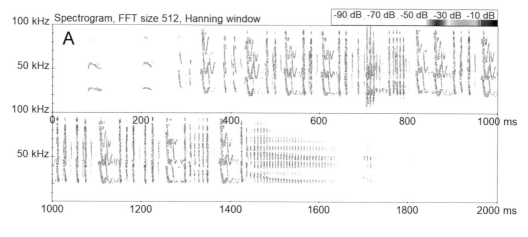

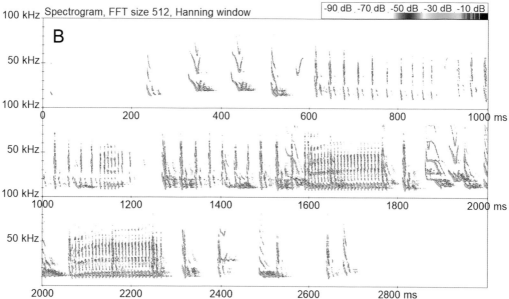

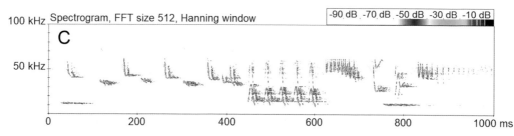

Figure 4.7 Courtship/territorial songs of (a) *Tadarida brasiliensis* recorded in Texas, USA (Michael Smotherman), (b) *T. teniotis* recorded in Portugal (Kirsten Bohn), and (c) *T. pumila* recorded in northern Tanzania (Grace Smarsh).

4.5 Environmental pressure

Environmental conditions and the distance to the target receiver can drive the use of certain types of vocalisations. Upon emission of sounds from the noseleaf or mouth, sound waves spread in a spherical manner, leading to a decrease in amplitude (Brenowitz 1986). This type of attenuation is frequency-independent, but attenuation from atmospheric conditions (high temperature or high humidity) is far stronger for higher than for lower frequencies (Griffin 1971, Lawrence and Simmons 1982). These principles form the basis of the sensory drive framework (and acoustic adaptation hypothesis), in which signals, signalling behaviour and signalling/receiving morphology should be evolutionarily intertwined such that the signal transmits best through a habitat to a receiver. Long-range communication signals, for example, should be longer and of lower frequency to reach another individual (Morton 1975, Endler 1992, Wilkins *et al.* 2013). High-frequency echolocation fails to be an effective communication signal in certain contexts, such as a pup calling to its mother in a noisy cave. Indeed, isolation calls are loud, low-frequency, and repetitive. During the courtship season, males in different species use vocalisations on the wing or perched to attract females to their roost (Smotherman *et al.* 2016). These advertisement songs are often loud and with low-frequency components, presumably to reach the ears of a passer-by. Zagmajster (2003) noted that the frequency-modulated syllable of *Vespertilio murinus* with a minimum frequency of approximately 10 kHz was easily recorded and analysed whereas the higher frequency syllables were not, and the congeneric pipistrelle social call motif was noted to be audible over 50 m away (Gerell-Lundberg and Gerell 1994). Knörnschild and colleagues estimated that the songs of *Saccopteryx bilineata* have a transmission distance of at least 120 m (Fulmer and Knörnschild 2012, Knörnschild *et al.* 2017). Territory-holding bats while foraging (e.g. megadermatid species) produce low-frequency songs and calls that can be heard 50–100 m or more away (Vaughan 1976, Vaughan and Vaughan 1986, Smarsh and Smotherman 2015). The frequency of communication vocalisations, however, may be ultimately constrained by the frequency of echolocation. Bohn *et al.* (2006) showed the correlation between peak frequency of echolocation pulses and peak frequency of the communication repertoire in bat species across families. This relationship might stem from sensory constraints in the auditory system of the bat. Her work suggests that we could reasonably predict how low-frequency the acoustic repertoire of a species might be, based on what we know about its echolocation pulses.

Less well understood is how more detailed aspects of the environment, such as forested or open areas, can influence bat vocal communication repertoires. The idea that certain syllable shapes transmit better through forest or fields has been investigated in bird song, suggesting that repetitive, tonal syllables with wide frequency bands would transmit better in reverberating, cluttered environments (Morton 1975, 1977, Wiley and Richards 1978). In a preliminary study, Russ (unpublished data) played back pure tones and sweeps of varying frequency and amplitude, as well as advertisement calls of *Pipistrellus nathusii, P. pipistrellus, P. pygmaeus, Nyctalus leisleri, N. noctula* and distress calls of *Barbastella barbastellus, Eptesicus serotinus, Myotis bechsteinii, M. brandtii, M. daubentonii, M. mystacinus, M. nattereri, P. nathusii, P. pipistrellus, P. pygmaeus, Plecotus auritus, Rhinolophus ferrumequinum* and *R. hipposideros* in woodland and grassland habitats. Attenuation of pure tones was higher in woodland than in grassland, and this effect was stronger for high-frequency tones, as expected (Figure 4.8). Degradation of pure tones and amplitude-modulated tones was similar across habitats, but frequency-modulated tones suffered the strongest degradation, especially in woodland (Figure 4.9). Downward sweeps with faster repetition rates (e.g. 1–2 ms) degraded more than lower repetition sweeps (e.g. 20 ms intervals), and degradation was overall stronger in woodland (Figure 4.10). In addition, distress calls, which contains lower frequencies, degrade less that advertisement calls which contain higher frequencies, with degradation again being higher in woodland (Figure 4.11).

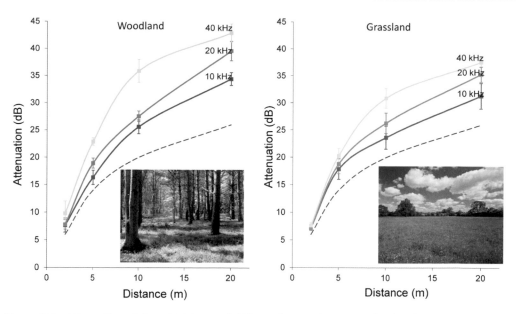

Figure 4.8   Attenuation of sinusoidal tones of different frequencies in woodland and over grassland. The dotted red line represents theoretical attenuation due to geometrical spreading alone. Error bars represent standard error. Overall attenuation was significantly higher in woodland than in grassland (ANOVA: F1, 120 = 9.091, $p < 0.05$) (J. Russ unpbl. data).

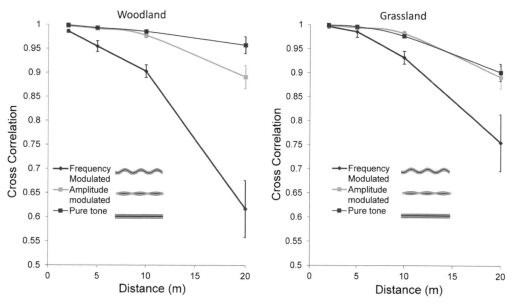

Figure 4.9   Comparison of cross-correlation values with distance for 20 kHz pure tone, amplitude-modulated and frequency-modulated pulses in woodland and over grassland. Error bars represent standard error. There was no difference between habitats except for frequency-modulated signals. There was a significant difference between the pure tone, amplitude-modulated and frequency-modulated tones (ANOVA: F2, 30 = 12.29, $p < 0.001$), post hoc tests – cross-correlation values were significantly lower for the frequency-modulated tones, no difference between pure tones and amplitude-modulated tones (J. Russ unpbl. data).

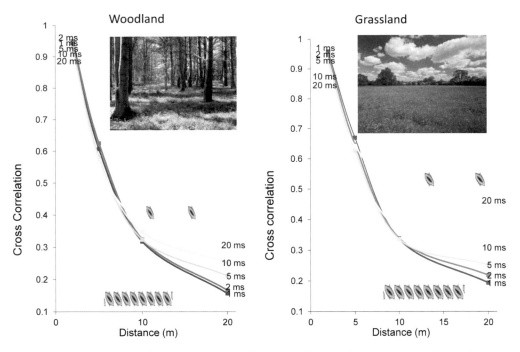

Figure 4.10   Comparison of cross-correlation values with distance for tones with different inter-pulse intervals (repetition rate) in woodland and over grassland. Error bars represent standard error. Increase in repetition rate results in higher degradation with distance. Call degradation is higher in woodland than in grassland (J. Russ unpbl. data).

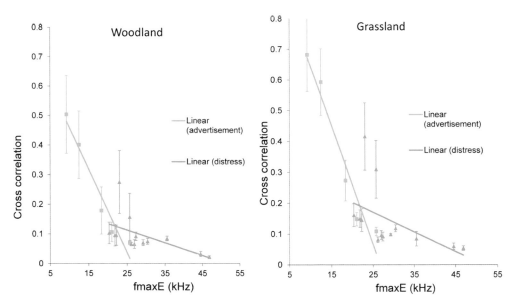

Figure 4.11   The relationship between cross-correlation values and the frequency containing maximum energy (FmaxE) for 'advertisement calls' and 'distress calls' in woodland and over grassland. Error bars represent standard error. Advertisement calls degrade at a much higher rate than distress calls in both habitats, with degradation being higher in woodland, owing to the higher frequencies contained in advertisement calls (J. Russ unpbl. data).

These results suggest that in cluttered habitats, long-range signals to reach individuals at a longer distance should not have very complex frequency modulations, nor consist of calls with high syllable repetition rate. Further field-based studies with multiple species and populations in mixed habitats can continue to test these predictions, including interactive playback studies to determine at what degradation threshold receivers respond to the signals. Additionally, while there is some recent evidence supporting the influence of climatic variables on echolocation frequency in rhinolophids (Mutumi *et al.* 2016, Maluleke *et al.* 2017), the relationship between temperature and humidity gradients on social-call evolution has yet to be demonstrated.

4.6 Energetic constraints

The energetic demands of communication can be an additional constraint on signalling behaviours (Gillooly and Ophir 2010). While acoustic signalling is generally considered to be energetically costly in animals, the costs of communication in bats is unclear (Smotherman *et al.* 2016). Evaluating the cost of signalling independent of contextual behaviours, such as flight, aggression or courtship displays, demands careful, targeted studies. Echolocation is not considered costly, because it is powered by the mechanical movement of the wings in flight (Speakman and Racey 1991, Voigt and Lewanzik 2012). Communication calls have also been observed on the wing in multiple types of bats, often produced within a sequence of echolocation pulses (Pfalzer and Kusch 2003).

In-flight calls have not been well studied, but they may facilitate a variety of social behaviours: contact calls to facilitate night-time social interaction between roost mates (Chaverri *et al.* 2010), between adult and juveniles during the learning phase of foraging (Wickler and Uhrig 1969), or to maintain contact among migrating bats (Reyes 2013); for courtship purposes (Barlow and Jones 1997a); or for foraging cooperation or competition (Springall *et al.* 2019). Depending upon the timing of emission, these communication calls may also benefit from the same energy-saving mechanism as sonar. To take advantage of the mechanism, we would expect social calls to be short, and produced between wingbeats. For species that have been noted to use the same call perched and in flight, we may expect to see modifications of the call in flight. *Pipistrellus nathusii* males, for example, produce complex advertisement songs composed of pipistrelle multisyllabic motifs, steep FM notes, trills and curved syllables (Jahelková *et al.* 2008). This species produces these songs both perched and on the wing, with most displays occurring while bats are sedentary at individual male roosts. Jahelková *et al.* (2008) found differences in songs by context: *P. nathusii* songs have a basic ABC motif structure, but perched bats produced the most complicated songs with additional D and E motifs, and flight songs were often fragmented, with one or two motifs. These observations could stem from the energetic or biomechanical constraint of flight song, or possibly from the male saving his 'fanciest' songs for a female approaching the roost entrance.

4.7 Signal variability and stability: acoustic signatures versus motivational cues

Beyond driving species-specific differences in communication signals, sexual and social selection can result in intraspecific variation. These types of selection drive acoustic repertoires with signatures, acoustic parameters varying between individuals in such a way as to give stable information to the receiver (Bradbury and Vehrencamp 2011). Because bats are long-lived and highly social, social selection likely has strong effects in many bat social-vocal behaviours. Acoustic signatures in bat communication include individual identity and group identity, and certain call parameters also include information on age and sex (Altringham and Fenton 2003). These signatures prevent costly mistakes during social interactions.

Individual identity plays a role in male advertisement during courtship displays, such that choosy females recognise preferred males. Age and sex information in vocal signals also helps individuals target appropriate mates, rather than wasting time, attention and energy on a juvenile or same-sex individual during courtship (Bradbury and Vehrencamp 2011). In aggressive contexts, such as within a territory network, the ability to discriminate and recognise individuals can prevent costly physical interactions (Tibbetts and Dale 2007). *Tadarida teniotis* recognise familiar individuals, as well as their age and sex, and accordingly display the appropriate levels of aggression in close interactions within their colonies (Ancillotto and Russo 2014). *Pipistrellus nathusii* male songs are individualistic and likely aid in individual identification for females (Russ and Racey 2007). *Saccopteryx bilineata* males engage in territorial counter-singing at roost sites, where multiple males may defend harems. Similarly, their songs encode individual signatures in the buzz phrase of the songs, as well as group signature (Eckenweber and Knörnschild 2013). These group signatures may be due to genetic closeness within the roost or from young males learning from harem males, and potentially allow related individuals to cooperatively defend a harem from stranger males. Contact calls have been shown to encode individual or group signatures for many types of bats, as discussed above. An interesting analysis by Prat *et al.* (2016) found that not only did colony social calls of *Rousettus aegyptiacus* encode behavioural context and signaller identity, they also encoded information on the intended receiver.

In addition to information content, vocalisations often contain emotional or motivational cues. Greater motivation in animals can be indicated by increasing call rate or amplitude, switching call types, or switching song type. Additionally, signals can be graded, with shifts in temporal patterns (syllable duration, inter-syllable interval), spectral shape and frequency, or composition (e.g. syllable type in complex songs) in heightened motivational situations while escalating other behaviours, such as in greater proximity to a mate or rival (Catchpole and Slater 2008, Bradbury and Vehrencamp 2011). An animal may call more in agonistic interactions with conspecifics, showing heightened aggressive motivation, before physically engaging. *Myotis myotis* switch from sweeps to noisy squawks during an escalation of aggression, and these squawks shift in frequency and tonality as well (Walter and Schnitzler 2019). Escalation of aggression between roost mates in *Hipposideros armiger* involves changes in visual and acoustic displays. Individuals switch from keeping their wings bent and baring their teeth to wing flapping and punching. These changes were accompanied by 'bent-upward frequency-modulated' calls that are lower in frequency and larger in bandwidth (Sun *et al.* 2018) (Figure 4.12). Heart-nosed bats *Cardioderma cor* sing lower-frequency, faster songs in response to perceived intruders on their foraging territories (Smarsh and Smotherman 2017).

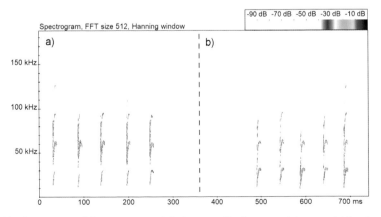

Figure 4.12 The bent-upward frequency-modulation vocalisations used by great Himalayan leaf-nosed bats *Hipposideros armiger* to defend their roosting territories, showing the transfer of energy from the second harmonic to the first harmonic with increased threat: (a) lower aggressive motivation; (b) higher aggressive motivation (Congnan Sun).

Observed variability of a call within individuals can thus be a result of changes in the social context (e.g. proximity of a conspecific), providing the basis for further investigation of the meaning and use of the signal through observation and playback experiments. Mayberry and Faure (2015) demonstrated that 13-day-old *Eptesicus fuscus* pups called more when isolated, but when provoked also produced calls that were longer, of larger bandwidths, and with more harmonics, similar to the calls of younger pups. When another individual passes by, male *Pipistrellus nathusii* alter their songs by increasing repetition rate of simple motifs in flight or increase the number of complex motifs while perching at the roost (Jahelková *et al.* 2008).

Morton's 'motivation-structural' (MS) rules (1977) suggest that the type of social-emotional behaviour a vocalisation is associated with is correlated with spectrotemporal parameters of the signal. The MS code describes a convergence of call structure in birds and mammals. Individuals should use broadband, noisy, low-frequency sounds in agonistic social situations, but tonal, higher-pitched (essentially, 'nicer') sounds when appeasing, fearful or friendly (Morton 1977, August and Anderson 1987). There has been some evidence of MS patterns in bats, including the *Myotis myotis* aggressive squawks, and the social-vocal repertoires of *Pteronotus parnellii*, *Murina leucogaster* and *Tadarida brasiliensis* (Bohn *et al.* 2008, Clement and Kanwal 2012, H.-J. Lin *et al.* 2015, Walter and Schnitzler 2019). Additionally, Luo *et al.* (2017) found that agonistic calls recorded from 31 species of bats were mostly noisy, low-frequency calls in accordance with MS rules. Russ *et al.* (2004) found that distress calls across pipistrelle species also supported the motivation-structural hypothesis. MS rules can thus provide insight into the possible functions of newly recorded social calls.

Morton's motivation-structural rules correlate with probably one of the best attempts to investigate patterns of communication signals across bat species. Pfalzer and Kusch (2003) recorded as many social calls as possible from 16 different vespertilionid species in Europe, both within and outside the roost. They found that they could statistically classify the calls into four general categories based on the spectral characteristics of the signals: type A (squawk), type B (repeated note, trill), type C (curved, cheep) and type D (complex, song). Interestingly, Pfalzer and Kusch (2003) also found that the calls tended to be used in certain social and spatial contexts: they hypothesised that type A is used in agonistic/combative interactions between individuals, observed in the roost; type C is used as an isolation call by pups in maternity roosts; type D is employed in mating contexts in which males compete for females, and also possibly for resource defence outside of the roost; and type B is used largely when in distress. Not all species had a call of each type, and specific studies of the social behaviour surrounding the different calls are still needed. Thus, when recording and describing 'new' vocalisations for a species, these categories must be applied with careful consideration. Some bats, such as *Rhinolophus ferrumequinum*, have many syllable types that cannot be easily categorised (Ma *et al.* 2006).

4.8 Conclusion

Bats remain an exciting group from which to learn more about mammalian social-vocal behaviour, and with improving technology, this is a promising area for discovery. We are still at the beginning of understanding the speech-like attributes in bat repertoires, including prosody (rhythm) and syntax, the intricacies of learning different sounds at different ages in variable social and natural environments, the influence of personality on repertoire use, and how females assess male displays. For the ecologist or wildlife specialist perusing this book, we hope that this information sparks some interest in the vast array of bat communication and social behaviour, and leads to closer consideration of the variety of calls observable in recordings.

Having a better understanding of the social calls we record or even hear by ear opens a window to useful information, including what species is nearby, the identity of the signaller and which receivers may be nearby, and what behaviours are occurring under the cover of darkness. In a still under-studied group of mammals, these behaviours provide clues into many aspects of a bat's life history, such as the timing and location of reproductive patterns, and provide further enlightenment on movement ecology and resource availability. For example, the presence of certain calls can signal the onset of a courtship period that correlates with the movement of individuals to additional roosts (e.g. *Nyctalus noctula*), the presence of a reproductive population by recordings of pup calls, the importance of particular trees associated with repeated use for social behaviours (e. g. *N. leisleri*), or decreasing food sources indicated by higher rates of competition calls (e.g. *Pipistrellus pipistrellus/Eptesicus fuscus*). Ongoing anthropogenic disturbance to bat habitats may have effects on bat fitness if the disturbance results in modification of areas needed for crucial social events in a bat's life, such as removal of roosts used for courtship displaying. The population-wide consequences of impeded social interactions from habitat disruption are not yet understood in the wide world of bats.

5 Equipment

Philip Briggs, Arjan Boonman, Jon Russ, Martijn Boonman, Jeremy Froidevaux and Kate Barlow

5.1 A history of bat-detector research

In 1938 in a laboratory at Harvard University Donald Griffin used a Pierce Sonic detector (Noyes and Pierce 1937) to listen to the echolocation calls of *Eptesicus fuscus* and *Myotis lucifugus*. This was the first detector ever to be used for such a purpose. Later, in the 1950s, Griffin took a self-made heterodyne bat detector to parks in New York to look for good locations to study the natural hunting behaviour of bats. For the actual studies on bat sonar, a high-speed camera filmed the oscilloscope images the echolocation sounds generated. The invention of the much more sensitive solid dielectric condenser microphone (Kuhl *et al.* 1954) was key to recording the sounds of bats more faithfully, especially outdoors. In the 1950s Franz Peter Moehres and Erwin Kulzer studied the echolocation of horseshoe bats (Moehres 1952) and tested the echolocation abilities of several other bat species (Moehres and Kulzer 1955). However, the main pioneer in the 1950s and 1960s was Alvin Novick. He first worked with Donald Griffin in Panama, recording fish-eating bats from canoes in the Chagres River using power generators on the river banks and filming oscilloscope images in the canoe (Griffin and Novick 1955). Novick went on to record bats throughout the 1950s in Mexico, Sri Lanka, the Philippines and the Belgian Congo (Novick 1958, 1963). Since batteries at this time were not yet up to the task, generators were used, or when possible, local power lines were tapped to power the oscilloscope and camera. In the early 1960s bats could be recorded with high-speed (152 cm·s^{-1}) tape recorders operated by batteries. The recordings could be analysed in the laboratory with a sonograph reading the slowed-down recordings. This advance made it much easier to record echolocation sounds.

By 1968 the echolocation of 54 bat species had been described by Alvin Novick, while David Pye added a further 54 species to the list of knowns, including a number of European bats (Pye 1968). By this time, concepts such as pulse compression and Doppler shift detection had been established through radar research, helping scientists such as James Simmons and Hans-Ulrich Schnitzler to interpret and study echolocation signals of bats. Still, progress in recording and cataloguing the sonar of bats throughout the world was slow and in the hands of a few scientists who had the technological means. Pye (1980) noted that about 200 of the known 650 echolocating bats in the world had been recorded. Brock Fenton and Gary Bell (1981), after recording much of the North American bat fauna, started to work on the concept of species recognition by sonar. Already back then, it was recognised that within-species variability of echolocation is highly task-dependent, which could render species recognition a difficult task.

The history of echolocation research always hinged on technological advances. Lazaro Spallanzani and Louis Jurine in the 1790s realised that the ear is key to the bat's ability to fly in complete darkness. How tantalisingly close they had come to discovering echolocation only became clear when the zoologist Sven Dijkgraaf (who discovered echolocation independently of Griffin during World War 2) went to a library in Reggio, Italy, to translate the old manuscripts he could find there. It turned out that, to repeat Jurine's findings, Spallanzani had put closable tubes in the ears of bats which clearly impeded performance only when closed, a key experiment to be repeated by Griffin and Robert Galambos 150 years later in Harvard. Spallanzani even reported on hearing 'the sounds of the bat's wing' when conducting a wire-avoidance experiment in a dark room (Dijkgraaf 1960). In reality, he probably heard the pressure envelope of the ultrasonic echolocation pulses, which had prompted Dijkgraaf to discover echolocation. Whereas Dijkgraaf and Griffin lived at a time when ultrasound and radar had been discovered, Spallanzani would have had to speculate about the existence of an concept unknown to physics to solve the mystery at hand. Although the eventual discovery of echolocation in bats was rejected in disbelief by some, the idea certainly did not come out of the blue. Not only had the physiologist Hamilton Hartridge (1920) suggested that bats use ultrasonic echolocation, in 1924 the Westinghouse Electric and Manufacturing Company had already built an ultrasound detector for biologist Frank Lutz to test the idea that insects communicate using ultrasound. Unfortunately, this ultrasound detector was not good enough, postponing the discovery of animal ultrasound to the late 1930s when the Pierce Sonic detector had been developed.

The much better mylar type of condenser microphones (the type Novick used on his worldwide expeditions) required appropriate plastics to be invented and applied, which did not take place until the 1950s. The large number of bat species that remained unrecorded and the huge swaths of the planet remaining acoustically unsurveyed was finally about to be remedied with the larger-scale production of cheap bat detectors, such as the QMC-mini (Queen Mary College, late 1970s) and the Pettersson D90 and D95 in the 1980s. Ingemar Ahlén (1981), Lee Miller and Jørgen Degn (1981), and Roland Weid and Otto von Helversen (1987) published descriptions of the European bat fauna and suggested survey methods. The availability of cheap detectors heralded a new era of amateur bat workers entering the field of acoustic surveying, as had been done for birds for many decades. A sound cassette with noisy frequency-division recordings of European bat species, each announced in Latin in Ahlén's Swedish accent, became an instant hit. By 1986, hundreds of volunteers learned how to use the new cheap heterodyne bat detectors to survey the whole of the Netherlands on a 5×5 km basis. The first European Bat Detector Workshop was organised in 1991 to instruct bat workers from all over Europe how to use bat detectors in field surveys. In the following years, more studies on species recognition were published, and enormous parts of Europe and the rest of the world have now been surveyed acoustically for bats.

In 1991, the first bat detector based on digitally sampling the microphone signal appeared (Pettersson D980), and this technique has become the method of choice. These digital recordings would be slowed down to be recorded on a cassette recorder in analogue form, to be digitised again on a computer at home. It was in the same year that Sandisk produced its first solid-state flash memory cards costing €40 per Mbyte, or €400 to store 17 seconds of ultrasound (currently €0.001 per Mbyte). Therefore, all the way from the 1960s into the 1990s, magnetic tape remained the main medium on which information was stored. Even digital audio tape (DAT) recorders first used videotape and later small cassettes to store digital information on. The 1990s saw the introduction of the minidisc recorder (data compression and moving disc), which became cheaper by the late 1990s and was often used to record transformed echolocation sounds of bats. Advances in nanotechnology led to ever-cheaper solid-state memory recorders and eventually to bat detectors directly recording onto flashcards.

Over the last two decades, a lot of effort has been put into developing software to identify bats by their echolocation. Artificial neural networks and support-vector machine learning – their real birth being in the 1990s – have developed rapidly and have been made easier to use.

The steady evolution of memory cards has spurred the development of stationary recording bat detectors over the last decade. These detectors are sufficiently weatherproof and have enough battery power and memory to record autonomously in the field for several nights early on in their development, up to several weeks at present. The ultrasonic vocalisations of bats trigger these detectors to store not only the trigger call but also some preceding and additional time to pick up possible consecutive calls. The advent of this type of detector in 2007 has pushed the idea of transects more to the background and has ushered in an era of long continuous acoustic monitoring.

In terms of state-of-the-art technology in 2021, miniaturisation in recording ultrasound now allows us to put tiny 'bat detectors' on bats themselves to let them carry out acoustic 'surveys'. This is possible as recordings also reveal other bats besides recording the calls of the bat to which the tag is glued. Combined Global Positioning System (GPS) and acoustic data from flying bats now provide researchers with an unprecedented information flow about what bats do from second to second.

A limiting technological factor, however, is still to be found in the sensitivity of small microphones. A detector using a small microphone still results in a very limited detection range, which in turn makes surveys inefficient, requiring more nights of surveying. Large condenser microphones have been available for a while, and their range extends much further than that of others. However, their fragility and price make them unsuitable for leaving in the field overnight. Again, we are waiting for the next technological advance, the graphene microphone (Zhou *et al.* 2015), to push yet another boundary by providing us with small yet quite sensitive microphones.

Within the next few years the prices of sampling devices, microphones and memory are expected to decrease further, making it easier to work with large numbers of them simultaneously during surveys. The future of acoustic bat surveys may involve attaching many recording devices to trees and other features, coming back weeks later to download all the data, which will automatically produce a map showing bat activity related to weather, geospatial data and other features readily available on the internet. This, in turn, means that biosoftware engineers will determine how data will be combined and compared.

5.2 Bat detectors

The three main systems for converting ultrasound produced by bats into a sound that we can hear are heterodyne, frequency division and time expansion. In addition, full-spectrum sampling (also known as direct sampling) enables the recording of ultrasound at a high sampling rate without converting frequencies to the audible range. The last three are all 'broadband' systems that simultaneously sample all frequencies in the bat calls, which means that all bat calls can be sampled whatever the frequencies, and that recordings from these systems are suitable for sonographic analysis. This enables assessment of call structure and measurement of call parameters to varying degrees of precision depending on the bat-detector system used, which can help to confirm species identity. Choice of a bat detector will be influenced by factors such as budget, what the bat detector is to be used for (for example for use on bat walks, to identify bats as a hobby, or to make systematic recordings of all bat species encountered during walked transect surveys or at fixed monitoring sites over a long period), and whether the bat detector is to be used in the hand or left unattended at monitoring sites.

5.2.1 Heterodyne

Heterodyne bat detectors tend to be relatively cheap, and they provide the quickest method for identifying bat species based on their calls, since identification is carried out in the field while the bat is present. Carrying out species call identification in the field is less time-consuming, though potentially less precise, than making recordings from broadband detectors for later identification using sound-analysis software. With experience, a range of species can be identified with a reasonable degree of confidence using this system.

In a simple heterodyne system, ultrasound is picked up by the microphone and mixed with a signal from a tuneable oscillator in the detector which the user can adjust, normally by turning a dial on the detector. The sum and the difference of these two signals are outputted through the speaker. For example, if a sound is coming in at 50 kHz and the detector is tuned to 49 kHz, the resultant sound will be 50 – 49 kHz = 1 kHz and 50 + 49 = 99 kHz. We can hear the 1 kHz, but the 99 kHz is in the ultrasound range so we can ignore this. Paradoxically, in theory, this means that when the tuned frequency is precisely the same as the incoming frequency we would hear nothing, as the subtraction equals zero. However, this would only be the case with an incoming sound that remains at the frequency to which the detector is tuned, whereas most bat calls sweep through a range of frequencies so there will nearly always be an audible difference between the two signals. Also most detectors 'listen' to frequencies within a certain bandwidth around the tuned frequency, so there will always be some output from the detector if it is tuned close to or at the frequency of the sound being picked up by the microphone.

Upgrades to manual heterodyning have been introduced, first in the Pettersson D980, which employs an automated scanning heterodyne system. The system is also used in a few other bat detectors such as the Elekon AG Batlogger, where the detector displays the frequency of maximum intensity that was found. Heterodyning being a superior detection technique, it is surprising that this technique is not very commonly used.

The key bat identification features provided by heterodyne detectors are that the resultant sound can have distinctive tonal qualities such as 'warbles', 'ticks', 'smacks', 'chips', 'chops' etc. (Figure 5.1). These sounds are heard in real time, which enables differences in rhythm and repetition rate to be discerned, and the 'approximate' frequencies of the sound can be determined in the field, all of which can provide clues to species identity.

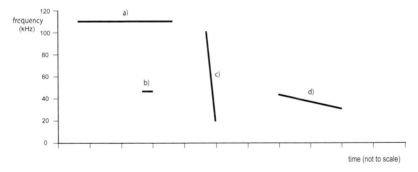

Figure 5.1   Diagrammatic representations of call types: (a) long CF signals ('warbles'); (b) short qCF signals ('slaps'); (c) steep FM signals ('ticks'); (d) shallow FM or qCF signals ('tocks').

When bats include CF components (horseshoe bats) or qCF components (pipistrelles, noctule, Leisler's bat, serotine) in their calls, this enables us to hear obvious changes in pitch as we tune around on the detector. In the example in Figure 5.2, as you tune closer to the qCF tail the sound from the speaker will become a deeper and more resonant 'smack', while tuning higher or lower will result in the sound becoming more high-pitched or 'tinny'. FM components (e.g. those largely used by *Myotis* species) tend to be too short in duration to enable the human ear to detect changes in pitch, so differences in tonal quality caused by tuning around will be less obvious. One technique for separating *Pipistrellus* from *Myotis* bats is to tune around to check for obvious changes in pitch. Pipistrelle calls will become richer in tone as you tune towards the qCF tail, whereas most *Myotis* calls lack a long qCF component and tend to sound like dry 'ticks' at whatever frequency you tune in to the calls.

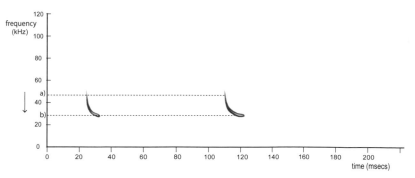

Figure 5.2   Tuning a heterodyne bat detector. At (a) the resultant sound is a 'tick', whereas as you tune towards (b) an increasingly deep 'smack' is heard. The 'smack' will sound deepest at (b) and get higher-pitched again as you continue tuning down below (b) before gradually fading away.

CF or qCF components contain the highest concentration of energy in the call, known as the peak frequency or frequency of maximum energy (FmaxE). Therefore the peak frequency can be determined by tuning to the frequency at which the sound from the detector is at its loudest and deepest. This is an important diagnostic characteristic used to separate species in groups that otherwise produce very similar calls, such as *Pipistrellus, Nyctalus* and *Rhinolophus*.

Of course, we are not just listening to one call. The bat is rapidly producing one call after another, so what we hear is a succession of ticks, smacks, chips or warbles, depending on the species and the tuned frequency. The speed at which these calls are emitted is known as the pulse repetition rate, and there are differences between species. For example, Natterer's bat has an extremely fast pulse repetition rate. In comparison, the repetition rate of the echolocation calls of noctule is very slow. Some species may have a very regular 'rhythm' to this repetition rate (e.g. Daubenton's bat) whereas others are erratic or irregular (e.g. soprano pipistrelle). Bats tend to couple wing beats with pulse emission, and therefore the rhythm at which bats flap their wings will strongly influence pulse repetition rate. In very open space they may skip pulses on wingbeat cycles. In its usual environment, a bat uses one pulse per wingbeat, in dense environment two pulses per wingbeat. In extremely dense environments the number of pulses per wingbeat will be much higher.

The repetition rate will vary depending on where and how the bat is flying, and it reaches a maximum when the bat homes in on an insect. Heterodyne bat detectors allow us to hear this: when the bat approaches and homes in on a prey item the pulse repetition rate increases dramatically and, on the detector, sounds like a rapid 'zzzzziippp' sound.

The ability to tune a heterodyne bat detector to different frequencies to learn more about the bat while it is still present makes for a rewarding experience, and visual clues (size, wing shape, flight pattern, behaviour, habitat etc.) can also be used to aid identification. These factors combine to make heterodyne bat detectors ideal for beginners, and they are an excellent tool for engaging people with bats on public bat walks. Heterodyne bat detectors also have a range of applications for professionals.

There are however several disadvantages with the heterodyne system. Identification is made by ear and therefore relies on the experience and abilities of the user, and recordings made from a heterodyne bat detector do not include useful frequency information that can be subsequently analysed, so it is not possible to verify records. This reduces the number of species that can be identified with confidence and makes identification more subjective. Furthermore, even easily recognisable species can elude identification if they fly by too briefly to enable the user to tune the heterodyne bat detector in to their calls. Finally, heterodyne bat detectors only output a narrow bandwidth of frequencies (typically around 3–5 kHz either side of the tuned frequency). Therefore it is not possible to survey for all species simultaneously and some bats may be missed.

The differences between some species heard on a heterodyne detector can be quite striking when listened to one after another for comparison, and there are several bat-call libraries available online, on CD and as smartphone apps that can be useful reference resources. However, it is not always possible to take these reference recordings into the field and it can be difficult for beginners to make judgements on the sounds they are hearing from their bat detector as a bat flies by. Therefore it helps to develop identification skills by committing a few reference sounds to memory before trying to use a bat detector in the field, and to keep referring back to these sounds. Pipistrelle calls are ideal for this purpose since they are frequently heard in the field and therefore easy to commit to memory, and some of the call characteristics are intermediate between other widely encountered species groups (e.g. *Myotis* species have a faster repetition rate and a drier tonal quality, while *Nyctalus/Eptesicus* species have a slower repetition rate and richer tonal quality) so make a useful reference against which to compare other sounds.

5.2.2 Frequency division (FD)

Frequency division has tended to be the cheapest of the 'broadband' systems that simultaneously monitor the full range of frequencies likely to be contained within bat calls. For this reason, it has long been popular with bat workers on a budget. However, the emergence of more competitively priced full-spectrum detectors which offer higher-resolution sampling is likely to cause a reduction in its popularity. FD detectors use a 'zero-crossing' circuit which produces a wave output with the same frequency as the fundamental of the incoming signal. The number of waves is counted, and for every 10 waves a simple wave of the same total duration is outputted (Figure 5.3).

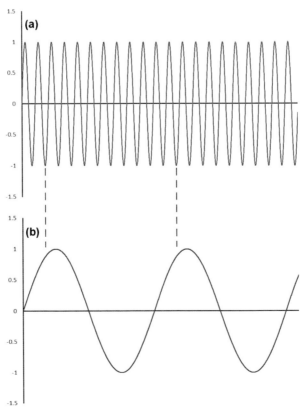

Figure 5.3   Diagrammatic representation of the frequency-division process showing (a) original sound wave and (b) wave outputted from a divide-by-10 frequency-division bat detector

Frequency division usually reduces frequencies by a factor of 10 (sometimes a different factor is used) and brings them within the audible range, but has no effect on time, so calls are heard in real time. Sufficient frequency information is preserved using this system to enable basic sonogram analysis: recordings can be recorded and analysed using software that processes the recordings to give us a visual image of the sound to represent frequency, time and amplitude. However, a lot of detail of the call structure is lost. For example, if the division factor is set to 10, it means that less than one-twentieth of the information is retained compared to a just sufficiently digitally sampled signal (two samples/period).

Moreover, only the loudest harmonic is preserved, usually the first harmonic or fundamental (other apparent harmonics are often seen in analysis of FD recordings, but these are artefacts of the way the calls are processed and should be ignored). In addition, quiet calls may not be detected, because the dynamic range is limited. One advantage that FD has over time expansion is that frequency-divided ultrasound is outputted in real time, allowing constant recordings, without the gaps inherent in the time-expansion system. Therefore, no bats will be missed that pass close enough to the bat detector for their calls to be picked up by the microphone. The data files of recordings from FD bat detectors are much smaller than from other broadband systems, making data storage easier.

The example sonograms given for each species described in Chapter 8 are from recordings made from time-expansion or full-spectrum bat detectors that retain the full call structure and detail. Sonograms from FD recordings provide a somewhat cruder representation of call shape as some of the detail of call structure is lost but, with practice and for some species, differences in structure can still be discerned. Example sonograms from FD detectors are shown in Figure 6.5 in Chapter 6, to help FD users relate their sonograms to those shown in Chapters 7 and 8.

5.2.3 Zero crossing (ZC)

Most FD detectors capture amplitude as well as frequency information, enabling frequency of maximum energy to be measured. Some detectors use zero crossing to generate pure data files (as opposed to the audio files generated by a conventional FD detector) which capture frequency but not amplitude information. For many years this has largely been the domain of the Anabat detectors produced by Titley Scientific, with their accompanying zero-crossing analysis software, Analook. Today, several models of full-spectrum bat detector also include the option to record in zero-crossing format. Software is available which converts full-spectrum files to zero crossing to reduce data storage capacity requirements and enable the files to be analysed using Analook software.

There are several advantages to this system. The economical approach to data capture (ZC files are tiny compared to audio files), combined with a triggering system that means that recording occurs only when ultrasound is detected, enables such detectors to be left unattended for very long periods before the CF or SD card fills up and needs changing. Also, analysis is relatively speedy, and call structure, though depicted as sequences of data points plotted on frequency/time axes, tends to be more clearly defined than with conventional FD systems. Since amplitude is not captured, the emphasis is more on analysing information on call structure and frequency parameters, though measurements analogous to FmaxE can be made. Zero-crossing analysis has also facilitated the measurement of slope (the rate at which a call component changes frequency over time, measured in octaves per second) which can provide an additional clue to narrowing down species identity within the *Myotis* genus. As zero-crossing analysis displays only the loudest part of a signal at any one time, this means that harmonics are often not visible (exceptions can include *Plecotus* species, which will often switch maximum energy from the first harmonic to the second harmonic part-way through emitting a call).

5.2.4 Time expansion (TE)

Along with full-spectrum sampling (see below) this method gives the most accurate repro-duction of bat calls. Basically, it stores the ultrasound signal digitally and replays it at a slower speed (usually ×10, but sometimes slower) so it can be recorded. The signal retains all the characteristics of the original signal (Figure 5.4). So we hear the entire call as it should sound except that it is 10 times lower in frequency and 10 times slower.

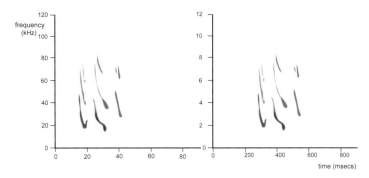

Figure 5.4   Ultrasonic call (left) and time-expanded call (right).

Time-expansion detectors also have a memory in which to store sound data temporarily. Some TE detectors can store up to 12 seconds of 'real' time, which when stretched out by a time expansion of 10 becomes 120 seconds of playback. Often TE detectors also include heterodyne and FD systems. Depending on the model of detector, signal capture can be triggered manually (by pressing a button) or automatically (when ultrasound is detected or at regular intervals). With most models, the time-expanded sounds will be played back in a continuous loop until this is interrupted by the triggering of signal capture. Some models allow pre-trigger data capture, which means that they are continually storing and discarding brief samples of sound. Manually hitting the capture button will play back the last stored sample of sound in TE mode. This enables you to record bat calls that occurred slightly before pressing the button.

As with FD and full-spectrum sampling (see below), a key advantage of TE is that, because no tuning is required, it is sampling all frequencies likely to be included in bat calls and therefore can simultaneously survey for all species. Because no information is discarded from the original calls, TE recordings provide excellent frequency information and produce much higher-quality sonograms than FD, provided of course that the recording device is set to a high enough sampling rate (see section 5.4.3). This allows a more accurate inter-pretation of call structure and measurement of call parameters from sonograms during sound analysis compared with recordings of calls from FD bat detectors. The disadvantage of TE is that during the period when the detector plays back the time-expanded sound, it is not capturing any new sounds. Thus with a time-expansion factor of 10, it may only be surveying for around 1 second out of every 11, and bats may be missed during these playback periods. Also, TE detectors tend to be more expensive than FD and heterodyne detectors. Newer models of bat detector have come on the market which provide TE functionality at a lower cost. One such model is the Batbox Baton XD (now discontinued), which uses a novel approach to TE that avoids the long gaps in sampling usually inherent in this system. It accomplishes this through a more 'intelligent' form of sampling where each sampled signal is played back in TE mode until another 'interesting' signal is detected; this means that a much higher proportion of call sequences from passing bats will be sampled either partially or in full.

5.2.5 Full-spectrum sampling

One of the biggest developments in bat-detector technology in recent years has been the growth in full-spectrum sampling (also known as direct sampling) as a standard way to carry out acoustic monitoring. These bat detectors record ultrasound in real time onto an internal high-speed data acquisition card. The increasing prevalence of this system has been facilitated by the availability of high-capacity, high-speed and low-cost CF and SD cards. The microphone captures sound at very high sample rates, thus enabling high-frequency sounds to be recorded directly. These enable the production of high-resolution spectrograms, as with TE, but also real-time continuous monitoring, as with FD, so you get the best features of both systems. One disadvantage is that the sounds outputted by the detector are not in the audible range so it is not usually possible to hear what you are recording in the field, though most models also include heterodyne, FD or TE systems, enabling you to listen to which species are around while you record their calls in full-spectrum mode. Some models are designed mainly for long-term unattended monitoring while others can also be used hand-held in the field and may display 'live' real-time sonograms. The high sampling rate means that the data files produced from these detectors are very large, and this needs to be considered both for recording and for storage. Most models can be set to trigger when they detect ultrasound above specified frequency and amplitude thresholds so that, when deployed remotely, wastage of data storage space on recordings without any bat calls is reduced. Most models also stamp files with the date and time as they are recorded, which assists subsequent analysis of large volumes of data. Full-spectrum detectors tend to be expensive, but cheaper models are now available, including the AudioMoth, which reduces this technology to the most basic components needed for passive recording and is among the cheapest bat detectors available.

5.2.6 Comparison between the output of detector types

A sequence of spectrograms showing frequency against time of the same sequence of bat calls recorded using each different system described above is shown in Figure 5.5.

5.3 Microphones, frequency ranges and detection ranges

5.3.1 Small, robust microphones

There are currently two types of commonly used small microphones (electret and MEMS), and there is another type coming soon (graphene microphone).

Electret microphones consist of a movable foil very close to a metal backplate. This unit is permanently charged and therefore has a certain capacitance. Sound waves hitting the foil cause it to move back and forth relative to the backplate, thus causing the output voltage to vary linearly with sound pressure. This voltage is the output of the microphone, and this is what the acquisition card will sample.

MEMS microphones (micro-electro-mechanical systems) are again based on the principle of a moving membrane and a backplate forming a capacitor. In this case, the membrane has tiny air holes while the whole unit consists of thin materials deposited on silicon wafers and etched away until extreme thinness. Charge is delivered externally. There are different MEMS designs, with various sizes of the so-called front chamber in front of the membrane. This front chamber works as a resonance chamber, amplifying ultrasonic frequencies. The microphone is integrated into the board that digitises the signal. The path from microphone to digitisers is therefore extremely short, meaning little signal loss. The initial digitisation is done with 1 bit. In this sense, the digital encoding of the system is like that of neurons (firing or not) with extremely short spiking intervals. It has to be taken into account that because the microphone air inlet is in a board of a certain size, the dimensions of this board will strongly influence the directional response of the microphone.

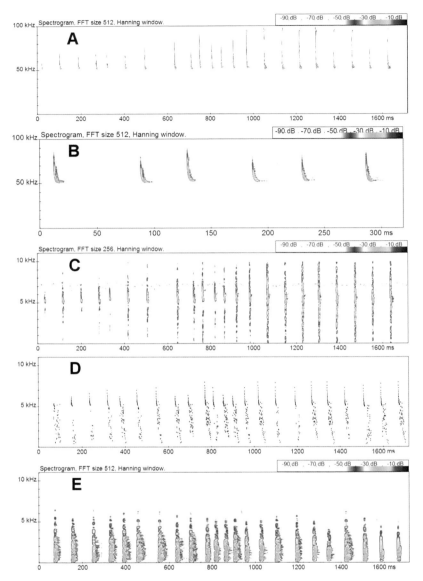

Figure 5.5   The same sequence of bat echolocation calls recorded using (A) a real-time ultrasound detector (high sample rate), (B) a time-expansion detector, (C) a frequency-division detector with amplitude retention, (D) a frequency-division detector without amplitude retention (zero crossing) and (E) a heterodyne detector. The grey shading shows the 'snapshot' of ultrasound recorded by the time-expansion detector.

The recent and not yet commercially available graphene microphone is made of a membrane consisting of graphene, which is nearly 'two-dimensional', consisting of a 20-nanometer layer of connected carbon atoms. Different from other microphones, in the prototype, the voltage of the membrane was kept constant while variations in current, caused by membrane motion, were picked up as the frequency of the sound. This means that the microphone did not measure displacement, but the velocity of the membrane. A bat detector with a 20-nanometer thick graphene microphone membrane has been made which was flat to within 10 dB from 20 Hz to 500 kHz. It was sensitive to even higher frequencies, which were however beyond the amplifier's reach (Zhou *et al.* 2015). The prototype still had a low absolute sensitivity (40 dB SPL), but many improvements can be expected. When being stretched, graphene is extremely strong. Its resilience to rupture due to perpendicular impacts depends on the perfection with which the atomic slice was made. All in all, miniaturisation holds a promising future for recording ultrasound, which requires ultra-thin membranes.

5.3.2 Typical bat-detection ranges of detectors with small microphones

If we take the Elekon Batcorder as an example, a *Rhinolophus hipposideros* emitting at 120 dB SPL will only trigger the detector within a distance of 5 m. A *Pipistrellus pipistrellus* emitting at 130 dB SPL will start to trigger the Batcorder within 18 m, *Eptesicus serotinus* within 38 m, and *Nyctalus noctula* at 130 dB SPL at 18 kHz at a distance of 60 m (but other detectors with small microphones at 70–80 m). At first sight, these distances seem shorter than expected, but we have to keep in mind that a beech forest is on average only 30 m high. Just picking up a noctule flying twice as high as the treetops seems reasonable. A heterodyne detector with an identical microphone would boost the range of low-frequency species, maybe to 80–90 m (with a Pettersson D240) in the case of *N. noctula*, but this means we can just hear a weak splash through the loudspeaker and we do not get an analysable recording. The Songmeter SM4 also performs better in the low frequencies, creating files for *N. noctula* up to 88 m and for the *E. serotinus* up to 47 m, but it is less sensitive than the Batcorder above 50 kHz. We must also take into account that the Songmeter is more 'trigger-happy' than the Batcorder, meaning a longer detection range at the cost of many false alarms. After processing all files and rejecting those which are not good enough for analysis, the detection ranges of detectors with small microphones will not differ much from each other, which is also logical because they use similar microphones.

5.3.3 Large condenser microphones

The condenser microphone works on the same principle as the electret microphone. The main difference here is that the charge is not permanent, but is delivered to the capacitor by an external source, becoming a very high voltage between membrane and backplate. Here the membrane is often plastic on one side and conductive on the other side, with the plastic side touching the backplate, which has many tiny grooves or holes. The pattern and size of the holes, the tension by which the membrane is stretched over the holes, the diameter of the capacitor (and of the surrounding area) and the material itself all affect the frequency response of the microphone. In general, in all microphones, the membrane is stretched tightly to ensure that the spring constant (which describes the stiffness of the membrane to movement) dominates all other effects. This is particularly important when recording ultrasonic sounds. The fact that the membrane diameter is large makes this type of microphone more sensitive than the smaller electret type. It also makes it more directional. In general, the potential for phase distortion when using large membranes is greater than with smaller membranes, but in the ultrasound, many small cheap electrets tend to have a highly fluctuating frequency response, so in practice this is not a disadvantage to worry about.

5.3.4 Typical bat-detection ranges of detectors with large condenser microphones

The numbers below apply to the Avisoft CM16 microphone and the trigger functions of Avisoft software. The results will be similar with other high-quality condenser microphones, such as on the D1000. When we use quite sensitive trigger settings on the Avisoft software, just above a very quiet noise level, a *Rhinolophus hipposideros* emitting at 120 dB SPL will trigger the system from a distance of 9 m, provided the microphone is perfectly aligned with the bat. A *Pipistrellus pipistrellus* emitting at 130 dB SPL will start to trigger the Avisoft within 32 m, *Eptesicus serotinus* within 57 m, and *Nyctalus noctula* at 130 dB SPL at 18 kHz at 93 m. We can see that with the much more sensitive condenser microphone, bat detection ranges extend much further than with the detectors using small microphones, particularly in the higher frequencies (Figure 5.6).

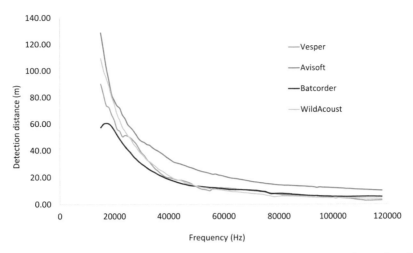

Figure 5.6 The maximum detection distances (the distance at which the signal can be analysed) of several detector types from 15 to 120 kHz, assuming an emission level of 130 dB SPL. Avisoft is a CM16 condenser microphone. Vesper is a miniature recording system to mount on bats using a MEMS microphone. The other two systems (Batcorder and Songmeter SM4) have small Knowles microphones. The shape of the curve is mainly dominated by the effects of atmospheric attenuation. The condenser microphone consistently detects bats at greater distances than all other types.

5.4 Recording sound

5.4.1 Recording formats

Digital recording devices may record to several different formats. The standard format is uncompressed, or lossless *.wav files which retain all information in the sound file but are also the largest file size for storage. Other file types compress the sound files in some way and are lossy, but the file sizes are generally smaller and therefore require less storage space. Lossy audio file types (*.mp3 is probably the most common; others include WMA, AAC and ATRAC) are designed to reduce storage space by discarding some of the information in the file, but in such a way that the effects on the human ear are imperceptible. However, because the information from these digital file types is discarded, there could be impacts on the details of bat calls that are stored using these methods and subsequently analysed using sound-analysis software. In the majority of cases, for example, if bats are being recorded for species identification from bat surveys, the effects are unlikely to be significant, as long as you are recording with the minimum sampling rate required for the calls of the species you are likely to detect (see Chapter 2). If, however, the aim of the study is to investigate bat acoustics in detail, it would be important to select equipment that would allow you to make the highest possible quality of recordings.

Different file types are coded in different ways, and this may affect compatibility with software. For example, *.wav files use the pulse-code modulation (PCM) codec, which is a generic format that can be read by most computers and software. Other file types such as *.mp3 use different codecs, and a separate piece of software might be required to convert files from one format to another for use with specific sound-analysis software. An internet search on 'audio converter' or similar will yield a range of downloadable sound-conversion programs, some of which are free of charge or have free trial periods. The following conversion settings are recommended for most sound-analysis software: 44.1 kHz, 16-bit, WAV PCM uncompressed. This recommendation applies only to recordings from bat detectors which convert ultrasound to the human audible range: FD and TE detectors. Full-spectrum detectors generate files with a far higher sampling rate recorded directly to an internal card; such files do not normally require conversion to open them using sound-analysis software, but should this be required then care should be taken not to reduce the sampling rate below twice the value of the maximum frequency of interest (see section 2.2.1, *Sampling rate*).

Data storage needs to be considered alongside the choice of recording device and file format. Different file formats will require different amounts of memory for storage, but for most audio files, external storage of files is likely to be necessary to avoid quickly filling up the hard drive of a computer or laptop. Options for file storage include external hard drives, servers or online services and cloud storage, each of which will have a cost attached.

5.4.2 Sound-recording devices

A variety of different devices are available for recording the output from bat detectors. These range from external plug-in devices usually recording in *.wav or *.mp3 format to built-in data cards which are found in many current models of bat detector. Recordings to external devices should be made using a cable (stereo if the output is stereo) with (usually) a 3.5 mm male jack to the output socket of the bat detector and a suitable connector to the line-in (where possible) input on the recording device. Factors to consider when selecting a recording device include the quality of recordings required, cost, ease of use in the field, storage type and storage capacity.

It is also possible to record directly from a bat detector to a laptop in the field, by using a high-speed data acquisition card to convert the analogue output to digital computer files. The bat detector is connected directly to the computer through the line-in socket, and output from the bat detector is recorded onto the laptop either into a sound file or fed directly into sound-analysis software to allow real-time viewing of spectrograms. At any stage when sound is being uploaded to a computer, it is always preferable to use the line-out on the device to the line-in on the computer for the best-quality recordings. It is possible to use the headphone socket, but this can result in a significant reduction in sound quality during the transfer. Bat detectors can also be connected to smartphones (see section 5.4.2.7).

5.4.2.1 Tape recorders

Tape recorders have become the old-fashioned way of recording calls from a bat detector, although some surveyors still use them. Good-quality tape recorders were required, as well as the best-quality tapes you could buy. The advantages were that, being mechanical, the tape recorders were long-lasting and reliable and worked well even in poor conditions. Disadvantages included the length of time it took to transfer recordings from a tape recorder onto the computer for analysis and search for specific sequences in the recordings. Tape recording of bat sounds has been made redundant by the wide availability of digital recording technology.

5.4.2.2 Digital audio tape recorders

DAT recorders allow uncompressed digital storage of sound files at high quality onto magnetic tapes which are smaller than standard portable tape. They are suitable for recording calls from bat detectors, but have not become widely popular, probably due to their high cost, and were quickly superseded by minidisc recorders.

5.4.2.3 Minidisc recorders

Minidisc recorders were for a time very popular for recording bat calls, as they were relatively cheap, compact, easier to get hold of than good-quality tape recorders, and normally enabled recordings of sufficiently high quality to be made from bat detectors. However, since they are not specifically designed for use in the field at night, some models can be fiddly to use on bat surveys. As with tape recorders, a disadvantage of conventional minidisc recorders is that the recordings need to be uploaded manually into a computer via an audio lead.

There was also Hi-MD minidisc, which was produced only by Sony. These recorders allow uploading of files directly to a computer via a USB lead from the device, thus avoiding the cumbersome process of manually recording the sounds into audio recording software. However, files from a Hi-MD recorder do need to be converted from the Sony format to *.wav format (using Sony's SonicStage software) before they can be opened in sound-analysis software. Hi-MD recorders tended to be more expensive than conventional minidisc recorders.

Minidisc recorders are no longer produced, and as a result tend to be quite expensive when offered for sale. They have been largely superseded by devices which record digitally to an internal data storage card, as these are widely available, can be relatively cheap, and make the process of transferring recordings to a computer considerably quicker.

5.4.2.4 MP3 recorders

There is currently a very wide range of these devices on the market, many of which are suitable for recording bat sounds. The advantage of these digital devices over minidisc is that sounds are recorded as files that can be quickly uploaded to a computer, usually via a USB lead. However, a key thing to bear in mind is that most models are designed to play music and may use high levels of compression to allow the storage of the maximum amount of music. This can result in poor reproduction of very-high-frequency sounds, which is not normally noticeable when listening to music but can be a disadvantage when recording bat calls. In common with minidisc recorders, since they are not designed with bat surveys in mind or even for use in demanding conditions outdoors, some models can be rather delicate or fiddly to use in the field at night. Therefore, if possible, it is worth getting recommendations from other users or viewing and handling a device before making a purchase.

There are two main types of MP3 recorders, hard drive and flash drive. Hard-drive devices usually have a large storage capacity, although flash-drive devices with smaller capacity are more common. Features to look for when selecting a device suitable for recording bat calls include good battery or charge life (for example, does it take external batteries or does it need to be recharged from a computer? – the former can be more convenient and longer-lasting); ease of use in the field (for example, is the display lit and the buttons accessible and large?); a line-in facility; the ability to make sufficiently high-quality recordings; as short a delay as possible in the device starting to record once the button is pressed (this can be quite common in MP3 recorders).

5.4.2.5 Solid-state digital recorders

These are the recording device of choice for many professional bat workers. They provide good-quality recordings to *.wav files (and often other file types), most record onto removable memory cards which can be replaced with different storage capacities, they take external batteries or have good battery life, they are generally robust and are often (though not always) well designed for use in the field. A range of these types of devices is available.

5.4.2.6 Built-in recording

Many bat detectors now available record directly onto memory cards built into the bat-detector unit. These have the advantage that only one device needs to be purchased, making use in the field less cumbersome and reducing the likelihood of user error, for example by incorrect linking of devices. There are different types of card including CF cards, SD cards and microSD cards, allowing storage of quite a range of file sizes. It is worth checking compatibility between card and device when purchasing additional cards. Cards can be slotted directly into a computer for transfer of files from the device, or a card reader can be used, normally connected to the computer using a USB cable. It is important to ensure that files are transferred and saved regularly, and labelled correctly and precisely to avoid later confusion and ensure that each file can be matched to a particular bat survey or night of data collection. However, many detectors automatically date- and time-stamp the recordings, often within the file name, making it easier to identify when and where the recordings were made, providing the date and time are set accurately before the survey and records are kept of survey dates and locations. Some detectors enable you to adjust the file naming settings so that the location is included in the file names. Several models of detector with built-in recording automatically display information about the calls as they are detected. This can range from displaying the FmaxE of each call to displaying live sonograms on the screen.

5.4.2.7 Smartphones

Bat detectors can be connected to smartphones for recording, though as with other recording devices it is important to ensure that the recording sample rate is high enough to accurately record the output from the detector. A variety of apps are available which provide extra functionality while making recordings. Features include tagging the audio files with geographic coordinates from the phone's built-in GPS module, displaying live sonograms and producing automated species identifications.

5.4.3 Sampling rate and bit resolution

See also section 2.2.1.

In time-expansion detectors, the microphone signal is sampled digitally at a very high rate, such as 350,000 times per second (350 kHz). Some detectors may do this at 8 bits or else at 12 bits. When using 8 bits it means that the intensity can be stored at 2^8 levels (voltages). If we only had 2 bits, we could only store at 4 levels: very weak, weak, strong and very strong, whereas with 8 bits we have 256 levels. The sampling rate of 350 kHz refers to how many times per second the intensity is measured. If you measure intensity 350,000 times per second, frequencies up to 175 kHz will be recorded correctly. Usually, a steep filter at half the sampling frequency (anti-aliasing filter) prevents higher frequencies from being recorded. In a strict classical sense, the higher one samples a signal, the more faithfully it will be recorded. A signal just below the upper limit would not be recorded very faithfully. Nowadays, however, sound is sampled using oversampling procedures. Oversampling means that we free up RAM by, for example, reducing 12 to 10 bits. This makes the amplitude resolution worse by a factor of 16, but the freed-up memory now allows 16 times the sampling frequency, so 16×350 kHz = 5.6 MHz. A sampling rate of 5.6 MHz means we have a fantastic time resolution and we don't have to worry about aliasing, but all at the expense of having a worse amplitude resolution. However, since we now track the outline of the sinus 16 times more accurately, but averaging this information eventually from 16 points to 1, this is like resampling our amplitude value 16 times, meaning a $1/\sqrt{16}$ improvement in the signal-to-noise ratio, which compensates exactly for our initial sacrifice of 2 bits. Oversampling works especially well if the signal of interest is a smooth sine function in fully random noise. Because of oversampling, we can assume that if we record a signal of 170 kHz with either 350 kHz or 500 kHz, the difference in quality would be negligible. 350 kHz is therefore fully sufficient, depending on the oversampling and interpolation the sound card uses.

An important consideration with any device used for recording bat sound is whether the sampling rate and bit resolution are high enough. The sampling rate should be twice that of the maximum frequency you expect to record. The number of bits is important in cases where one expects to record very weak, but also very strong signals that should all be analysable.

In some older detector types, the initial output is a slowed-down analogue version of the ultrasound, and we must again sample it in order for it to be recorded. For example, the FmaxE of *Rhinolophus hipposideros* at *c.*110 kHz. When this sound is output by a TE or FD detector with an expansion or division factor of 10, these frequencies will be reduced to around 11 kHz. A standard sampling rate of 22.05 kHz may be just about adequate (as 2×11 kHz = 22 kHz) but would not allow for fluctuations in frequency above 11 kHz. Therefore, the standard sampling rate needed to accurately record *R. hipposideros* calls with frequencies around 110 kHz would be 44.1 kHz (as many recorders offer fixed sampling rates of 11.025 kHz, 22.05 kHz, 44.1 kHz, 48 kHz, 96 kHz etc). Also, some species that produce FM calls can use a starting frequency well above 110 kHz (e.g. *Myotis nattereri* at around 145 kHz) so a sampling rate of 44.1 kHz is more than adequate to accurately record these frequencies when converted by a TE or FD bat detector.

Sampling frequency can also be expressed in bit-rate or kilobytes per second (kbps). It is determined by the sampling rate multiplied by the resolution and the number of channels, so the ideal bit-rate if recording a stereo signal (e.g. from a detector which outputs FD in one channel and heterodyne in the other) would be $44.1 \times 16 \times 2 = 1,411.2$ kbps. The maximum bit-rate available on MP3 recorders is usually 320 kbps, however. It is also worth being aware that different MP3 recorders use different codecs to compress sound files, and these can also affect the quality of recordings, so recordings from two different devices with the same bit-rate may not be identical. If the bit-rate is low, there is an increased risk of missing information, compression artefacts (sounds that were not in the original signal) and poorer signal-to-noise ratio. It is, therefore, best to set the bit-rate as high as possible and aim for a device that has bit-rates higher than 192 kbps, although, for most sound-recording require-ments, the effect on recording quality would not be sufficient to cause problems for species identification.

5.4.4 Recording levels

Whatever combination of bat detector and recording device is used, the quality of the recordings made will at least in part depend on setting the correct recording level. Where possible, choose a recording device that allows manual adjustment of the input level that can be seen easily when using the device in the field. Some cheaper recording devices may only have automatic gain control, which is less useful as it is best to adjust the input to maximise the signal-to-noise ratio of the bat calls and to maximise the use of the full dynamic range of the bat detector. When using a new recording device, it might take a bit of time to get used to the input levels and get the recording levels set correctly. Recording levels may also need to be adjusted during a survey, depending on where the recordings are being made, the distance to the bats being recorded, the level of background noise or other factors. The aim is to maximise the signal-to-noise ratio of the recorded calls while ensuring that recordings are not overloaded, which will result in clipped calls.

If recordings are made to a tape recorder or minidisc and subsequently need to be recorded onto the computer, it is also important to set the recording level during input of the calls. The recording levels can be adjusted on the computer, for example by adjusting the line-in settings when recording directly to a sound file or using sound-analysis software.

5.5 Automated identification of bat calls

Bat detectors that automatically record bat calls have been widely used for more than a decade. They are used for monitoring bat activity along transects or from fixed points such as (potential) roost sites, highway underpasses or the nacelles of wind turbines. A huge amount of data can be collected in a single night using these detectors. Analysing all these files manually is very time-consuming and monotonous work. To save time, automated identification of bat calls is possible by using several software packages. The consensus of most authors is that automated identification is not completely reliable (e.g. Lemen *et al.* 2015, Russo and Voigt 2016, Rydell *et al.* 2017). Without automated identification, however, long-term monitoring would be nearly impossible and many important questions relating to the ecology and protection of bats could not be answered within a reasonable time. The purpose of this section is to describe how automated identification of bat calls can be done responsibly. So how does automated identification generally work?

5.5.1 Triggering by the detector

The first selection stage in detecting bats of certain frequencies is performed in the field by the detector. Selection depends on a combination of microphone sensitivity, atmospheric attenuation and the 'trigger-happiness' of the detector (see Figure 5.6). The detector can also determine whether the sound is produced by a bat or something else, thereby potentially reducing false alarms and excessive data due to insects and car noise. Some detectors (e.g. Elekon and Batcorder) are only triggered by a sinusoidal frequency of a certain minimum duration above a threshold value, ensuring the recorded pulse will be loud enough to be analysed in a meaningful way.

5.5.2 First data-processing step

Within each recording, bat calls are detected by selecting high-energy 'events' of a certain duration (usually somewhere between 1 and 50 ms) within a predetermined frequency range. All calls are separated by splitting the sound recording into pieces so that several small spectrograms show only a single call. Filtering is applied to remove noise before extracting dozens of call parameters from the spectrogram. These call parameters (duration, bandwidth, shape etc.) are compared to those of spectrograms of calls from identified bats (library of calls). Some software packages (e.g. SonoChiro, BatIdent) use a random forest classification (Skowronski and Harris 2006). Identification using these packages takes place in several stages. A call is first classified as belonging to a certain class of bats, say '*Rhinolophus*' or the '*Myotis* group', and after this pre-selection stage call features are matched to a library of this group of bats (see section 5.5.4). The software then identifies the species the calls are most similar to. Usually, the software calculates a probability of a match for several species.

The following lists some of the more frequently used software packages currently available for automatically identifying European bat species.

- SonoChiro (Windows; Biotope, France)
- Kaleidoscope (Windows/Linux/MacOS; Wildlife Acoustics, USA, with a self-made library)
- BatClassify (Windows; University of Leeds, UK)
- Batscope (MacOS, Windows; WSL, Switzerland)
- BatExplorer (Windows; Elekon, Switzerland)
- BatIdent (MacOS; EcoObs, Germany)
- iBatsID (Europe) (Walters *et al.* 2012)

Some programs are designed around a detector (e.g. Batscope for Batlogger, BatIdent for Batcorder) and are probably more reliable when used in this combination. This is because detector microphones do not have a flat frequency response (or the way the microphone is embedded creates typical angular filtering), so a recording made by detector A side by side to detector B will not result in totally identical recordings. Many other (semi-automatic) software packages exist. In Analook, for example (Windows; Titley Scientific, Australia), zero-crossing files can be scanned by using self-made filters.

None of the automated identification programs is flawless (e.g. Russo and Voigt 2016), so visual validation of certain files is always necessary. For this reason, automated programs will only become useful when more than a few hundred files are to be analysed. It is also critical that these programs are employed efficiently and a margin of error is used that is appropriate for the study. Otherwise, the purpose of using these programs, to save time, is defeated.

5.5.3 Removing noise

Detectors are triggered not just by bat calls but also by other acoustic sources such as grasshoppers, rain, car brakes or phones (hereafter called noise). A major advantage of automated identification programs is to get rid of files that contain only noise. Although this is probably the most important step in the automated identification process, it is hard to find out how this is done by each program, and settings can usually not be adjusted. Most programs seem to use duration, energy and bandwidth to identify the short high-energy structured sounds of bats in the right frequency range (Andreassen *et al.* 2014). In zero-crossing files, 'smoothness' seems to be a key factor in filters, which determines to what extent separated points are lumped together. This process of eliminating noise is usually done quite well. Rydell *et al.* (2017) reported a 100% success rate in distinguishing rain from bat calls. Thousands of files resulting from an electrical loop (which can occur in wind turbines) or rain can be separated from bat calls in a single step, saving you days or even weeks of work. Bat calls may occasionally still be present in the files that were labelled as noise. These discarded files have a poor signal-to-noise ratio, and in many cases any calls that are present cannot reliably be identified to species or species group. Although it might seem alarming that some bat calls are not picked up by the program, it is worth remembering that a similar selection process has already taken place, with the choice of a particular detector sensitivity setting. Browsing through the discarded files should be avoided, as this might result in a bias towards easily identifiable bat species.

The remaining files that the program identifies as bat calls can be false positives (non-bat sounds that are retained). Several randomly selected files should always be checked to make sure that there are not too many false positives present (a few per cent seems acceptable, depending on your study). Since all the rare species suggestions need to be checked anyway (see section 5.5.4), a large number of false positives will be quickly detected. In recordings made from wind turbines, false positives are usually present in files that are made during windy nights when bat activity is very low. Once such a problem is detected it will be necessary to manually remove all these false positives.

5.5.4 Species identification

Identification of bat calls is performed by different processes, as discussed in depth by Russo and Voigt (2016) and others. Examples include algorithms of pattern recognition (Obrist *et al.* 2004), support-vector machines (Redgwell *et al.* 2009), hierarchical ensembles of neural networks (Redgwell *et al.* 2009, Walters *et al.* 2012) and random forest classification and machine learning (Skowronski and Harris 2006). These processes (such as artificial neural networks) are now successfully used in speech recognition (transforming spoken language into text by computer), automated detection of tumours for medical diagnosis, and automated human face recognition. It is safe to say that compared to these tasks the recognition of bat calls seems rather simple. Nonetheless, all the software packages using these processes fail to accurately identify all bat calls. Indeed, error rates exceed 50% in species with less characteristic echolocation calls (Rydell *et al.* 2017).

Given the state-of-the-art technology, why is automated bat identification still such a daunting task? Since identifying specific shapes or patterns can be done very accurately, the reasons for this failure must lie elsewhere: bats adapt their echolocation calls to the environment they fly in. The calls they use in open areas are very different from those used in cluttered environments. This results in considerable overlap between species. Species identi-fication using the above-mentioned processes usually involves a learning process based on a library of calls. The reliability of the analyses is therefore dependent on the library of calls that is used. If the library is poor, so are the analyses. The library only contains recordings that belong with 100% certainty to a specific species. To ensure this, many recordings of bats are of hand-released individuals or bats emerging from their roost, whereas calls from bats in extreme open conditions tend to be rarer in confirmed call libraries. The recorded species, however, may generally prefer to fly in open habitats and thus use long and narrowband pulses that form a minor part of the library on which the neural network is trained. Such a biased library therefore actively reduces the probability of correct identification. There are two possible solutions to this problem. The first is to expand the library to contain thousands of calls from a large range of habitats and locations so that it becomes representative of the species. This may be impossible to achieve, however, particularly when dealing with rare species or high fliers. The second solution would be to make a library that contains an equal number of calls for each pulse duration so that the library is deliberately unbiased towards a possible preference for the use of a certain pulse duration, reflecting the openness of the habitat.

Other problems with call libraries may be that they often lack social calls and/or feeding buzzes, with the result that the neural network performs poorly when the library only contains a limited number of recordings. The library should also contain both high- and lower-quality recordings to ensure that a match is possible with recordings made during suboptimal (field) conditions. Yet this should be done with caution, because for some species degradation of recording can minimise differences between species. Typically, the software produces very rare species identifications when analysing poor-quality recordings, suggesting that high-quality recordings of these rare species (and/or poor-quality recordings of common species) are missing in the library. When datasets are limited, even recording quality itself can bias the software.

Another flaw in automated identification is that the analysis focuses on individual pulses without information about the context. For example, a social call of common pipistrelle consists of four pulses with very short inter-pulse intervals. Although the frequency range and shape are slightly different, one pulse is sometimes mistaken for a search-phase call of a long-eared bat. In visual analysis, the surrounding pulses are seen, so it is immediately clear that this is a social call. A noctule approaching an object such as the nacelle of a wind turbine uses calls that are similar to calls made by serotines. The beginning of the recording or even the previous recording might show the characteristic low-frequency qCF pulses of the noctule when it was still in search phase. For reliable identification, it is therefore much better to take several calls or even several recordings into account. In this way harmonics can be seen, search-phase calls can be separated from approach-phase calls, alternating calls such as used by noctules can be identified, and it can easily be seen when calls belonging to different individuals overlap. The reason why human analysts outperform existing software (Rydell *et al.* 2017) is probably strongly related to the benefit of context.

The identification software applies filtering in order to separate calls from noise and to get a clear view of the shape of bat calls used in automated identification. During filtering, some important information is lost. For identifying *Myotis* species, for example, key measurements are the highest and lowest frequency. Natterer's bats use pulses with extreme low end frequencies (below 20 kHz) and Geoffroy's bats have a start frequency of more than 130 kHz. Since the filtering shifts most energy towards the middle frequencies, bandwidth is drastically reduced. Automated identification of *Myotis* species is therefore mostly based on call shape, and could be much more reliable if accurate information about the frequency range were used.

Several species might also call simultaneously in a recording; therefore, algorithms should be designed that retain only one individual (based on call-level and pulse repetition rate, of course with a margin of error).

Because of these shortcomings, automated identification is not fully reliable. It is important to realise that the confidence/probability of identification as presented by the software package is the result of internal validation. A 100% match with the library does not mean that a concrete identification has been made. Clement *et al.* (2014) demonstrated the inherent bias of each program by looking at classification agreement across different packages, and found only 40% agreement.

5.5.5 How can we use automated identification?

Technological developments should soon overcome or reduce the impact of many of the limitations mentioned above, leading to improvements in the reliability of automated identification. Species with structurally distinct calls such as soprano or common pipistrelles are already accurately identified (83–99%; Rydell *et al.* 2017). Using automated identification to identify these species seems acceptable as long as it is done with care. Careful use means for instance still checking questionable identifications such as soprano pipistrelles in regions where the species is scarce, eliminating species in areas far removed from its geographical distribution (e.g. Kuhl's pipistrelle or Schreiber's bat in northern Europe) and realising that bats flying in confined areas such as highway underpasses will alter their calls (common pipistrelle calls will become more similar to those of Schreiber's bats and soprano pipistrelles). In many areas of northwestern Europe, more than 80% of all bat recordings belong to species with structurally distinct calls. If automated identification is only used to set aside these recordings, the manual workload is reduced to less than 20% of all recordings simply by discarding pipistrelles and noise. Combining automatic and manual identification optimises bat-call classification (López-Baucells *et al.* 2019). For the time being, we recommend that automated identification is used only for these species, and that visual validation of all other species (nyctaloids, *Myotis* and *Plecotus*) is still necessary.

5.5.6 The magic human ear

A discussion sometimes arises that pits the human ear against the machine. It is then argued that the human ear can do more with the sound being played (slowed down) through the loudspeaker than with the 'limited information' contained in a spectrogram. Sometimes, this is even argued for heterodyne sounds. We would like to stress that recorded bat calls, either listened to or in spectrographic form, are usually filtered versions of the call the bat emitted. None of the microphones bat detectors use has a perfectly flat frequency response, and some have a large surface. Recordings of *Rousettus* clicks with any of these detectors will for example show 4–5 periods lasting 100 microseconds. Using an instrumentation microphone shows that the actual clicks only have three periods and are 60 microseconds, often with a different phase than recorded with the bat detector. Playing a bat-detector recording back slowed down does not counter the information loss that has taken place during recording, which will affect the sound no matter how it is analysed. In heterodyne recordings the original sound is nearly totally 'destroyed' and we are listening to a proxy signal of what is happening in a certain frequency band (see Chapter 3). When a time-expanded sound is played back, this sound is no more than what was recorded, digitised in sampling rate and bits. A sound file is no more than a vector of numbers (voltage values over time). This series of numbers is what either the ear or the computer will analyse.

The argument that 'the ear does the best job' would therefore only hold if the spectrogram (or any other computer-assisted analysis) threw away vital information. The spectrogram throws out phase information, while the waveform representation still contains all information that was recorded. Yet looking at a picture of a waveform barely helps us to identify a bat, whereas playing it back is much more helpful. The best computer analysis is as good as the programmer of the identification software, or the person looking at the spectrogram, and either may overlook certain tiny details that the human ear would pick up. Still, the point that *all* information about the sound is contained within the sound file (or waveform) is an important one. It implies that if the human ear hears something essential that a computer-aided analysis overlooked, this essential information is present in the sound file – which leads to the conclusion that, given enough programming talent, it must be possible to extract it, describe it and plot it. As echolocation pulses are very simply structured, it is unlikely that there is anything a programmer could not extract.

With complex social calls, figuring out the essential features may be more challenging, but the bottom line is that 'golden ears' should always be in dialogue with programmers to describe their sensations in physical terms. If automated identification software is still inadequate, this does not mean that the human ear can hear things the original sound file does not contain, but simply that the programmers have not chosen the right acoustic features to analyse.

5.6 Passive recording

As the statistical power of bat-detector transects has still not been fully resolved, attention has focused on how to conduct statistically powerful surveys with stationary autonomous detectors (e.g. Skalak *et al.* 2012, Froidevaux *et al.* 2014, Obrist and Giavi 2016, Braun de Torrez *et al.* 2017, Newson 2017, Richardson *et al.* 2019).

Temporal variation in bat occurrence and activity occurs both within and between nights (Hayes 1997). Taking the former into account, Skalak *et al.* (2012) and Froidevaux *et al.* (2014) concluded from their studies that sampling the entire night (dusk to dawn) outperformed other sampling periods (e.g. at dusk only) as it captured species-specific activity patterns throughout the night. To tackle the latter, it is highly recommended to sample for several nights at the same locations to increase the likelihood of detecting rare and elusive species. In Nevada, Skalak *et al.* (2012) highlighted that on average 20–30 nights were required to reach 80–90% of the estimated species richness at a given site. Froidevaux *et al.* (2014) found similar results in the Swiss forests, with 12–33 full nights required to reach 90% of all species, depending on the way microhabitats were sampled. Interestingly, Braun de Torrez *et al.* (2017) and Bruckner (2016) both found that sites of great importance for bats, such as water bodies, required much less sampling effort (e.g. 8 nights to reach 90% of all species in Everglades National Park). Sampling for a repeated number of nights also allows us to reliably identify sites of high activity. Based on results from England, Richardson *et al.* (2019) recommended at least 12 nights of surveying to correctly estimate *Nyctalus noctula* activity at wind turbines.

Regardless of the scale of the study, spatial variation in bat occurrence and activity should also be accounted for by sampling several locations simultaneously. At the local scale, within-site variability of detection probability is likely to be high (Kubista and Bruckner 2017), especially in structurally complex habitats such as forests. The deployment of detectors in different microhabitats important for bats is therefore crucial (Froidevaux *et al.* 2014). Likewise, when assessing bat activity and species richness near wind turbines, detectors should be installed at both ground and above-ground levels to consider the vertical stratification of bat activity (Collins and Jones 2009, Wellig *et al.* 2018). Regarding large-scale studies, bat surveys conducted in heterogeneous mosaic landscapes will require greater effort than in homogeneous ones. Nevertheless, the number of detectors required can be reduced by implementing a strategic sampling design, focusing on landscape elements favoured by bats such as water sites and woody linear features (Newson 2017, Braun de Torrez *et al.* 2017). Sampling near lit areas (e.g. close to streetlights and light traps) must be avoided, as photophobic species are likely to be missed from the inventory (Stone and Jones 2009, Azam *et al.* 2018, Froidevaux *et al.* 2018).

Overall, optimising the sampling design will involve finding the best trade-off between the number of detectors to be deployed, the number of nights to sample, and the sampling costs in terms of resources and time (Froidevaux *et al.* 2014, Obrist and Giavi 2016). The optimal combinations of these three factors may vary considerably depending on the region and sites being sampled, the type of detector used, and the study objectives.

6 Call Analysis

Jon Russ, Kate Barlow and Philip Briggs

6.1 Sound-analysis software

The idea of representing sound visually using sonograms, oscillograms and power spectra, and how variation in frequency, amplitude and duration of bat calls affects the appearance of these diagrams, was introduced in Chapter 2. Sound-analysis software processes the calls recorded from bat detectors, allowing us to visualise call shapes, measure call parameters and identify the calls of different bat species. Various software packages are available, ranging from freeware that can be downloaded from the internet and which provides basic call analysis and reporting functionality, through to programs with enhanced features such as automated call analysis that come at an additional cost or are compatible only with specific models of bat detector. Selection of the most appropriate software should be based on the level of call analysis to be undertaken, budget and functions required – for example, whether there is a requirement for some degree of automated analysis (e.g. for processing large quantities of files) or more sophisticated reporting tools.

The majority of sound-analysis packages are designed for analysing audio files (e.g. in *.wav format) using a Fast Fourier Transform (FFT) algorithm to produce sonograms of bat calls as described in Chapter 2. This enables assessment of call structure, and measurement of frequency, time and amplitude parameters, often to a high level of detail, depending on the quality of the recording and the type of bat-detector system used (time-expansion and full-spectrum bat detectors produce much higher-resolution recordings than frequency division). Most sound-analysis software can be used to carry out manual analysis of recordings from any bat detector that produces files in a compatible file format, but some of the enhanced features may only work on recordings from bat detectors produced by the same manufacturer, for example extracting date, time and GPS data from the files.

Some packages are also exclusively designed to analyse zero-crossing files (*.zc format) (see section 5.2.3). Zero-crossing analysis (ZCA) is applied to the frequency-divided output from the detector to produce graphical representations of the recorded sound: call structure is displayed in the form of data points plotted on frequency/time axes. ZC files do not retain amplitude information and therefore it is not possible to measure the frequency of maximum energy (FmaxE), which can be a diagnostic feature in species identification; however, approximately analogous measurements can be taken in the form of 'cycles' and 'characteristic frequency' (Fc), both of which indicate the frequency in the flattest part of the call – which is usually where the peak in amplitude is located. A wide range of other measurements can be taken, including 'slope', which measures the rate of change in frequency over time (measured in octaves per second), while the AnalookW and Anabat Insight programs (Titley Scientific) enable the user to create species filters that automatically identify and label files containing call characteristics that fall within specified parameters (see section 6.5). ZC detectors have the advantage of producing small file sizes, reducing data storage needs and speeding up data processing.

This chapter focuses on the analysis of audio recordings using software based on FFT; section 6.5 also briefly describes AnalookW, the dedicated ZCA software that has developed the range of techniques widely used for analysing zero-crossing files.

6.2 Automatic recognition software

There is now a growing number of automatic identification systems available on the market that automatically perform some or all of the following: processing of recordings, extraction of calls, and assignment of calls to a species or species group for identification based on a set of parameters programmed into the software. In some cases, the software comes with the species identification parameters and algorithms pre-installed, based on a large call library collated for a particular geographical area by the manufacturer of the program. Alternatively, users can set up their own algorithms based on their call libraries or known call parameters.

Automatic recognition and identification can significantly increase the speed with which recordings of bat calls can be analysed (see Chapter 5). However, there is always some error and uncertainty attached to using an automated system. The two main factors that contribute to the uncertainty are the nature of the echolocation calls themselves (variation within a species, and similarities between species or groups of species) and how effective the program is at extracting the calls. The accuracy of the output from an automatic recognition system will always depend on what datasets have been used to develop and train the identification system. A very wide range and a number of calls for each species would increase the likelihood of call classification being correct, whereas systems trained using only limited numbers of bat calls, or calls collected from known species in limited environments (which are less likely to be representative of the full range of natural call variation) may misclassify a greater percentage of inputted calls.

6.3 Using sound-analysis software

6.3.1 Views

Sound files are opened in the chosen sound-analysis software to view a recording from a bat detector and then inspected for the presence of bat calls, and selected calls are measured. If a single recording has been made over a long period in the field, the resulting file is likely to be very large, may take a long time to open in the sound-analysis software, and will be unwieldy to inspect for bat calls and analyse. It is preferable to split recordings in the field into manageable files, perhaps by recording for set periods of time onto separate files; alternatively, large sound files can be split into shorter sections of a specified length. An internet search will reveal several programs that will carry out this function.

Bat detectors which record directly to an internal CompactFlash (CF) or Secure Digital (SD) card typically enable you to set maximum file length to ensure that the resulting files are more manageable to analyse in terms of file length and number of files. For example, files which are 5 seconds in length have the advantage that, when opening each file by clicking through files in sequence with 'zoom to fill screen' set, any sounds shown in the spectrogram will already be zoomed in to the extent that a very quick assessment can be made as to whether any bat calls are present, and even which species or species group. The downside of setting a file length of 5 seconds can be a large number of files to click through, particularly if the bat-detector trigger setting was set to a high sensitivity, resulting in a wearying number of files to click through containing non-bat noise, such as surveyors trudging through a woodland. Files which are 15 seconds long will reduce the number of files to examine, but more information will be squeezed into the displayed spectrogram so it is more likely that portions of the file will need to be zoomed in to confirm the presence of bat calls and identification of species.

Most sound-analysis programs provide options to view the oscillogram (showing amplitude against time) or sonogram/spectrogram (showing frequency against time with amplitude shown as a colour variation), or both simultaneously. If possible, it is helpful initially to view both to allow assessment of whether bats are present in a recording, to check signal-to-noise ratios and recording quality, and to select a section of the recording for analysis. Some software also displays a power spectrum (showing amplitude against frequency), or this may need to be selected for a specific call or section of recording during analysis.

It is important to check the axis scales and confirm that options are set correctly where possible for the type of recordings being analysed. For example, time-expanded (TE) recordings are slowed down in time by a factor of 10 (for ×10 TE) and this should be taken into account when looking at the time axis scale, or set accordingly. Similarly, the frequency range of frequency-division (FD) recordings will be one-tenth (for FD by 10) of the original call frequency, which again needs to be accounted for when reading the frequency axis, or set accordingly.

The range of frequency and time axes can usually be adjusted within the software to improve the viewing of bat calls. Bat species found in Europe can produce echolocation and social calls between 6 and 180 kHz, and this is a good starting range for axes when looking at bat recordings.

Some software packages also allow the user to choose different options that control how the recordings are processed. Most programs use FFT to process sound files into sonograms, and the settings used will have an impact on the sonograms produced. For example, the FFT size will affect how the calls are viewed in the sonogram: as size increases, the resolution on the frequency axis increases, but resolution decreases on the time axis, and vice versa (Figure 6.1). Most programs use a default setting that generally gives a good compromise between frequency and time resolution.

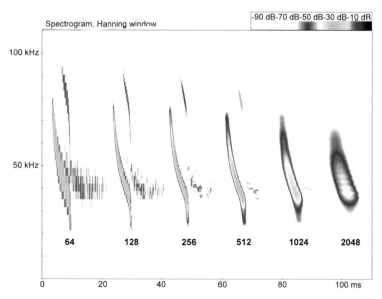

Figure 6.1   The effect of differing Fast Fourier Transform (FFT) sizes on the sonogram of a single bat echolocation call (*Myotis* sp.).

6.3.2 Threshold/gain settings

In most sound-analysis programs there is an option to adjust the 'threshold' or 'gain'. This is a feature that allows the user to have control over the amount of unwanted or 'background' noise in the analysis and also to establish measurement criteria. Attenuation can be selected to any value offered by the particular software (e.g. 0–40 dB, or an index of amplitude level). The portions of the power spectra below the threshold will not be displayed in the spectrogram, so the threshold sets the weakest signal to be displayed on the sonogram (Figures 6.2 and 6.3). It is advisable when comparing calls between species to choose a threshold level and stick to it for all calls.

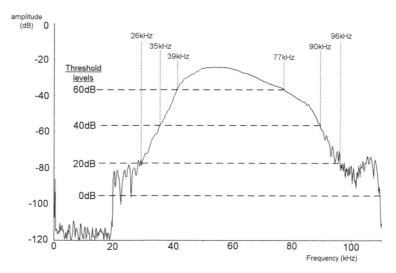

Figure 6.2   Diagram of threshold levels. A threshold level of 20 dB results in a sonogram with minimum and maximum frequencies of 26 kHz and 96 kHz, a threshold level of 40 dB results in a sonogram with minimum and maximum frequencies of 35 kHz and 90 kHz, and a threshold level of 60 dB results in a sonogram with minimum and maximum frequencies of 39 kHz and 77 kHz (see Figure 6.3 for sonogram).

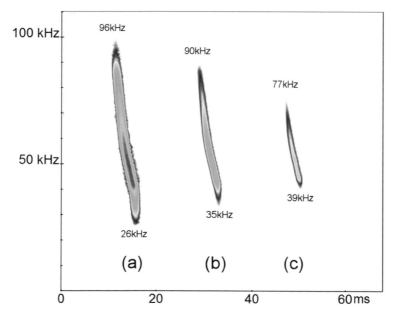

Figure 6.3   Sonographic results from threshold settings of (a) 20 dB, (b) 40 dB, (c) 60 dB.

The threshold may also need to be adjusted depending on the recording level and the signal-to-noise ratio of the recording. A weak signal (low signal-to-noise ratio) requires a lower threshold, while a strong signal (high signal-to-noise ratio) may look best with a higher threshold.

6.4 Measuring call parameters

The process of analysing a recorded sequence of bat calls manually using sound-analysis software is best approached using the series of steps described below.

6.4.1 Play the sequence

If the sound-analysis software allows recordings to be played back through the computer's speakers, this can help to identify the presence of bat calls. If the recordings were made from a full-spectrum bat detector, then the bat sounds will still be in the ultrasonic range and therefore mostly inaudible to human ears. Sound-analysis software often includes a function to slow down the playback of recordings so that they can be listened to in time-expansion mode.

A bat call will be relatively tonal, and when played back will sound like a series of chirps or whistles of varying length and pitch depending on the species recorded (if recorded from a TE detector, or a slowed-down recording from a full-spectrum detector) or like a rhythmic series of high-pitched clicks or smacks (if recorded from an FD detector). Noise is highly variable: it may be of very short or very long duration or be repetitive, and it may be harsh, or sound like a hiss, crackle or whining. In general, a bat call will be longer than 2.5 ms and shorter than 70 ms. Bat calls are also produced as multiple pulses in a rapid sequence, although it is possible that only a few more spaced-out calls may be present instead, particularly from species that produce low-frequency calls with long inter-pulse intervals. Occasionally a bat may fly within the range of the microphone only very fleetingly, resulting in a single pulse being detected. Such instances should be treated with great caution when carrying out species identification and are often best ignored, since analysis of at least three pulses is recommended to confirm species identification; a single pulse could potentially be an anomalous call from the bat species that produced it, or a non-bat sound that happens to resemble a bat call (Middleton 2020). By listening to a range of call sequences, it is quite easy to build up sufficient experience to become confident in identifying bat calls from other noise in recordings.

6.4.2 Select suitable calls for analysis

Choose around three to five calls with a good signal-to-noise ratio where possible (Figure 6.4). It is important to look at a sequence of calls rather than a single call, to help ensure that species identification is accurate.

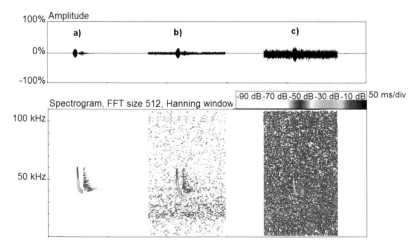

Figure 6.4   Oscillogram and sonogram of a single bat echolocation call (*Pipistrellus* sp.), showing recordings with (a) high, (b) moderate and (c) low signal-to-noise ratio.

6.4.3 Look at call shape

Call shape can help provide an initial suggestion of the species or species group of the recorded bat. There are several different call shapes produced by European bats, and it is worth becoming familiar with these shapes, how they typically appear on sonograms from different recording types, and how they vary across the species and habitat in which the bat is flying (see Chapter 3).

The main call structures used by European bats and how these typically appear as sonograms from recordings made with the different broadband detector systems is illustrated in Figure 6.5. It is provided to enable ZC and FD users to relate their recordings to the TE/full-spectrum sonograms shown in Chapter 8. It should be noted that recordings from all systems can vary considerably in quality. For example, a poor TE sonogram may look no clearer than a typical FD sonogram. Conversely, a very good-quality FD recording may display a higher level of detail than suggested below.

For FM calls, such as those of the *Myotis* species, it may be necessary to increase the threshold to reveal higher frequencies, or lower FFT to see clear call shape (see sections 6.3.1 and 6.3.2)

6.4.4 Measure frequency parameters

Measuring parameters from the series of three to five selected calls is good practice and ensures that any variation between calls is taken into account; if there are alternating call types, parameters from both are measured. The commonly measured frequency and time parameters are shown in Table 6.1. Frequency parameters are normally reported in kilohertz (kHz), time parameters in milliseconds (ms). When carrying out more detailed call analysis, both minimum and maximum frequency (or in some instances start and end frequency) are measured. For most call types these will not be different, but for some call shapes, such as those from horseshoe bats where there is a tail at the start and end of the call, it may be helpful to record all these frequency parameters for completeness.

Table 6.1   Commonly used call parameters.

Abbreviation	Parameter	Description
FmaxE	Frequency containing maximum energy	Frequency of the maximum amplitude of the spectrum
Fmin	Minimum frequency	Minimum frequency of the call
Fmax	Maximum frequency	Maximum frequency of the call
FStart	Start frequency	Frequency at the start of the call
FEnd	End frequency	Frequency at the end of the call
IPI	Inter-pulse interval	Duration between the start of two adjacent calls
Dur	Duration	Duration of the call

Frequency parameters can be measured from the spectrogram and/or power spectrum. When comparing measurements from a series of calls, it is best to ensure they are all measured in a consistent way. Frequency containing maximum energy (FmaxE) is often the key parameter used to identify many species, in conjunction with call shape. It can usually be measured from the power spectrum, although for some species that produce FM calls there may be no clear peak (Figure 6.6). For species that produce these call types, in particular, end or minimum frequency can also be very useful for distinguishing between species. Start or maximum frequency can be very difficult to measure, depending on the level of background noise and the quality of the recordings, and the value will often depend on the threshold

Call type	Species	Time expansion/ Full spectrum	Anabat (Zero crossing)	Frequency Division
FM-CF-FM	*Rhinolophus* species	 "Staple" shape. Start/end FM sweeps not always clear.	 "Staple" shape. Start/end FM sweeps not always clear.	 Clear CF portion. Start/end FM sweeps often unclear or not visible.
FM	*Myotis* species	 Normally long sweep, sometimes with a slight kink.	 Normally long sweep, sometimes with a slight kink.	 Normally long sweep. Kink in structure rarely discernible.
FM-QCF	*Pipistrellus* species *Nyctalus* species *Eptesicus serotinus*	 "Hockey stick/ reverse j" shape.	 "Hockey stick/ reverse j" shape.	 "Hockey stick" shape occasionally discernible but typically a "tear drop" shape.

Figure 6.5   Examples of sonograms of typical echolocation calls produced by European bats recorded using different detector systems.

Call type	Species	Time expansion/ Full spectrum	Anabat (Zero crossing)	Frequency Division
QCF	*Pipistrellus* species in open habitat *Nyctalus* species in open habitat (often alternating with FM-QCF calls) *Eptesicus serotinus* in open habitat	*Almost flat line with slight curve.*	*Almost flat line with slight curve.*	*Almost flat line, slight curve not always discernible.*
FM with strong harmonics	*Plecotus* species	*FM sweep often with slight curve and strong harmonic.*	*FM sweep often with slight curve. Harmonic not always captured.*	*FM sweep. Harmonic not captured and curve in structure not usually discernible.*
QCF-FM (type 1) and/or short FM (type 2) *Sometimes alternates type 1 and type 2 calls in a sequence*	*Barbastelle barbastellus*	*Type 1: FM sweep often with slight curve visible at start of call. Often very quiet. Type 2: Very short FM sweep.*	*Type 1: FM sweep. Slight curve at start of call not always visible. Often very quiet. Type 2: Very short FM sweep.*	*Type 1: FM sweep. Slight curve at start of call not always visible. Often very quiet. Type 2: Very short FM sweep, more "blobby" in appearance.*

settings for the spectrogram and therefore may vary from user to user. Both start and end frequency (or maximum and minimum frequency) can be particularly difficult to distinguish on spectrograms from FD recordings because of the poorer quality of FD compared with TE or full-spectrum sampling and may, therefore, be less useful for species identification when obtained using FD rather than higher-quality recordings. Regardless of the quality of the recording, higher frequencies are more prone to attenuation (see Chapter 2), and the highest frequencies in the calls may not always be picked up by the detector. Therefore, measurements of start or maximum frequency often need to be treated with caution.

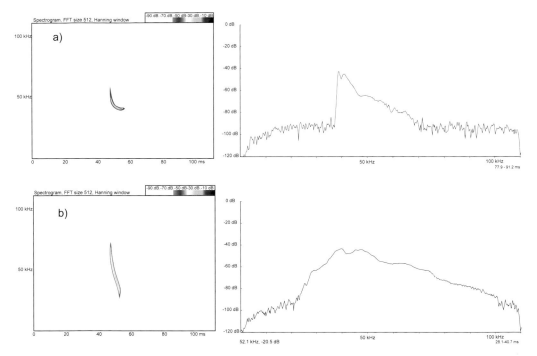

Figure 6.6   Power spectra, showing typical calls of bats of the genera *Pipistrellus* and *Myotis*.

6.4.5 Measure time parameters

As with frequency parameters, measuring time parameters from the series of three to five selected calls is good practice. The time parameters duration and inter-pulse interval are rarely diagnostic but often measured to help confirm likely species identification. Figures 6.7 and 6.8 show how various call parameters can be measured from the spectrogram and power spectrum. Ideally, call duration should be measured from the oscillogram in preference to the spectrogram, as the FFT setting selected can have a significant impact on call duration, whereas this does not vary on the oscillogram. Inter-pulse interval is less likely to be affected by the method of measurement and can be measured either from the oscillogram or from the spectrogram.

It should be noted that there is a large amount of variation in all call parameters within species; finding a parameter measurement outside the described range for a species does not necessarily mean that species identification is incorrect. Call parameter measurement ranges described in Chapter 8 should be used as a guide to the normal range for each species.

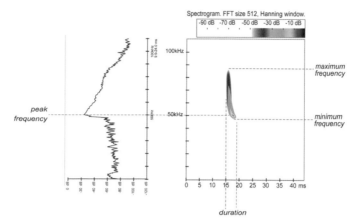

Figure 6.7   Example of measurements taken from a single echolocation call, shown as a power spectrum and a sonogram.

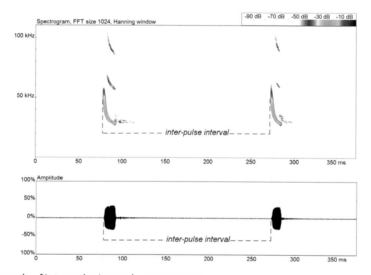

Figure 6.8   Example of inter-pulse interval measurement.

6.4.6 Identification

Once the call shape and call parameters have been determined, refer to Chapter 8 and compare the observed call shape and measured call parameters to allow identification to species or species group. However, a basic echolocation key is provided in Chapter 7.

6.5 Zero-crossing analysis

Some sound-analysis programs, for example Kaleidoscope (Wildlife Acoustics), include zero-crossing analysis (ZCA) functionality alongside the standard FFT analysis used for audio files. For many years the most widely used dedicated ZCA program has been AnalookW (Titley Scientific), which developed a range of techniques for analysing zero-crossing files. Many of these techniques have been incorporated into newer programs which offer both FFT-based and zero-crossing analysis, such as Kaleidoscope and Titley Scientific's Anabat Insight. Here we will describe the functionality of AnalookW, as it is still widely used and free to download, and this provides an overview of key ZCA techniques, many of which are common to other programs with ZCA functionality.

As with standard FFT-based sound-analysis software, the main view in AnalookW is an on-screen sonogram showing frequency against time. All sound files can be labelled with location, notes and the species identification in dedicated boxes at the bottom of the screen (this facility is now included in several other sound-analysis programs). Magnification of calls on the screen can be changed across a range selected by using the F1 to F10 keys on the computer keyboard. The F6 option is widely used. Simple call parameters can be measured on the sonogram view in a similar way to those described for standard sound-analysis software. An instant measure of 15 call parameters from the calls displayed on the screen is also available in tabular form on the AnalookW screen using the *measures* function (selected from the *view* dropdown list or by pressing M on the keyboard).

Filters can be applied to files or a series of files to reduce 'noise', enabling bat calls to be selected in a particular frequency band. Species-specific filters can also be developed, based on known call parameter ranges, and then applied to individual files or series of files. The user needs good species identification skills and call parameters reference information to create the filters, which often require much testing and modifying before they produce satisfactory results. Filters will not give 100% accurate identifications, but the aim should be to reduce false positives and false negatives to a level considered acceptable for the survey.

Additionally, scans enable one or more filters to be applied to large amounts of data, from whole folders or folder 'trees'. Several filters run together in one scan can target different call types from the same species, resulting in a text file that summarises either the results of all bat pulses which have passed the filters or all the 'measures' sonogram parameters of the pulses that have passed the filters. Scans can enable objective, consistent, repeatable and much speedier analysis in this way.

A wide range of viewing tools is available to assist in the identification of bat calls in AnalookW, including split-screen tools. These display graphic representations of *cycles*, *slope* and *time between calls* (TBC) – see below for explanations of each of these. The split-screen options can be selected from the *view* menu to create a split-screen with the right-hand half of the screen displaying the selected option.

The *cycles* display is roughly analogous to the power-spectrum display in FFT-based sound analysis. The difference is that rather than showing a graph of frequency (in kHz) against amplitude (in dB) it plots frequency (in kHz) against how much time the call spends at that frequency, so the peak shows the frequency at which the call spends the most time, i.e. the frequency of the flattest part of the call. In the bottom right-hand corner of the screen that frequency is shown, measured from the series of calls displayed on the left-hand side of the screen. This will be approximately equivalent to published FmaxE values for each species, since the FmaxE is typically located in the flattest part of the call. In the *measures* and *filters* functions, this is measured as Fc (characteristic frequency), the frequency at the right-hand end of the portion of the call with the lowest absolute slope, which gives a similar value to cycles and FmaxE.

The TBC option simply shows the time (typically measured in milliseconds) between the start of one call and the start of the next call. The right-hand side of the screen shows a histogram of TBC for the series of calls displayed on the left of the screen.

The *slope* option provides a measure of how vertical or horizontal the call sonogram appears, measured in octaves per second (OPS); it is the speed at which a call changes in pitch over time. Slope can be positive (starts high in pitch and descends) or negative (starts low in pitch and ascends) or have zero slope (remains at the same pitch over time), as illustrated in Figure 6.9. The right-hand side of the screen shows the slope of the series of calls displayed on the left of the screen and whether it is positive, negative or if there is little or zero slope. Looking at the slope can be a useful additional clue in narrowing down species identity with difficult groups such as the genus *Myotis*. However, it should be noted that, like other call parameters, slope varies according to the habitat in which the bat is flying. For example, the slope of calls is likely to be steeper when a bat is flying in a cluttered environment compared with an open environment, and this should be taken into account when using slope to guide species identification.

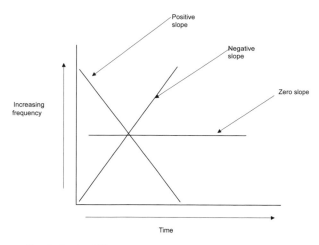

Figure 6.9   The slope option in AnalookW.

6.6 Common analysis problems

6.6.1 Selecting the correct channel

If recordings are made from a bat detector that outputs in stereo, for example recording a heterodyne output to one channel and TE or FD output to the other, it is vital to ensure that the TE or FD channel is being viewed and analysed in the sound-analysis software. Most programs have options for viewing stereo or single channels, so it is worth checking these if unfamiliar with viewing bat-call spectrograms. In the heterodyne channel, calls are often squashed down at the bottom of the screen (Figure 6.10) if you were tuned near to the FmaxE when listening to the bats in the field. Alternatively, depending on what frequency you were tuned to, heterodyne recordings can sometimes resemble FD recordings, except that the call sequences will tend to veer up and down in frequency quite markedly (corresponding to tuning around on the detector in the field), whereas FD or TE recordings are likely to vary much less over a sequence of calls.

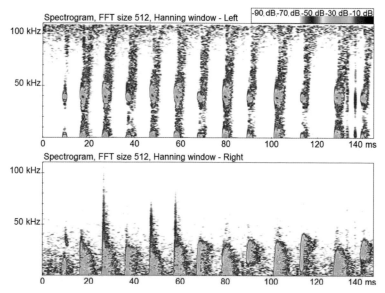

Figure 6.10   A comparison of a typical sequence of bat calls recorded from a frequency-division output (top sonogram) and a heterodyne output (bottom sonogram) from a bat detector.

6.6.2 Overloaded calls and clipping

It is possible on a sonogram to obtain harmonics that were not contained in the original signal. This is a result of the incoming signal being of greater amplitude than the maximum allowed amplitude of the equipment. The amplitude peak is in effect 'clipped' from the original wave. This can be seen in Figure 6.11, where (a) shows the original wave and the amplitude floor and ceiling of the equipment and (b) shows the result of clipping. Clipping can easily be seen on inspection of the oscillogram: the pulses will look square with a flat top and bottom instead of rising to clear peaks. Figure 6.12 shows the spurious harmonics produced as a result of clipping. The obvious features are that the harmonics are very intense and are many. Most bats do not produce many harmonics. If recordings are made from FD detectors, any harmonics visualised in the sound-analysis software are spurious harmonics and artefacts either of clipped calls or of the processing of the square waves by the software, and should be ignored.

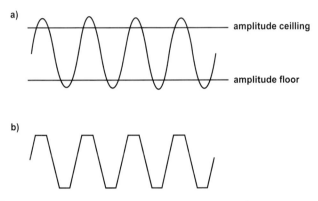

Figure 6.11   The effect of clipping on a wave. Part (a) shows the limits of the recording device and (b) shows the resultant waveform.

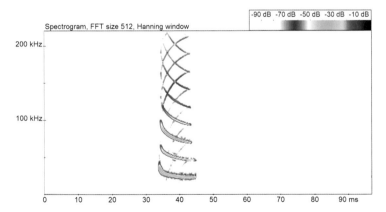

Figure 6.12   Spurious harmonics as a result of clipping (with aliasing also occurring).

Clipping can occur from overloading the signal into the bat detector, for example, if the bat flies very close to the detector, from setting the input level on the recording device too high, or during the recording of calls into the computer if input levels are set too high. It may be possible to restore the clipped signals to their original full signal by adjusting the input levels at the time of recording, although if the calls have been recorded onto the device at an excessively high amplitude, it will not be possible to adjust for this at the sound-analysis stage.

6.6.3 Interference: 'missing frequencies'

Calls from bats flying close to water, for example Daubenton's bats, have sonograms that appear to have 'missing' frequencies (Figures 6.13 and 6.14). This is not an effect the bats themselves are producing but is due to interference between the sound produced by the bat and the sound reflected off the water (Figure 6.15). For a bat flying at height x above the water, sound waves arrive at the point of the bat detector's microphone (A) directly from the bat (y) and also from the flat water surface, where they have been reflected from the original source (z). These waves usually arrive out of synchronisation with each other, and combining these signals produces some reinforcement of the original sound and some cancelling out, resulting in the 'missing' frequencies through the call. Analysis of these calls should be avoided, as the frequency structure will not accurately reflect that of the original call.

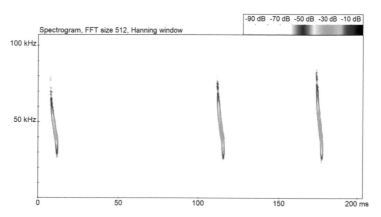

Figure 6.13   Echolocation calls produced by a Daubenton's bat.

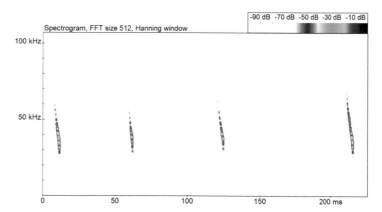

Figure 6.14   'Missing' frequencies as a result of interference when bats are flying close to the water.

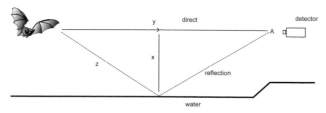

Figure 6.15   The interference paths of sound waves produced by a single bat and received by a detector. See text for explanation.

6.6.4 Echoes or more than one bat?

Occasionally, calls may be recorded which appear to have repeated or 'double' calls (Figure 6.16). In most cases, this appears on a sonogram as a distorted sound that looks like noise directly after the recorded call, although sometimes the call can be repeated almost in its entirety, as seen in the example here. These echoes are due to the emitted sound from the bat being picked up directly by the detector but also bouncing off an object such as a wall and the resulting echo also being picked up by the detector. The first call is the call produced by the bat whereas the second, usually quieter, call is the echo.

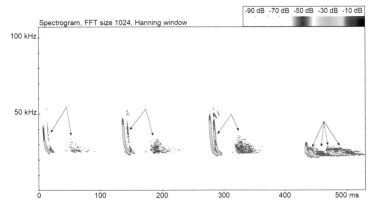

Figure 6.16   Repeated or 'double' calls produced by the original call bouncing off a nearby object plus noisier echoes from a less flat and more distant surface. The echoes are arrowed.

If there are several similar calls in one sequence that look similar to echoes but which are not seen as regular repeats of call patterns, it is more likely that there is more than one bat of the same species in the recorded sequence. It is usually possible to pick out each bat's call sequence from differences in call strength, frequency and inter-pulse interval for up to around three bats (Figure 6.17). If there are more than three bats it is often not easy either to determine how many bats make up the call sequence or to separate each individual's calls. This needs to be taken into account during analysis.

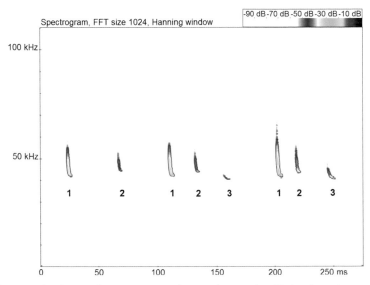

Figure 6.17   An example of a recording sequence with more than one bat (*Pipistrellus* sp.) present (labelled 1–3 on the sonogram). The frequency and amplitude differences between the bats can be seen on the sonogram.

7 A Basic Echolocation Guide to Species

Jon Russ, Yves Bas, Charlotte Roemer, Arjan Boonman, Alex Lefevre and Marc Van De Sijpe

This chapter contains a relatively simple guide to echolocation for beginners. It provides information designed to assist in the analysis of sequences of calls in order to arrive at the most likely species or species group. It is based on call shape, frequency and duration, with detailed notes providing information on individual species and groups of species that produce similar echolocation calls.

The chapter is divided into three main parts: separation of calls by call shape, then broad separation into species or species groups using measured parameters, and finally fine separation to species between those species with similar calls.

First of all, select the **call shape** that most resembles your call(s) (categories A–E). Once call-shape category has been selected, go to the graph under the relevant **category** and determine which species, or group of species, best fits your measured parameters. Finally, look at the **species notes** for distinguishing characteristics. Species are presented in the order of call-shape categories (A–E) with species that overlap most in call shape and measured parameters located next to each other, arranged in order of decreasing call frequency.

More detailed information on each of the 44 species occurring in Europe is contained in Chapter 8. In Chapter 7 the order of species is based on call shape and parameters, whereas the order of species in Chapter 8 is taxonomic.

The figures in this chapter use abbreviated forms of the species names, and these are listed in Table 7.1. This table also provides a cross-reference to the relevant section for each species in Chapter 8.

Table 7.1 Species abbreviations used in Chapter 7, and locations of species accounts in Chapter 8.

Abbreviation	Species	Common name	Section number
Rhip	*Rhinolophus hipposideros*	Lesser horseshoe bat	8.1
Rfer	*Rhinolophus ferrumequinum*	Greater horseshoe bat	8.2
Reur	*Rhinolophus euryale*	Mediterranean horseshoe bat	8.3
Rmeh	*Rhinolophus mehelyi*	Mehely's horseshoe bat	8.4
Rbla	*Rhinolophus blasii*	Blasius's horseshoe bat	8.5
Mdau	*Myotis daubentonii*	Daubenton's bat	8.6
Mdas	*Myotis dasycneme*	Pond bat	8.7
Mcap	*Myotis capaccinii*	Long-fingered bat	8.8
Mbra	*Myotis brandtii*	Brandt's bat	8.9

Abbreviation	Species	Common name	Section number
Mmys	*Myotis mystacinus*	Whiskered bat	8.10
Mdav	*Myotis davidii*	David's myotis	8.11
Malc	*Myotis alcathoe*	Alcathoe whiskered bat	8.12
Mema	*Myotis emarginatus*	Geoffroy's bat	8.13
Mnat	*Myotis nattereri*	Natterer's bat	8.14
Mcry	*Myotis crypticus*	Cryptic myotis	8.15
Mesc	*Myotis escalerai*	Iberian Natterer's bat	8.16
Mbec	*Myotis bechsteinii*	Bechstein's bat	8.17
Mmyo	*Myotis myotis*	Greater mouse-eared bat	8.18
Mbly	*Myotis blythii*	Lesser mouse-eared bat	8.19
Mpun	*Myotis punicus*	Maghreb mouse-eared bat	8.20
Nnoc	*Nyctalus noctula*	Noctule	8.21
Nlas	*Nyctalus lasiopterus*	Greater noctule	8.22
Blei	*Nyctalus leisleri*	Leisler's bat	8.23
Nazo	*Nyctalus azoreum*	Azorean noctule	8.24
Eser	*Eptesicus serotinus*	Serotine	8.25
Eisa	*Eptesicus isabellinus*	Meridional serotine	8.26
Eana	*Eptesicus anatolicus*	Anatolian serotine	8.27
Enil	*Eptesicus nilssonii*	Northern bat	8.28
Vmur	*Vespertilio murinus*	Parti-coloured bat	8.29
Ppip	*Pipistrellus pipistrellus*	Common pipistrelle	8.30
Ppyg	*Pipistrellus pygmaeus*	Soprano pipistrelle	8.31
Phan	*Pipistrellus hanaki*	Hanak's pipistrelle	8.32
Pnat	*Pipistrellus nathusii*	Nathusius's pipistrelle	8.33
Pkuh	*Pipistrellus kuhlii*	Kuhl's pipistrelle	8.34
Pmad	*Pipistrellus maderensis*	Madeira pipistrelle	8.35
Hsav	*Hypsugo savii*	Savi's pipistrelle	8.36
Bbar	*Barbastella barbastellus*	Western barbastelle	8.37
Paur	*Plecotus auritus*	Brown long-eared bat	8.38
Paus	*Plecotus macrobullaris*	Alpine long-eared bat	8.39
Psar	*Plecotus sardus*	Sardinian long-eared bat	8.40
Paus	*Plecotus austriacus*	Grey long-eared bat	8.41
Pkol	*Plecotus kolombatovici*	Mediterranean long-eared bat	8.42
Msch	*Miniopterus schreibersii*	Schreiber's bent-winged bat	8.43
Tten	*Tadarida teniotis*	European free-tailed bat	8.44

See Figure 2.25 in Chapter 2 for descriptions of components of call shape, and Table 6.1 in Chapter 6 for descriptions of commonly used call parameters.

Call shape

In Figure 7.1, sonograms are presented showing calls in a cluttered environment (closed habitat) on the left, moving to an uncluttered (open habitat) on the right. Note that call shape can vary depending on screen size and the degree to which you have zoomed in or out of the echolocation call, and therefore it is important to pay special attention to the frequency and time scales in the sonograms shown here. Note also the harmonics used by different species. Finally, note that the meaning of 'clutter' varies between species. For example, an environment that is cluttered for a noctule (e.g. foraging between two treelines above a canal) is not the same as a cluttered environment for a brown long-eared bat (foraging in among the branches of a tree).

Category A

Category A contains echolocation calls that are primarily an FM-CF-FM shape, typically produced by European bats of the genus *Rhinolophus*. For this group FmaxE is the most useful parameter for separating the species. In Figure 7.2, the ranges of values of the frequency containing maximum energy (FmaxE) are represented by vertical bars. Mean values and the spread around the mean are shown by the shading (darkest = mean, fading to white at the ends of the bars where these values rarely occur). Species with a restricted geographical range are shown in red.

Category B

Category B contains echolocation calls that are either FM-qCF or qCF in shape, typically produced by European bats of the genera *Pipistrellus*, *Miniopterus*, *Hypsugo*, *Nyctalus*, *Eptesicus*, *Vespertilio* and *Tadarida*. For this group Fmin and duration are the most useful parameters for initially separating the species. Figure 7.3 shows the end frequency of the qCF part of the call. It does not include the end hook, which is only occasionally produced. The ranges of values of the minimum frequency (Fmin) are represented by the vertical bars, with the horizontal spread representing duration range at specific frequencies. For FmaxE, mean values and the spread around the mean are shown by the density of shading (darkest = mean, fading to white at the ends of the bars where these values rarely occur). This is not shown for duration. Species with a restricted geographical range are shown in red. Fmin is chosen over FmaxE, as the overlap between species is smaller and it is not biased by recording conditions (distance and angle from the microphone).

Note that for identification, very short calls of less than 3 ms duration, such as those in extreme clutter, have an unreliable Fmin (and an even more unreliable FmaxE).

Category C

Category C contains echolocation calls that are generally FM in shape, which are typically produced by European bats of the genus *Myotis*. For this group Fmin, Fmax and duration are the most useful parameters for initially separating the species. In Figure 7.4, the ranges of maximum frequencies (Fmax) and minimum frequencies (Fmin) are represented by the vertical bars, with the horizontal spread representing duration range at specific frequencies. Mean values and the spread around the mean are shown by the density of shading (darkest = mean, fading to white at the ends of the bars where these values rarely occur). This is not shown for duration. Species with a restricted geographical range are shown in red, and species for which parameters are assumed are shown in green.

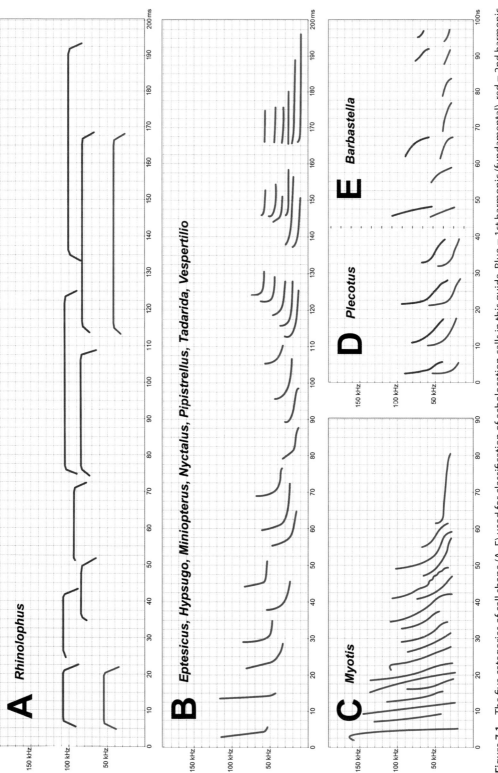

Figure 7.1 The five categories of call shape (A–E) used for classification of echolocation calls in this guide. Blue = 1st harmonic (fundamental), red = 2nd harmonic.

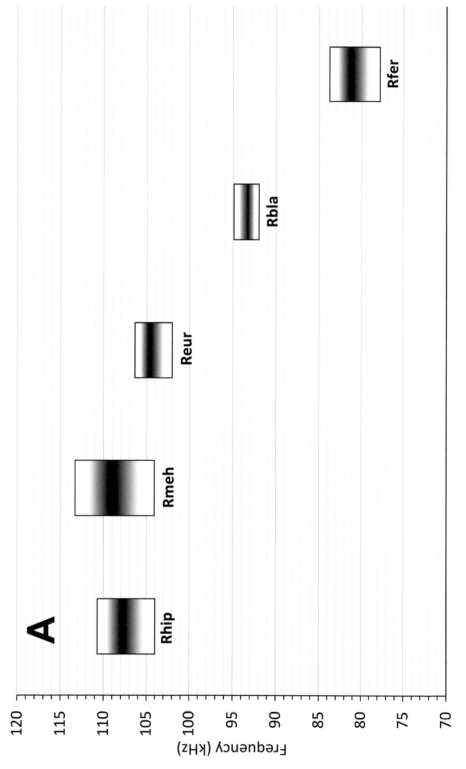

Figure 7.2 Category A FmaxE ranges around the mean (second harmonic). Red = restricted geographical range. For species abbreviations, see Table 7.1.

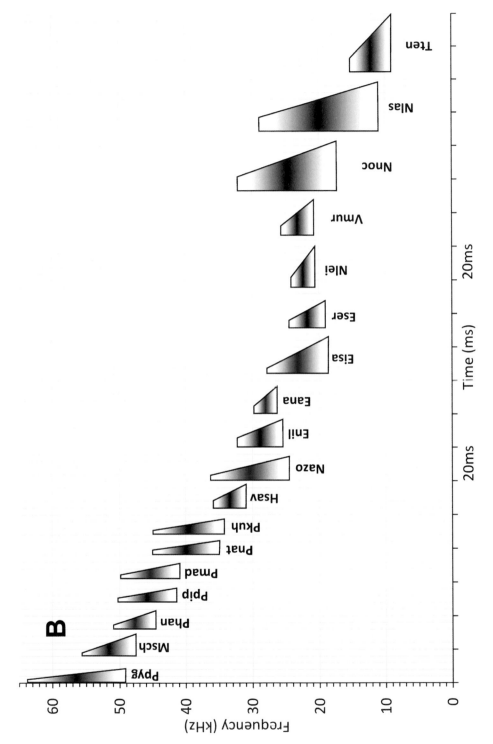

Figure 7.3 Category B Fmin ranges around the mean, with duration ranges also shown. Red = restricted geographical range. For species abbreviations, see Table 7.1.

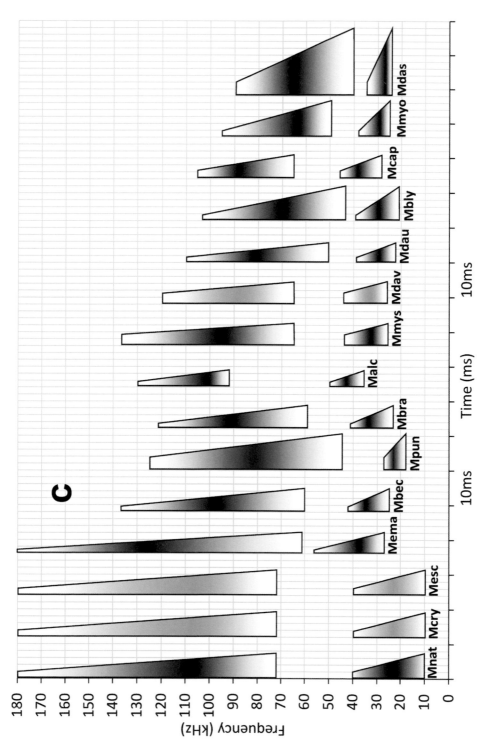

Figure 7.4 Category C Fmax and Fmin ranges around the mean, with duration ranges also shown. Red = restricted geographical range, green = assumed parameters. For species abbreviations, see Table 7.1.

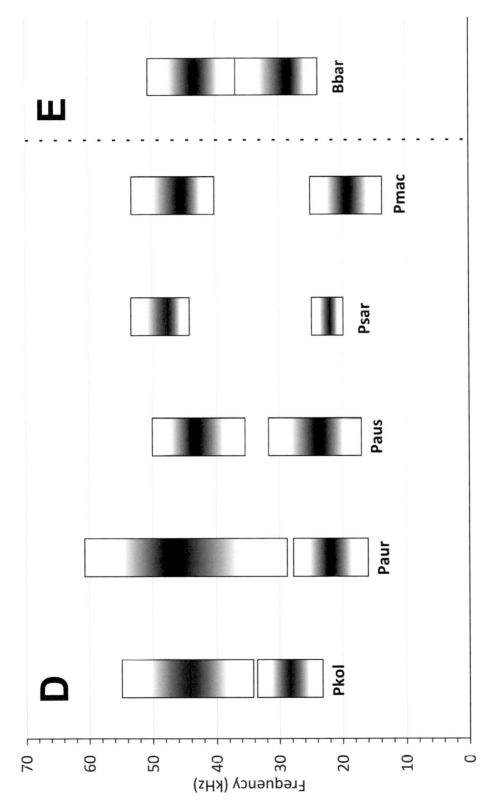

Figure 7.5 Categories D and E first harmonic (fundamental) Fmax and Fmin ranges around the mean. Red = restricted geographical range. For species abbreviations, see Table 7.1.

Categories D and E

Categories D and E contain echolocation calls that are generally low-frequency and relatively short duration, often with harmonics, which are typically produced by European bats of the genera *Plecotus* and *Barbastella*. For this group Fmin and Fmax of the first harmonic (fundamental) are the most useful parameters for initially separating the species. In Figure 7.5, the ranges of maximum frequencies (Fmax) and minimum frequencies (Fmin) are represented by the vertical bars. Mean values and the spread around the mean are shown by the density of shading (darkest = mean, fading to white at the ends of the bars where these values rarely occur). Species with a restricted geographical range are shown in red.

Species notes
Category A – *Rhinolophus*
Lesser horseshoe bat *Rhinolophus hipposideros*

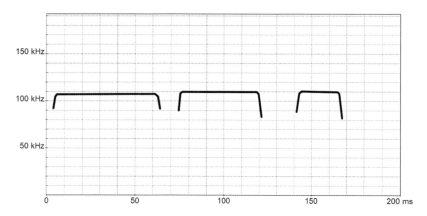

- See Chapter 8, section 8.1
- To avoid the Doppler effect, use only the highest-frequency calls of long sequences from static recordings or when facing the individual from active recordings
- Overlaps with *R. mehelyi* and *R. euryale*
- FmaxE can go down to 105 kHz on mainland Europe, mean 110 kHz in the UK, mean 115 kHz in Crete

Mehely's horseshoe bat *Rhinolophus mehelyi*

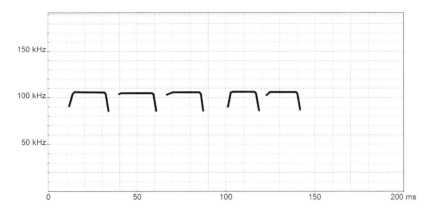

- See Chapter 8, section 8.4
- To avoid the Doppler effect, use only the highest-frequency calls of long sequences from static recordings or when facing the individual from active recordings
- Overlaps with *R. hipposideros* and *R. euryale*
- FmaxE mean in Sicily 112 kHz, in Sardinia 107 kHz, in Greece 109.5 kHz, in Spain 106.5 kHz

Mediterranean horseshoe bat *Rhinolophus euryale*

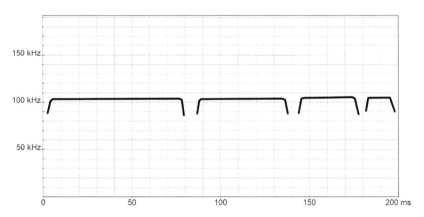

- See Chapter 8, section 8.3
- To avoid the Doppler effect, use only the highest-frequency calls of long sequences from static recordings or when facing the individual from active recordings
- Overlaps with *R. hipposideros* and *R. mehelyi*
- FmaxE mean in eastern and central Europe 104 kHz, in western Europe 102 kHz, in southern Europe (Greece) 105–108 kHz

Blasius's horseshoe bat *Rhinolophus blasii*

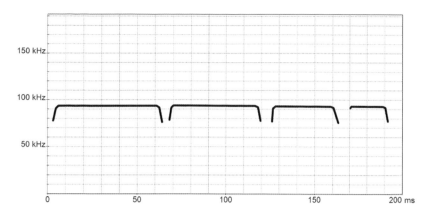

- See Chapter 8, section 8.5
- To avoid the Doppler effect, use only the highest-frequency calls of long sequences from static recordings or when facing the individual from active recordings
- No overlap with other rhinolophids

Greater horseshoe bat *Rhinolophus ferrumequinum*

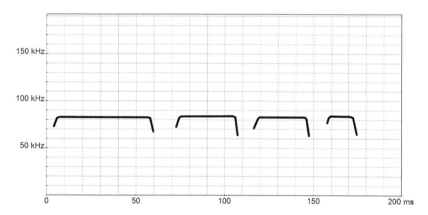

- See Chapter 8, section 8.2
- To avoid the Doppler effect, use only the highest-frequency calls of long sequences from static recordings or when facing the individual from active recordings
- No overlap with other rhinolophids

Category B – *Eptesicus, Hypsugo, Miniopterus, Nyctalus, Pipistrellus, Tadarida, Vespertilio*

Soprano pipistrelle *Pipistrellus pygmaeus*

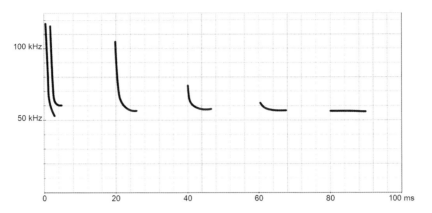

- See Chapter 8, section 8.31
- Overlap with *Miniopterus schreibersii* (and *P. pipistrellus* in southern Europe)
- Social calls may be used to separate the species
- Calls never start with a hook in clutter, whereas those of *M. schreibersii* do
- Feeding buzzes include a clear approach phase, whereas those of *M. schreibersii* do not
- FM calls sometimes end with an upward hook, whereas those of *M. schreibersii* do not
- The curve of qCF calls can show an abrupt angle, whereas the curve of *M. schreibersii* is always a smooth transition
- Calls are rarely longer than 8 ms, whereas calls of *M. schreibersii* are often longer than 10 ms
- In southern Europe, end frequency >54 kHz, whereas *P. pipistrellus* is lower

Schreiber's bent-winged bat *Miniopterus schreibersii*

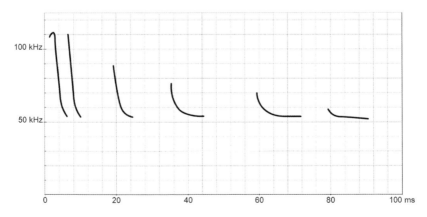

- See Chapter 8, section 8.43
- Sometimes has a starting hook in clutter
- Sometimes has a downward end hook but never an upward end hook

- Feeding buzzes lack a clear approach phase compared to *Pipistrellus pygmaeus* (and other species)
- End frequency 50–52 kHz in >90% of cases, very rarely above 53 kHz
- Except in clutter, many calls are more than 8 ms long (rare in *Pipistrellus*) and can be as long as 16 ms
- Nearly always a smooth transition from FM to qCF
- Can be a very irregular IPI (e.g. short IPI 60–80 ms often mixed with much longer ones (~200 ms))
- In open environments, often alternates between short calls with high energy and long calls with low energy (almost inaudible), which might be due to alternating head movements
- The curve of qCF calls is always gradual
- Length of sequence (from the first call to the last call) often very short (about 2.5 seconds) due to its high flight speed, whereas sequences of *Pipistrellus* often last about 4–5 seconds

Hanak's pipistrelle *Pipistrellus hanaki*

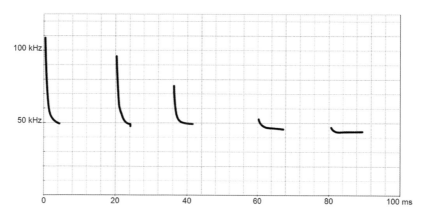

- See Chapter 8, section 8.32
- Only found on the island of Crete (plus Cyrenaica peninsula)
- May be confused with *P. pipistrellus*, which also occurs on the island (mean FmaxE *P. hanaki* = 48 kHz, *P. pipistrellus* = 46 kHz). However, social calls may be used to separate these species

Common pipistrelle *Pipistrellus pipistrellus*

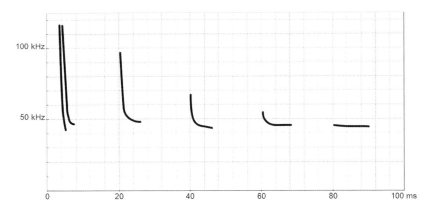

- See Chapter 8, section 8.30
- On the island of Crete may be confused with *P. hanaki* (mean FmaxE *P. hanaki* = 48 kHz, *P. pipistrellus* = 46 kHz)
- Can be confused with *Miniopterus schreibersii* and *P. pygmaeus* in southern Europe
- Social calls may be used to separate the species
- In southern Europe, end frequency < 54 kHz, whereas *P. pygmaeus* is higher
- FM calls sometimes end with an upward hook, whereas those of *M. schreibersii* do not
- Feeding buzzes include a clear approach phase, whereas those of *M. schreibersii* do not
- The curve of qCF calls can show an abrupt angle, whereas the curve of *M. schreibersii* is always a smooth transition
- Calls are rarely longer than 8 ms, whereas calls of *M. schreibersii* are often longer than 10 ms

Madeira pipistrelle *Pipistrellus maderensis*

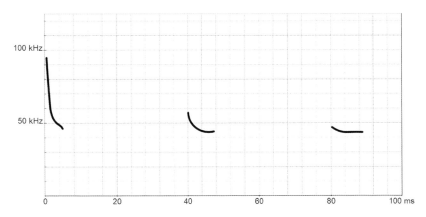

- See Chapter 8, section 8.35
- Restricted to Madeira and the Canary Islands. No other similar species present

Nathusius's pipistrelle *Pipistrellus nathusii*

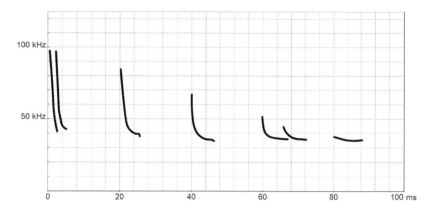

- See Chapter 8, section 8.33
- Overlap with *P. kuhlii*
- Produces few social calls (except in autumn, when they are frequently produced) but they can be used to easily separate the species from *P. kuhlii*
- When foraging, produces qCF calls frequently, whereas *P. kuhlii* produces few qCF calls
- Very flat qCF calls around 39–40 kHz are more common in *P. nathusii* but may occasionally be found in *P. kuhlii*

Kuhl's pipistrelle *Pipistrellus kuhlii*

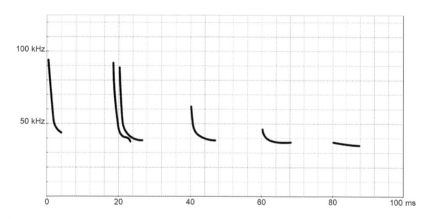

- See Chapter 8, section 8.34
- Overlap with *P. nathusii* and may be confused with *Hypsugo savii* in clutter
- Frequently emits social calls, which can be used to separate the species from *H. savii*. In addition, structure of social call very different from those of *H. savii* (which are emitted rarely) and *P. nathusii*
- Fmin goes down often to 35 kHz, occasionally 33 kHz during social behaviour
- When foraging, produces few qCF calls, whereas *P. nathusii* and *H. savii* often produce qCF calls
- FM calls often display a downward hook with a bandwidth that can be larger than 5 kHz, whereas *P. nathusii* and *H. savii* have end hooks with a bandwidth less than 5 kHz

Savi's pipistrelle *Hypsugo savii*

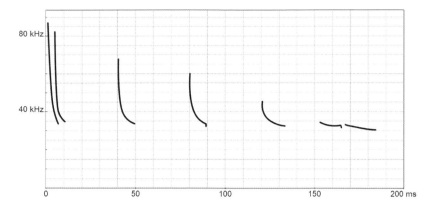

- See Chapter 8, section 8.36
- May be confused with *Pipistrellus kuhlii*
- When foraging frequently produces qCF calls, whereas *P. kuhlii* rarely produces qCF calls
- Fmin of qCF calls can be as low as 30 kHz (sometimes 29 kHz with Doppler effect, which may be confused with *Eptesicus nilssonii*), whereas Fmin of *P. kuhlii* lowest qCF only as low as 33 kHz
- In the open *H. savii* does not overlap with *P. kuhlii* and *P. nathusii*: FmaxE of calls is 30–33 kHz and call duration of qCF is longer than both *P. kuhlii* and *P. nathusii*, up to 18 ms
- Fmin can go to 40 kHz in clutter (lots of calls like this when near roost)
- Produces few social calls away from the roost, which makes it difficult to confirm in the absence of low qCF calls. However, many social calls in the vicinity of roosts (around cliffs and buildings, for example)

Azorean noctule *Nyctalus azoreum*

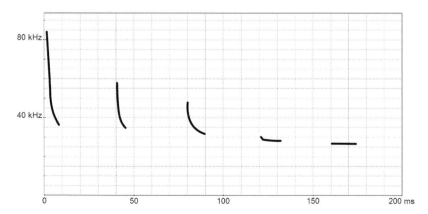

- See Chapter 8, section 8.24
- No other bats on the Azores that can be confused with this species

Northern bat *Eptesicus nilssonii*

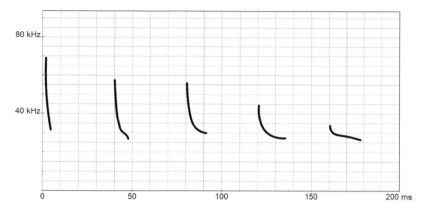

- See Chapter 8, section 8.28
- *Eptesicus* are like big *Pipistrellus* – frequency does not fluctuate (jump about) as much as *Nyctalus* over a regular sequence of calls
- May be confused with *Hypsugo savii* when *H. savii* produces flat qCF calls with a Doppler effect making them as low as 29 kHz
- May be confused with *N. leisleri* but can produce calls with Fmin > 25 kHz and call length > 18 ms, whereas *N. leisleri* can not
- qCF rarely shorter than 10 ms, whereas *N. leisleri* can be longer
- May be confused in clutter with *E. serotinus* and *Vespertilio murinus*, but qCF calls are higher than in those species

Anatolian serotine *Eptesicus anatolicus*

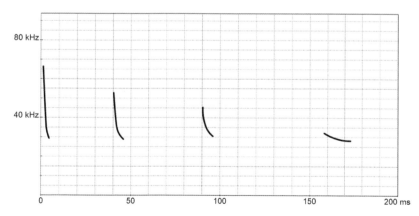

- See Chapter 8, section 8.27
- *Eptesicus* are like big *Pipistrellus* – frequency does not fluctuate (jump about) as much as *Nyctalus* over a regular sequence of calls
- In Europe restricted to the Greek islands off the Anatolian coast, including Rhodes, and possibly Samos and Cyprus. No overlap with other similar species in this area

Meridional serotine *Eptesicus isabellinus*

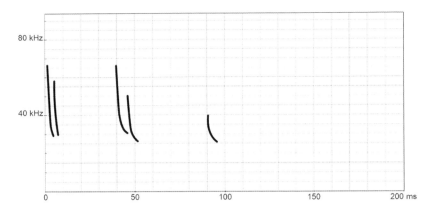

- See Chapter 8, section 8.26
- *Eptesicus* are like big *Pipistrellus* – frequency does not fluctuate (jump about) as much as *Nyctalus* over a regular sequence of calls
- Restricted to just over the southern half of the Iberian peninsula. In this area, there is overlap with *Nyctalus leisleri*, although frequencies are generally lower for *N. leisleri* (mean qCF = 29.5 kHz for *E. isabellinus*, compared to mean qCF = 23 kHz and mean FM-qCF = 27 kHz for *N. leisleri*)
- Fmin can go down to 22 kHz

Serotine *Eptesicus serotinus*

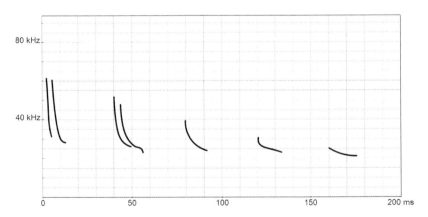

- See Chapter 8, section 8.25
- *Eptesicus* are like big *Pipistrellus* – frequency does not fluctuate (jump about) as much as *Nyctalus* over a regular sequence of calls
- Overlap in clutter with *E. nilssonii* and *Vespertilio murinus*, but qCF calls are lower than in those species
- Overlap with *N. noctula* and *N. leisleri*, especially in clutter
- When producing a series of FM calls, one call is often missing in the series although the rest of the calls display very regular IPI. *Nyctalus* do not show such regular IPI

- When flying in open environments, displays series of very typical long calls with low bandwidth
- qCF calls are rarely shorter than 10 ms, whereas *N. leisleri* can be longer
- Fmax is no higher than 64 kHz
- For qCF calls usually maintains a bandwidth over 2 kHz, whereas a bandwidth below 0.5 kHz regularly occurs for other species

Leisler's bat *Nyctalus leisleri*

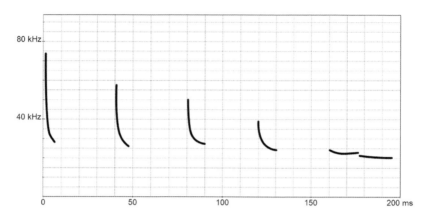

- See Chapter 8, section 8.23
- *Nyctalus* calls tend to jump between call types over a long sequence of calls (e.g. FM-qCF and qCF) compared to *Eptesicus*
- May be confused with *E. nilssonii* but cannot produce calls with Fmin > 25 kHz and duration > 18 ms, whereas *E. nilssonii*, in general, has higher Fmin with longer call lengths > 18 ms
- Only bat in the *Nyctalus/Eptesicus/Vespertilio* group capable of producing very short qCF calls (< 8 ms), but it can also produce very long calls (up to 26 ms, as long as *V. murinus*)
- Can be very plastic in frequency (Fmin of qCF varies from 20 to 26 kHz)
- In areas where *V. murinus* is absent, only bat to produce very 'flat' calls at 22–23 kHz
- Only long qCF (> 10 ms) are below 25 kHz
- May be confused with *N. noctula* when strong Doppler effect reduces call frequency down to 20–21 kHz, but calls will be completely flat in *N. leisleri* at these frequencies, whereas *N. noctula* calls will not be

Parti-coloured bat *Vespertilio murinus*

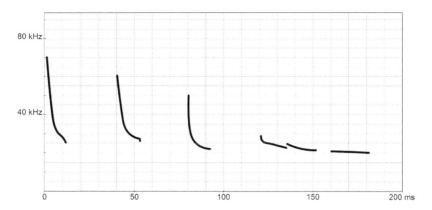

- See Chapter 8, section 8.29
- Similar to a big *Pipistrellus*. No alternating calls
- Overlap with *Eptesicus* and *Nyctalus* in clutter
- Very difficult to distinguish from *N. leisleri* because although *N. leisleri* displays several unique acoustic criteria, *V. murinus* does not
- Duration is generally higher in *V. murinus* but there is some overlap
- For the same call length, Fmin higher than *E. serotinus*
- Produces long series of diagonal (no knee) qCF without alternating calls

Noctule *Nyctalus noctula*

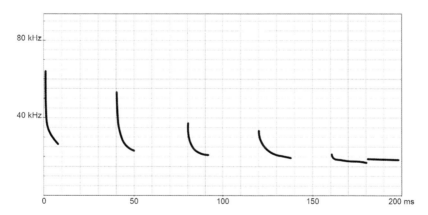

- See Chapter 8, section 8.21
- *Nyctalus* calls tend to jump between call types over a long sequence of calls (e.g. FM-qCF and qCF) compared to *Eptesicus*
- Overlap with *Eptesicus*, *Vespertilio* and *N. leisleri* in clutter
- The only bat to produce flat qCF at 17–18 kHz
- May be confused with *N. leisleri* when strong Doppler effect reduces call frequency of *N. leisleri* down to 20–21 kHz, but calls will be completely flat in *N. leisleri* at these frequencies, whereas *N. noctula* calls will not be
- May be confused with *N. lasiopterus* when strong Doppler effect reduces call frequency down to 16 kHz, but calls will be completely flat in *N. noctula* at these frequencies, whereas *N. lasiopterus* calls will not be

Greater noctule *Nyctalus lasiopterus*

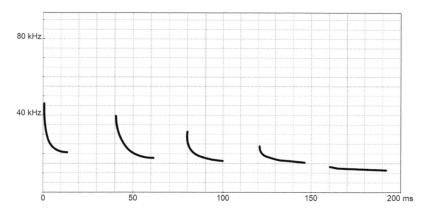

- See Chapter 8, section 8.22
- *Nyctalus* calls tend to jump between call types over a long sequence of calls (e.g. FM-qCF and qCF) compared to *Eptesicus*
- Frequencies are generally higher than *Tadarida teniotis*. For same bandwidth, Fmin of *N. lasiopterus* FM calls are often > 17 kHz and/or longer than 19 ms; Fmin of *T. teniotis* qCF calls are often > 13 kHz and/or longer than 23 ms
- May be confused with *N. noctula* when strong Doppler effect reduces call frequency of *N. noctula* down to 16 kHz, but calls will be completely flat in *N. noctula* at these frequencies, whereas *N. lasiopterus* calls will not be
- May be confused with *N. leisleri* social calls (especially common in autumn). However, these social calls start with a qCF component, then an FM and qCF again, whereas *N. lasiopterus* starts with an FM. In addition, social calls are emitted while stationary

European free-tailed bat *Tadarida teniotis*

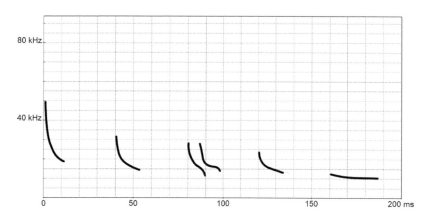

- See Chapter 8, section 8.44
- When flying in the open calls FmaxE < 13 kHz, *Nyctalus lasiopterus* > 13 kHz
- Calls can be very variable – the S-shaped calls are often used in semi-open habitats
- *T. teniotis* FmaxE frequently goes down to 10 kHz and occasionally 9 kHz
- The most common FmaxE when commuting is 11 kHz

- Calls with FmaxE 14–15 kHz are generally emitted only when actively foraging or in social interaction
- May be confused with *N. leisleri* social calls (especially common in autumn). However, these social calls start with a qCF component, then an FM and qCF again, whereas *T. teniotis* starts with an FM. In addition, social calls are produced while stationary

Category C – *Myotis*

Natterer's bat *Myotis nattereri*/Cryptic myotis *M. crypticus*/ Iberian Natterer's bat *M. escalerai*

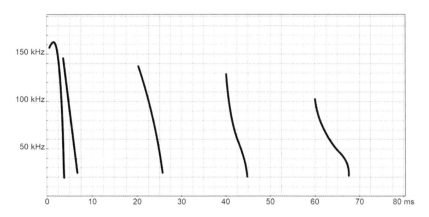

- See Chapter 8, sections 8.14, 8.15 and 8.16
- Very high bandwidth
- End frequency down into the audible range (< 20 kHz)
- Able to scan more than 100 kHz in 1 ms
- In closed and semi-closed environments, no knees and heels or only slightly visible
- Start frequency up to 155 kHz
- Warbles are common in commuting *M. nattereri*

Geoffroy's bat *Myotis emarginatus*

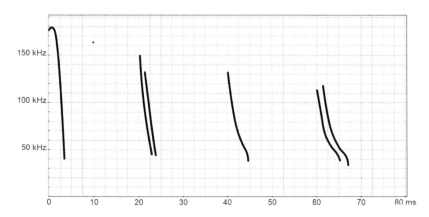

- See Chapter 8, section 8.13
- Frequently the calls are very straight

- Often the start frequency is very high – up to 175khz
- End frequency is very high, usually around 40 kHz in most cases
- The heel is sometimes visible but very rarely the knee (which makes this unique among the commuting calls >4 ms)
- Signals rather short (1–3 ms in most cases)

Maghreb mouse-eared bat *Myotis punicus*

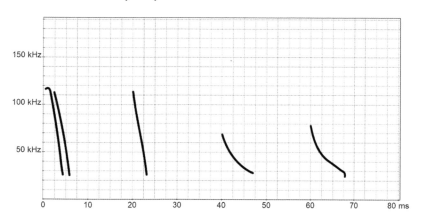

- See Chapter 8, section 8.20
- In Europe restricted to islands of Malta, Corsica (France), Sardinia and Sicily

Alcathoe whiskered bat *Myotis alcathoe*

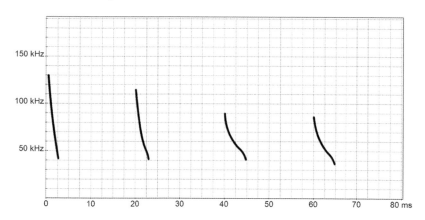

- See Chapter 8, section 8.12
- High end frequency, which is usually above 43 kHz (compared to *M. emarginatus*, which is usually above 40 kHz)
- Knee visible but no heel on short-duration calls
- The call starts very abruptly with lots of energy between start and knee
- Heel around 45 kHz

David's myotis *Myotis davidii*

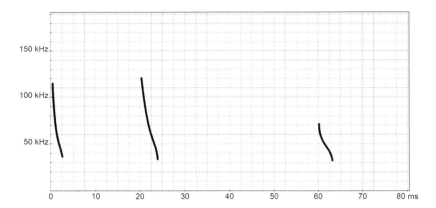

- See Chapter 8, section 8.11
- Restricted to far east of Europe: northeast Bulgaria, eastern Romania, Moldova, southern Ukraine
- Little information

Brandt's bat *Myotis brandtii*/Whiskered bat *M. mystacinus*

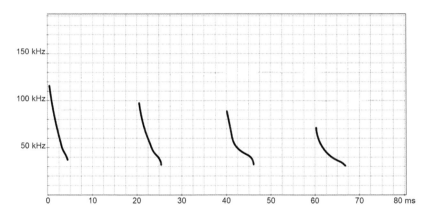

- See Chapter 8, sections 8.9 and 8.10
- Knee often more pronounced than heel
- The section between start and knee abrupt except during commuting in the open
- In semi-closed environments, signals close to those of *M. alcathoe* but lower frequency
- In semi-open environments, signals very close to those of *M. daubentonii*
- Unlike *M. daubentonii*, end frequency decreases with a lengthening of the calls

Daubenton's bat *Myotis daubentonii*

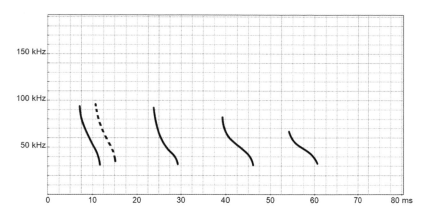

- See Chapter 8, section 8.6
- Signal regularly S-shaped
- Start frequency very rarely higher than 100 kHz
- Heel around 38–40 kHz, knee often around 60 kHz
- Bandwidth between knee and heel (around 20 kHz) often larger than in *M. capaccinii* (around 10 kHz)
- End frequency around 25 kHz. End frequency increases with increasing call duration
- Interference when flying over the water – 'dashed'
- Pulses > 2.5 ms curved with clear final hook
- Similar to *M. capaccinii* but with slightly lower frequencies (heel around 38–40 kHz and end frequency around 25 kHz)
- End frequencies do not alternate in frequency within a sequence
- End frequency variable, often lower than *M. dasycneme*
- Bandwidth between the knee and end frequency is fairly large (15–20 kHz) compared to *M. dasycneme*
- Social behaviour of *P. kuhlii* can resemble echolocation calls of *M. daubentonii*, although they will never be produced during the whole sequence in *P. kuhlii*, and 'usual' echolocation calls will eventually appear

Long-fingered bat *Myotis capaccinii*

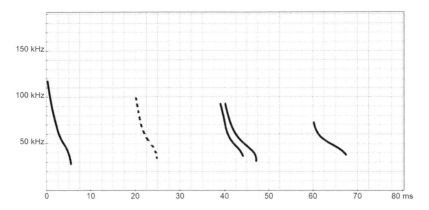

- See Chapter 8, section 8.8
- Similar to *M. daubentonii* but with slightly higher frequencies – heel around 45 kHz and end frequency around 35 kHz
- Among the bats with end frequencies around 35 kHz, only *M. capaccinii* can have signals of duration >5 ms
- Interference when flying over the water – 'dashed'
- End frequencies can alternate in frequency within a sequence – not known in *M. daubentonii*

Pond bat *Myotis dasycneme*

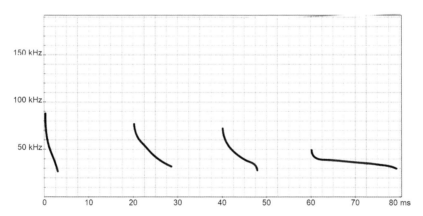

- See Chapter 8, section 8.7
- Pulses >2.5 ms curved with a clear final hook, positioned after qCF part
- Heel usually 30–35 kHz
- Able to emit very long calls (>20 ms)
- FmaxE generally lower than *M. daubentonii*
- Usually a concave curve with no obvious knee
- Duration over 8 ms is common, compared to *M. daubentonii*, which uses calls over 8 ms very rarely
- End frequency often more stable and higher than *M. daubentonii*
- Bandwidth between the knee and end frequency is small (10 kHz) compared to *M. daubentonii*

Bechstein's bat *Myotis bechsteinii*

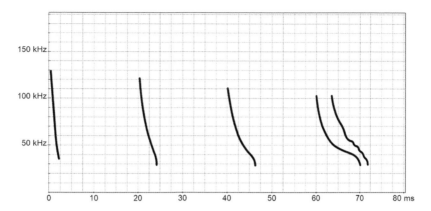

- See Chapter 8, section 8.17
- In closed environments, signals are similar to those of *M. emarginatus*
- In semi-open environments, signals similar to those of *M. daubentonii*, but may lengthen the calls beyond 8 ms in the open
- Unlike *M. daubentonii*, end frequency decreases with lengthening of calls
- May have warble like *M. myotis*
- In a semi-open environment, FmaxE is higher than *M. myotis* (> 35 kHz)
- In cluttered environments, soft short FM; in open environments, louder diagonal sweep with qCF at 50 kHz

Greater mouse-eared bat *Myotis myotis*/Lesser mouse-eared bat *M. blythii*

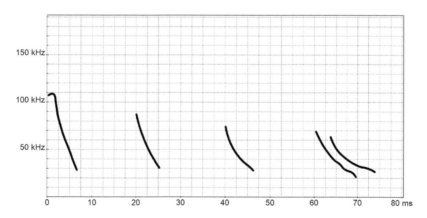

- See Chapter 8, sections 8.18 and 8.19
- End frequency 12–30 kHz
- Able to emit very long calls (> 10 ms)
- FmaxE down to 30–35 kHz
- Warbling shape with long-duration calls (> 6 ms)
- This is the only *Myotis* species for which the frequency of maximum energy is close to the end frequency
- Possible confusion with *M. nattereri* when the repetition rate is high (i.e. approach phase)

- When longer than 4 ms the calls are curved; when shorter than 4 ms the calls are linear with no clear terminal hook
- The heel can be below 30 kHz

Category D – *Plecotus*
Brown long-eared bat *Plecotus auritus*

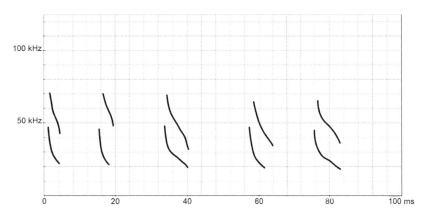

- See Chapter 8, section 8.38
- *P. auritus* start frequency is often above 48 kHz, whereas for *P. austriacus* it is often below 48 kHz
- *P. auritus* has a higher start frequency and FmaxE and larger bandwidth of the first harmonic
- *P. auritus* bandwidth of the first harmonic (fundamental) generally above 27 kHz; below 27 kHz for *P. austriacus*
- *P. auritus* – reported that the two first harmonics can overlap with signals of less than 4 ms

Grey long-eared bat *Plecotus austriacus*

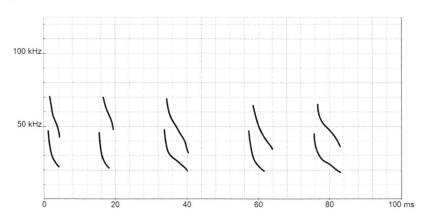

- See Chapter 8, section 8.41
- Start frequency on first harmonic not usually above 50 kHz
- *P. austriacus* start frequency is often below 48 kHz, whereas for *P. auritus* it is often above 48 kHz

- *P. austriacus* has a lower start frequency and FmaxE and smaller bandwidth of the first harmonic
- *P. austriacus* bandwidth of the 1st harmonic (fundamental) generally below 27 kHz; above 27 kHz for *P. auritus*
- *P. austriacus* – reported that two first harmonics do not overlap with signals of less than 4 ms

Mediterranean long-eared bat *Plecotus kolombatovici*

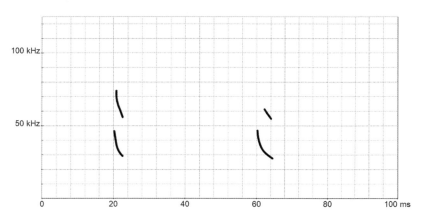

- See Chapter 8, section 8.42
- Little information
- Restricted to Croatia through the Adriatic islands and narrow coastal strip along the Adriatic Sea, an area near the Greek coast and on some islands in the Aegean Sea. Also on Crete, Rhodes and Cyprus

Alpine long-eared bat *Plecotus macrobullaris*

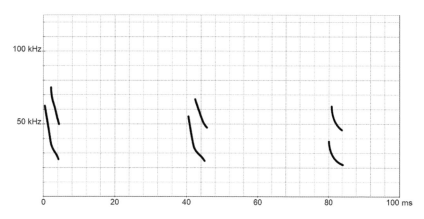

- See Chapter 8, section 8.39
- Calls similar to those of *P. austriacus*
- FmaxE of *P. macrobullaris* is lower (30 kHz) than that of *P. auritus* (34 kHz)

Sardinian long-eared bat *Plecotus sardus*

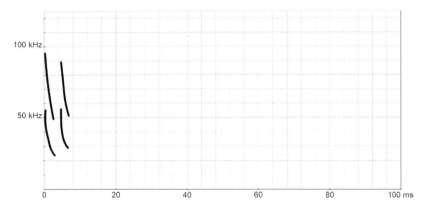

- See Chapter 8, section 8.40
- Little information
- Restricted to the central part of Sardinia

Category E – *Barbastella*
Western barbastelle *Barbastella barbastellus*

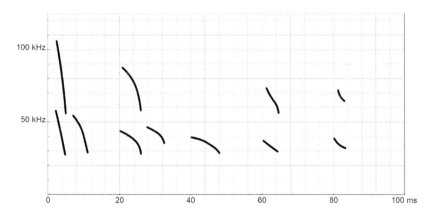

- See Chapter 8, section 8.37
- Uses alternate call types when hunting and commuting
- In cluttered and semi-cluttered environments FM calls tend to be convex compared to *Plecotus* calls, which are concave

8 The Bat Species

The species sections provide details of all 44 species of European bat and are arranged as follows:

DISTRIBUTION

Maps are based on those in Dietz and Kiefer (2016) but have been modified where appropriate to reflect current knowledge.

EMERGENCE

Emergence times are from a variety of sources and show the range of emergence for each species. The main period of the species' emergence is shown in red, fading from the central mean. Ranges are shown above a bar showing the range for all species combined. This bar is shaded to show periods of daylight, dusk and night.

FLIGHT AND FORAGING BEHAVIOUR

General characteristics of flight for the species, and behaviour when foraging.

HABITAT

The habitats that the species are typically found in are described. These are usually the foraging habitats of the species, but other habitats are also mentioned. Also included are typical roosts used throughout the year.

ECHOLOCATION

Tables of measured parameters are presented. Where more than one call type (e.g. FM-CF-FM, qCF, FM) is recognised, these are included in separate columns. Descriptions of echolocation calls are divided into heterodyne and time expansion/full spectrum. Sonograms have been fixed in the time and frequency scales to facilitate comparison with similar species. For time-expansion/full-spectrum calls, the features of echolocation pulses presented in Figure 2.25 are used in descriptions.

SOCIAL CALLS

Descriptions of social calls are divided into heterodyne and time expansion/full spectrum. Sonograms vary in time and frequency scales to best present the calls. The social calls include only those that could be obtained at the time of writing. A bat's vocal repertoire is likely to be broader than that presented. No attempt has been made to categorise calls, owing to the difficulty in doing so, but more importantly to avoid the assumption of function that would then arise. Where function has been established by research, this is included, but more often than not no research is available.

SPECIES WITH SIMILAR OR OVERLAPPING ECHOLOCATION CALLS

A list of those species with similar or overlapping calls is included, with reference to the basic echolocation guide to species presented in Chapter 7.

NOTES

Any further information not included in the main text.

8.1 Lesser horseshoe bat

Rhinolophus hipposideros (Bechstein, 1800)

Maggie Andrews, Amelia Hodnett and Peter Andrews

© James Shipman

DISTRIBUTION

Originally present up to 51–52°N, but now largely absent from much of Germany and the adjacent countries. Widely distributed in the Mediterranean. Southwest England and south, north and west Wales, Wirral, Cheshire and Staffordshire, and the west of Ireland.

EMERGENCE

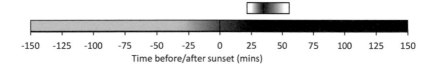

-150   -125   -100   -75   -50   -25   0   25   50   75   100   125   150
Time before/after sunset (mins)

FLIGHT AND FORAGING BEHAVIOUR

Skilful and fairly fast, wing movements almost whirring. Flight is low, near the ground and up to 5 m above. Extremely manoeuvrable and capable of hovering. Spends most of the time flying close to clutter among deciduous trees. Insects are caught by aerial hawking, gleaning and perch feeding (Schofield and McAney 2008), and by pouncing on prey close to the ground. When perched, regularly scans the surroundings by continuously turning its body around its legs. Often uses a night-roost perch, where it eats its captured prey. During social interactions outside the nursery roost, the bats are very agile and can avoid each other at close range near the exit hole (Andrews *et al.* 2017).

HABITAT

Mainly deciduous woodland, but also pasture, woodland edge and hedgerows. Also, over water and in farmyards. Patrols river edges next to luxuriant riparian vegetation. Broadleaf woodland generally used more than any other habitat.

Nursery roosts are typically located in the attic of a roof in a stone-built house, church or farm outbuilding. Individuals hang from beams in a nursery roost and form clusters. Night roosts in porches, outbuildings and cellars are often used to perch and eat their prey (Schofield and McAney 2008). Females often carry infants to the night roosts and forage in the deciduous woods (Amelia Hodnett, personal observation). Social behaviour occurs outside nursery roosts in passageways or covered areas when communal flights take place in a circle or figure of eight after the dusk emergence and when bats return to the roost during the night to feed infants (Andrews *et al.* 2017). Hibernation takes place in caves, mines or cellars, and individuals hang separately where the ambient temperature is 5–11 °C and the air is moist (Schofield and McAney 2008). Greater horseshoe bats may hibernate nearby, but nursery roosts are typically in different parts of the roof in a building (Schofield 2008).

ECHOLOCATION

Measured parameters	Mean (range)	
	Adult FM-CF-FM call	Infant FM-CF-FM call
Inter-pulse interval (ms)	70.4 (14.1–113.7)	n/a
Call duration (ms)	43.6 (11.9–61.4)	24.3 (14.0–33.5)
Frequency of maximum energy (kHz)	111.1 (107.3–114.0)	101.5 (97.0–106.0)
Start frequency (kHz)	99.0 (92.3–107.8)	94.0 (87.0–101.8)
End frequency (kHz)	96.6 (83.4–110.3)	91.5 (78.0–104.3)

Heterodyne
When a detector is tuned to around 110 kHz a continuous warbling sound is heard. The echolocation calls are distinctive from other species except for other rhinolophids. Calls overlap with those of *R. mehelyi* and *R. euryale*. Near the roost or in a confined space the first harmonic (fundamental) can be heard at around 55 kHz.

Time expansion/full spectrum
Adult echolocation calls are composed of a short FM upward sweep up, followed by a CF pulse 40.8 ± 10.5 ms long at 110–114 kHz, then a short downward FM sweep (Vaughan *et al.* 1997). The echolocation call is the second harmonic emitted through the nose. In the roost and sometimes in the field the first harmonic or fundamental sound can be heard in the frequency range 55–57 kHz.

Calls show the typical FM-CF-FM structure found in all rhinolophids (Figures 8.1.1–8.1.4). The component with the most energy (FmaxE), peaking at around 111 kHz (Vaughan *et al.* 1997), is actually the second harmonic emitted through the nose. The first harmonic is the fundamental sound produced in the larynx (Neuweiler 2000). Call frequency may vary slightly as the individual adjusts its echolocation to compensate for the Doppler effect during flight. Also, when many bats are flying at once the echolocation frequency is variable. Echolocation can vary according to age, sex and body size (Vaughan *et al.* 1997, Jones and Rayner 1989), so differences in these signals offer the potential for intraspecific communication (Jones and Siemers 2011). The short CF pulses and large FM sweeps in rhinolophid echolocation calls provide maximum information about the distance of an object in front of the bat (Simmons *et al.* 1979, Andrews 1995). Call duration in the nursery roost is relatively short, between 15 and 30 ms, the FM components are pronounced and the fundamental is often recorded. In cluttered environments, the inter-pulse interval and duration decrease and the bandwidth of the FM components may increase.

Echolocation calls overlap with those of *R. mehelyi* and *R. euryale*.

In mainland Europe, mean FmaxE can go down to 105 kHz, and in Crete it can go up to 115 kHz.

Infant echolocation calls recorded at 10–20 days of age are similar to the adult FM-CF-FM calls but are made at lower frequencies and are often shorter (Figures 8.1.1 and 8.1.5). In the example shown there are short echolocation calls and long calls at 97 kHz. They are produced initially as the third or fourth harmonic but develop so that the second harmonic is produced through the nose (Konstantinov 1973, Konstantinov *et al.* 1990).

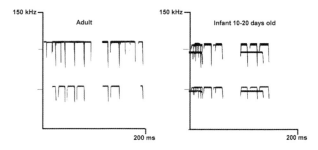

Figure 8.1.1   Echolocation calls of adult lesser horseshoe bats and infant bats.

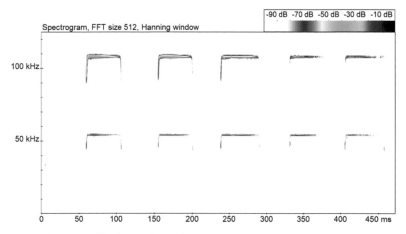

Figure 8.1.2   Echolocation calls of lesser horseshoe bats emerging from a roost (Jon Russ).

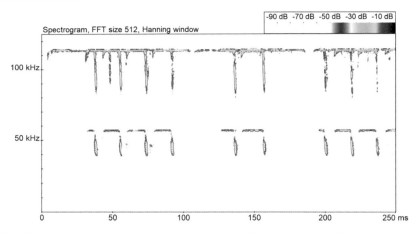

Figure 8.1.3　Echolocation calls of adult lesser horseshoe bats in a nursery roost (Amelia Hodnett).

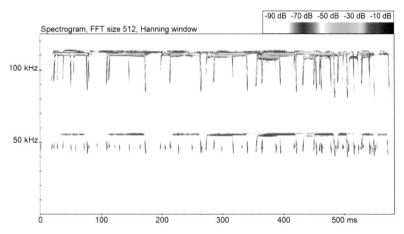

Figure 8.1.4　Echolocation calls of lesser horseshoe bats flying within the branches of an oak tree (Jon Russ).

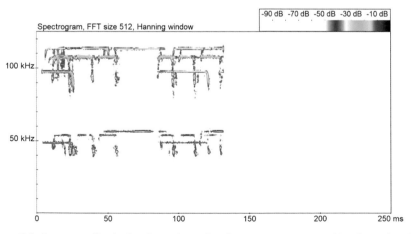

Figure 8.1.5　Echolocation calls of infant lesser horseshoe bats in a nursery roost (Amelia Hodnett).

SOCIAL CALLS

Heterodyne
The fundamental frequencies of the social calls are in the range 15–49 kHz, but it is not advisable to try to identify these calls using heterodyne detectors because the identification of each type of call requires broadband recordings and analysis with software.

Time expansion/full spectrum
The types of ultrasonic social calls made by lesser horseshoe bats are identified according to their parameters in three main groups: single-component, multiple-component and modified echolocation calls. The fundamental frequencies in the range 15–42 kHz are the predominant sounds, and many calls have harmonics, but generally the social calls are below the echolocation frequency (Andrews *et al.* 2017).

Single-component calls
The frequency of single-component calls may be constant, or it may rise or fall during the call. Andrews *et al.* (2017) describe parameters of these calls in detail.

Constant-frequency calls
Adult lesser horseshoe bats use constant-frequency calls more than any other social calls inside and outside the nursery roost and in a hibernaculum (Figures 8.1.6 and 8.1.7). The fundamental frequency is low (21.6 ± 4.5 kHz) and generally very long but the duration is variable (107.7 ± 68.4 ms).

Isolation calls of infant lesser horseshoe bats are polyharmonic constant frequency calls (Figures 8.1.6 and 8.1.8). These calls are characteristic of newborn to day-old infant bats in a nursery roost in June and July and occur in double or triple syllables. The fundamental frequencies are higher than adult calls, ranging from 21 to 33 kHz, and the duration is shorter (27.2 ± 8.5 ms).

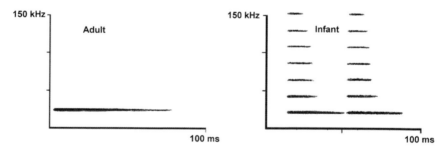

Figure 8.1.6 Single-component CF social calls made by an adult lesser horseshoe bat and infant isolation calls from the same species.

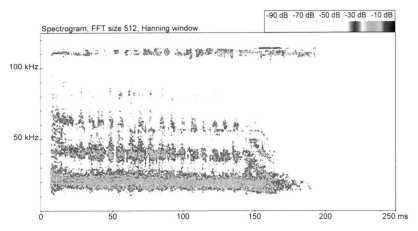

Figure 8.1.7   Single-component CF social calls of adult lesser horseshoe bats in a nursery roost (Amelia Hodnett).

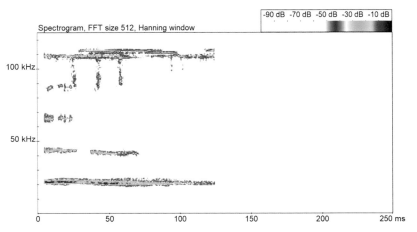

Figure 8.1.8   Infant lesser horseshoe bat isolation call in a nursery roost with adult echolocation calls (Amelia Hodnett).

Rising- or falling-frequency FM calls

Rising-frequency single-component calls made by adult lesser horseshoe bats are often observed in nursery roost recordings (Figures 8.1.9 and 8.1.10). The duration of these calls is variable but often short (21.9 ± 11.0 ms) and the fundamental frequency ranges from 18 to 27 kHz, sometimes with a second harmonic. The first harmonic of falling-frequency calls ranges from 13 to 16 kHz, and these calls are shorter (6.6 ± 2.8 ms) and less frequent than the rising-frequency calls.

Infant lesser horseshoe bats develop the rising- or falling-frequency call soon after birth, and these calls are observed as a double-syllable call with the constant-frequency isolation calls (Figures 8.1.9 and 8.1.11). Typically, the fundamental frequencies are 34–37 kHz and they are polyharmonic short calls (17.3 ± 4.0 ms). Rising-frequency calls also occur in double-syllable calls.

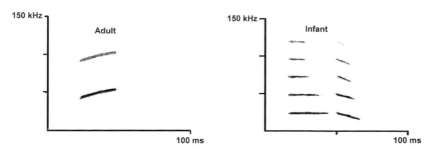

Figure 8.1.9 Single-component FM social calls made by an adult lesser horseshoe bat and infant development calls.

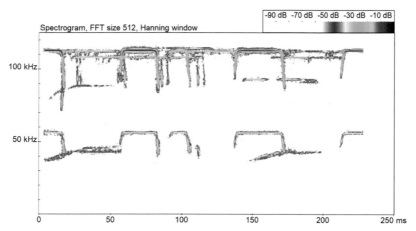

Figure 8.1.10 Single-component FM social calls of adult lesser horseshoe bats in a nursery roost (Amelia Hodnett).

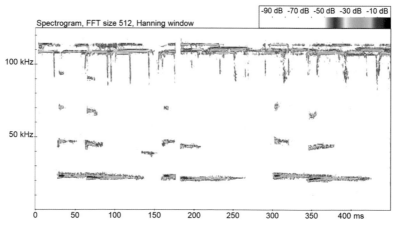

Figure 8.1.11 Infant lesser horseshoe bat FM development call in a nursery roost with adult echolocation calls (Amelia Hodnett).

Modified echolocation calls

Modified echolocation calls are similar to echolocation calls but their fundamental (first harmonic) is much lower, and therefore the second harmonic is not in the range suitable for echolocation. These calls are found in a sequence or set of calls in which the fundamental frequency of each call is 5 kHz below the previous call. There are often five calls in the sequence and they look like echolocation calls but are lower in frequency.

Modified echolocation calls are identified by the central part of the call as constant-frequency (CF) or frequency-modulated (FM) modified echolocation calls depending on the central part of the call.

Constant-frequency modified echolocation calls

The adult CF modified echolocation calls are measured as a cascade of short calls, each of which is 10 ms in duration (Figures 8.1.12 and 8.1.13). Measurements of the second harmonic of the first calls in the cascade are in the range 110–112 kHz and the last calls in the sequence of five calls range from 70 to 79 kHz. The cascades of modified echolocation calls are associated with flight inside the roost.

The infant CF modified echolocation calls are longer than adult calls (10–21 ms) and occur in a random sequence of 3–5 calls (Figures 8.1.12 and 8.1.14). These calls are attempts at echolocation; the frequencies of the second harmonic of the first call in a sequence range from 60 to 69 kHz, and the last call is 100–105 kHz. The first harmonic is always observed.

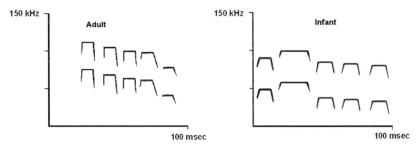

Figure 8.1.12   Adult lesser horseshoe bat CF modified echolocation social calls and infant development calls.

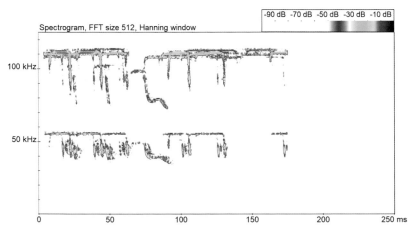

Figure 8.1.13   Adult lesser horseshoe bat CF modified echolocation social calls in a nursery roost with adult echolocation calls (Amelia Hodnett).

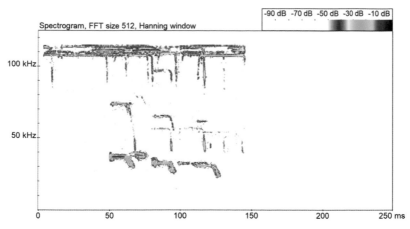

Figure 8.1.14   Infant lesser horseshoe bat CF development calls in a nursery roost with echolocation calls (Amelia Hodnett).

Frequency-modulated modified echolocation calls

The adult FM modified echolocation calls are measured as a sequence of short calls (6.9±0.5 ms) (Figures 8.1.15 and 8.1.16). The second harmonics of calls in a sequence of three rising frequencies range from 80–89 kHz to 100–109 kHz and are followed by a series of echolocation calls at 110–114 kHz. The sequence can also be observed as a cascade of calls of decreasing frequency. Often a cascade of decreasing-frequency calls is followed by a sequence of increasing-frequency calls.

Infant FM modified echolocation calls are attempts at echolocation, but the frequency containing maximum energy (FmaxE) of the call is not sustained and the echolocation calls that follow at 100–106 kHz are below the adult frequency (Figures 8.1.15 and 8.1.17). The frequency range of the calls is similar to the adult FM modified echolocation calls but the calls are longer (7.7±2.5 ms).

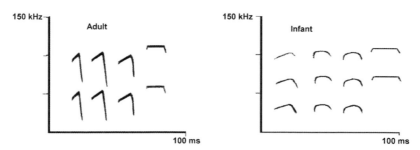

Figure 8.1.15   Adult lesser horseshoe bat FM modified echolocation social calls and infant development calls.

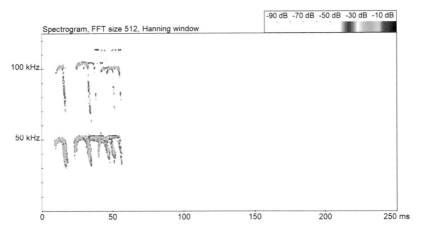

Figure 8.1.16 Adult lesser horseshoe bat FM modified echolocation social calls in a nursery roost (Amelia Hodnett).

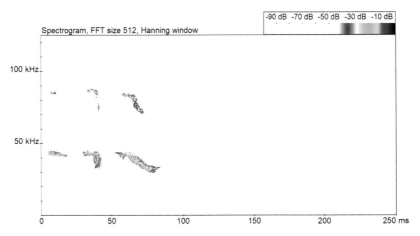

Figure 8.1.17 Infant lesser horseshoe bat FM development calls in a nursery roost (Amelia Hodnett).

Multiple-component FM calls

If the frequency rises then falls several times the call is identified as a complex FM multiple-component call, and each variation in frequency is identified as a component (Figure 8.1.18). There is sufficient variation in these calls to provide enough information for bats to recognise individuals or groups. If there are oscillations during the rising and falling components the call is identified as a 'trill advertisement call'. The presence of oscillations is sufficient to identify these calls inside and outside nursery roosts but also in a hibernaculum.

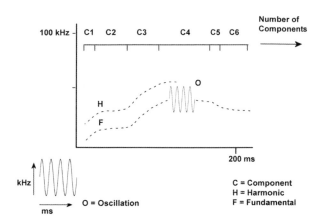

Figure 8.1.18   Measurement of lesser horseshoe bat multiple-component FM social calls.

The fundamental frequency ranges from 20 to 52 kHz and the durations are variable, depending on the length of each component (Figures 8.1.19 and 8.1.20). The duration of complex FM calls with four components ranges from 38 to 52 ms (44.8 ± 6.7 ms) but the call with five components ranges from 32 to 42 ms (36.8 ± 4.3 ms).

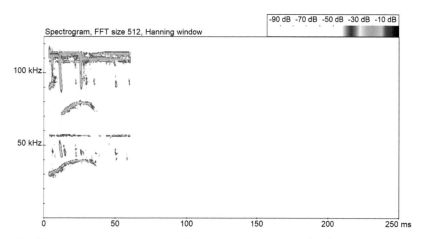

Figure 8.1.19   Adult lesser horseshoe bat multiple-component FM social calls in a nursery roost (Amelia Hodnett).

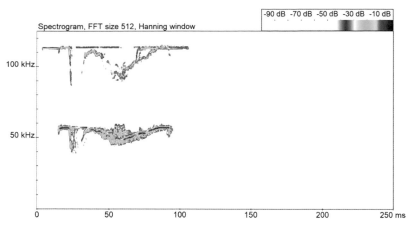

Figure 8.1.20 Adult lesser horseshoe bat multiple-component social call in a nursery roost (Amelia Hodnett).

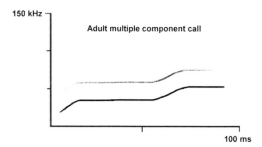

Figure 8.1.21 Complex multiple-component social calls of an adult lesser horseshoe bat outside a nursery roost.

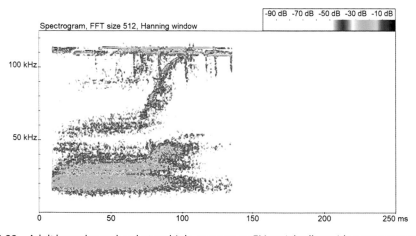

Figure 8.1.22 Adult lesser horseshoe bat multiple-component FM social call outside a nursery roost (Maggie Andrews).

The relatively low frequencies of trill advertisement calls carry a longer distance in a cave hibernaculum than echolocation calls and are made at the roosting site away from the hibernaculum entrance. Trill calls occur most frequently between 23.00 and 01.00 h, and up to 05.00 h in a hibernaculum, but trill calls are also observed during social interactions after emergence from the nursery roost (Andrews *et al.* 2017).

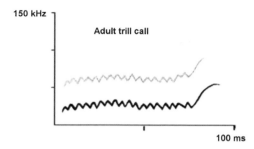

Figure 8.1.23   Multiple-component trill social call made by a lesser horseshoe bat outside a nursery roost.

Adult lesser horseshoe bats produce trill calls during social behaviour outside nursery roosts especially from May to July (Figure 8.1.24). The best time to record them is during emergence when the bats fly around the exit hole, or in a covered area near the exit hole before they disperse to forage (Andrews *et al.* 2017).

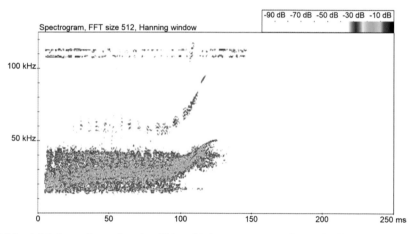

Figure 8.1.24   Adult lesser horseshoe bat FM multiple-component trill social call outside a nursery roost (Maggie Andrews).

Distress calls

Distress calls from bats held in the hand are very harsh and consist of a very loud trill (Figure 8.1.25)

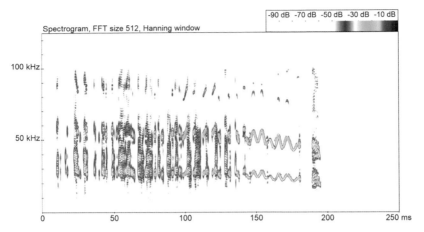

Figure 8.1.25 Distress calls produced by an adult lesser horseshoe bat held in the hand (Jon Russ).

SPECIES WITH SIMILAR OR OVERLAPPING ECHOLOCATION CALLS

Rhinolophus mehelyi, Rhinolophus euryale (see Chapter 7 species notes).

NOTES

Peter Andrews set up modified time-expansion bat detectors, which enabled remote ultrasonic recordings to be made in the nursery roost and made a major contribution to the scientific analysis of *R. hipposideros* ultrasonic social calls.

When identifying infant development calls and adult social calls it is necessary to record simultaneous adult echolocation calls, because greater horseshoe bat social calls are made in a similar range of frequencies. Varying the intensity of the call using the sound-analysis software is often needed to visualise the parameters, because some calls have a lot of noise associated with them and it is sometimes easier to recognise the call from the second harmonic.

Remote time-expansion recording for 24 hours in October is recommended for identification of trill calls in a hibernaculum and at 1 hour after emergence from a nursery roost. Placement of a time-expansion detector with external batteries and a battery-operated digital recorder between 10.00 and 11.00 h is advisable, when the lesser horseshoe bats are torpid. The number of calls per hour can be used to establish periods when the bats are of torpid in a 24-hour period (Andrews *et al.* 2017).

8.2 Greater horseshoe bat

Rhinolophus ferrumequinum (Schreber, 1774)

Maggie Andrews and Peter Andrews

© James Shipman

DISTRIBUTION

Mainly central and southern Europe, also southwest England and south and west Wales.

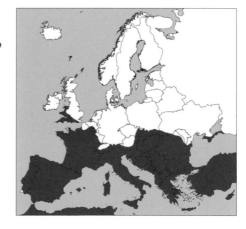

EMERGENCE

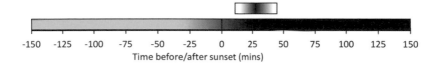

-150 -125 -100 -75 -50 -25 0 25 50 75 100 125 150

Time before/after sunset (mins)

FLIGHT AND FORAGING BEHAVIOUR

Flight is low and swift on emergence but slow and fluttering with short glides while foraging 0.3–6 m from the ground. When perched turns its body with the head up and alternately moves its ears backwards and forwards to focus on hearing insects in the clutter (Andrews 1995). Very manoeuvrable and able to avoid other bats at close range near the exit hole of a nursery roost. Insects caught by aerial hawking, gleaning and perch feeding. Can pick up insects, such as beetles, from the ground (Ransome 2008).

HABITAT

Flies under cover where possible after emergence from a nursery roost. Forages along hedgerows, over pasture and scrubland or at the edge of deciduous woodland near lakes, streams or ponds. Also forages in trees where there are moths and other insects, and flies over open country. Dispersal 40 km from nursery roost (Ransome 2008).

Typically, nursery roosts are located in the attic space of a warm roof of a large house or stable and in warm underground sites (Ransome 2008). Individuals hang from beams in a nursery roost and often form small clusters of 5–7 bats or larger clusters that may number in the hundreds. Social behaviour occurs outside nursery roosts when flight excursions are made around the exit/entrance hole after dusk and when females return to the roost to feed infants during the night in June and July. Adult males occupy separate sites where they attract females for mating. Hibernation usually takes place in a cave, mine or cellar where the ambient temperature is approximately 7 °C or less and the air is moist (Ransome 1971). Lesser horseshoe bats may hibernate nearby, but nursery roosts are typically in different parts of the roof of a building.

ECHOLOCATION

Measured parameters	Mean (range) FM-CF-FM call
Inter-pulse interval (ms)	90 (24.9–186.6)
Call duration (ms)	50.5 (16.3–73.8)
Frequency of maximum energy (kHz)	81.3 (77.8–83.8)
Start frequency (kHz)	70.2 (62.2–78.5)
End frequency (kHz)	67.3 (58.1–80.9)

Heterodyne
A continuous warbling sound is heard when the detector is tuned to around 82 kHz. Heterodyne detectors, used with headphones, are useful for exit counts at nursery roosts because there is no delay in the sound.

Time expansion/full spectrum
Calls show the typical FM-CF-FM structure found in all rhinolophids (Figures 8.2.1–8.2.3). The frequency containing maximum energy (FmaxE), which peaks at around 82 kHz (Vaughan *et al.* 1997), is actually the second harmonic emitted through the nose. The first is much weaker at around 41 kHz (e.g. Figure 8.2.2). The first harmonic is the fundamental sound produced in the larynx (Neuweiler 2000). Call frequency may vary slightly as the individual adjusts its echolocation to compensate for the Doppler effect during flight and when many bats fly together in and around the roost. Echolocation can vary according to age, sex and body size (Vaughan *et al.* 1997, Jones and Rayner 1989), so differences in these

signals offer the potential for intraspecific communication (Jones and Siemers 2011). The short CF pulses and large FM sweeps in rhinolophid echolocation calls provide maximum information about the distance of an object in front of the bat (Simmons *et al.* 1979, Andrews 1995). When the short CF calls are longer than 10 ms greater horseshoe bats are also able to detect small changes in a bat approach velocity (Simmons *et al.* 1979). Call duration may vary between 15 and 75 ms. In the nursery roost, the duration is short and the FM components are pronounced, and the fundamental is often recorded. In cluttered environments, the inter-pulse interval and duration decrease and the bandwidth of the FM components may increase.

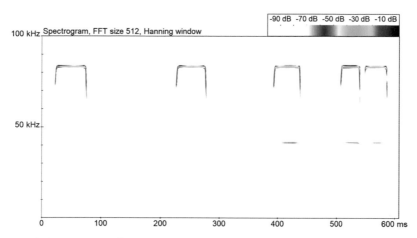

Figure 8.2.1   Echolocation calls of an adult greater horseshoe bat emerging from a cave (Jon Russ).

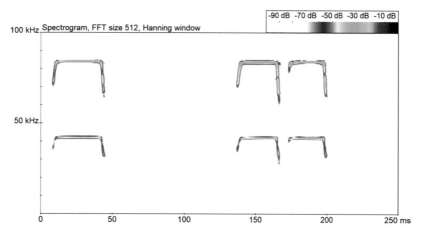

Figure 8.2.2   Echolocation calls of an adult greater horseshoe bat flying at an exposed cave entrance (Jon Russ).

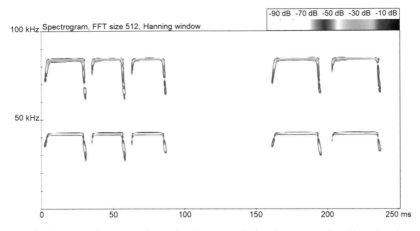

Figure 8.2.3   Echolocation of a greater horseshoe bat recorded in dense woodland (Jon Russ).

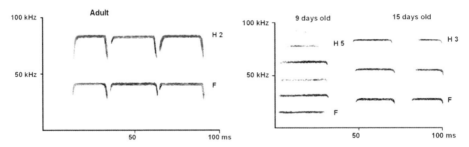

Figure 8.2.4   Adult and infant greater horseshoe bat echolocation calls. F – fundamental, H = harmonic, CF = constant-frequency, FM = frequency-modulated.

Infant echolocation calls can be recorded inside or at the exit hole of a nursery roost from June to July (Figures 8.2.4 and 8.2.5). They are distinct from the adult echolocation calls because the fundamental is lower and there are up to five harmonics. The echolocation frequency is achieved by using either the fifth or third harmonic. The infant echolocation develops through its ability to produce the CF component at higher frequencies. These calls are shorter than the adult echolocation because the infant cannot sustain the CF component. Infants as young as 9 days old can produce echolocation sufficient to start flying in the roost. Infants aged 15–21 days produce echolocation similar to adult bats but the frequency is lower.

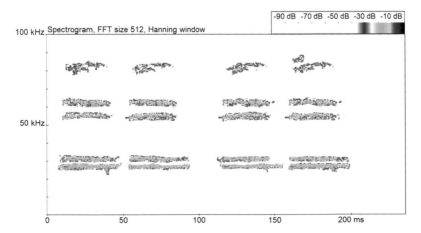

Figure 8.2.5   Echolocation of an infant greater horseshoe bat recorded in a nursery roost (Maggie Andrews).

SOCIAL CALLS

Heterodyne
The fundamental frequencies of the social calls are in the range 15–39 kHz but it is not advisable to try to identify these calls using heterodyne detectors, because the identification of each type of call requires broadband recordings and analysis with software.

Time expansion/full spectrum
The types of ultrasonic social calls made by greater horseshoe bats are identified according to their parameters in three main groups: single-component, multiple-component and modified echolocation calls. The fundamental frequencies in the range 11–39 kHz are the predominant sounds, and many calls have harmonics, but generally the social calls are below the echolocation frequency (Andrews and Andrews 2003).

Single-component calls
The frequency of single-component calls is either constant, or it rises or falls during the call. Andrews and Andrews (2003) and Andrews *et al.* (2006, 2011) describe parameters of these calls in detail.

Constant-frequency calls
Adult greater horseshoe bats use constant frequency calls more than any other social calls inside and outside the nursery roost and in a hibernaculum (Figures 8.2.6 and 8.2.7). The fundamental frequency is low (20.1 ± 5.9 kHz) and generally long but the duration is very variable (43.9 ± 41.7 ms).

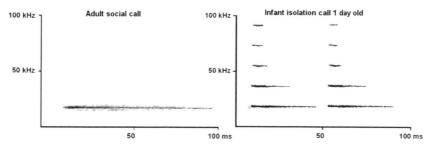

Figure 8.2.6   Adult greater horseshoe bat single-component CF ultrasound social calls and infant isolation calls.

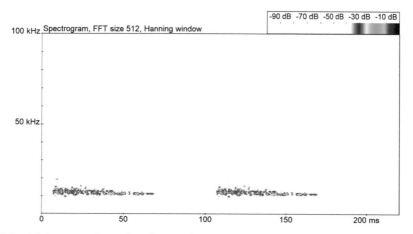

Figure 8.2.7   Adult greater horseshoe bat single-component CF social calls in a nursery roost (Maggie Andrews).

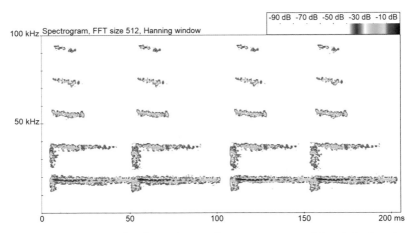

Figure 8.2.8   Infant greater horseshoe bat isolation calls in a nursery roost (Maggie Andrews).

Isolation calls of infant greater horseshoe bats are polyharmonic constant frequency calls (Figure 8.2.8). These calls are characteristic of newborn to day-old infant bats in a nursery roost in June and July and occur in double syllables. The fundamental frequencies are lower than adult calls ranging from 15–19 kHz and the duration is shorter (15.5 ± 7.2 ms).

Rising- or falling-frequency FM calls

Rising-frequency single-component calls made by adult greater horseshoe bats are often observed in nursery roost recordings (Figures 8.2.9 and 8.2.10). The duration of these calls is variable but often short (29.9 ± 28.4 ms) and the fundamental frequency ranges from 10 to 38 kHz, sometimes with a second harmonic. Falling-frequency calls, similar in frequency and duration, are less frequent.

Infant greater horseshoe bats develop the ability to make a rising-frequency call at 3 days of age, and the fundamental frequencies are typically 20–24 kHz. These short calls (15.2 ± 7.4) are polyharmonic and often occur in double syllables. Few falling-frequency calls occur (Figure 8.2.11).

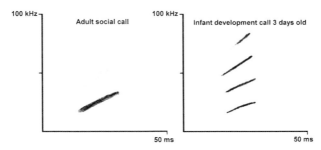

Figure 8.2.9   Adult greater horseshoe bat single-component FM ultrasound social call and an infant development call.

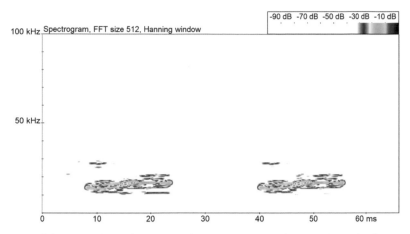

Figure 8.2.10   Adult greater horseshoe bat single-component FM ultrasound social call in a nursery roost (Maggie Andrews).

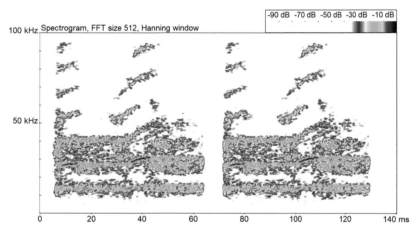

Figure 8.2.11   Infant greater horseshoe bat, 3 days old, FM development call (Maggie Andrews).

Modified echolocation calls

Modified echolocation calls are similar to echolocation calls but their fundamental (first harmonic) is much lower, and therefore the second harmonic is not in the range suitable for echolocation. These calls are found in a sequence or set of calls in which the fundamental frequency of each call is 5 kHz above or below the previous call. There are often 10 calls in the sequence, and they look like echolocation calls but are lower in frequency.

Modified echolocation calls are identified as constant-frequency (CF) or frequency-modulated (FM) modified echolocation calls depending on the central part of the call. The sequence of calls is made with each call at a lower or higher frequency than the call before it.

Constant-frequency modified echolocation calls

The adult CF modified echolocation calls are measured as a cascade of short calls (4.5 ± 2.3 ms) (Figures 8.2.12 and 8.2.13). Measurements of the second harmonic of the first calls in the cascade are in the range 80–68 kHz and the last calls, ranging from 20 to 29 kHz, are the fundamental only. The second harmonic is a nasal call and the fundamental is uttered through the mouth. The cascades of modified echolocation calls are observed most often 1 hour before emergence from a nursery roost and within 1 hour after they return. They are associated with flight inside the roost.

The infant CF modified echolocation calls are attempts at echolocation and occur in a random sequence, and the second harmonic is always observed (Figure 8.2.14).

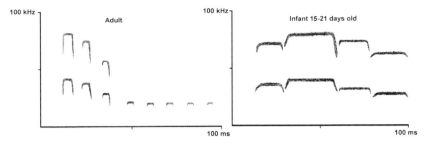

Figure 8.2.12   Adult greater horseshoe bat CF modified echolocation social calls and infant development calls.

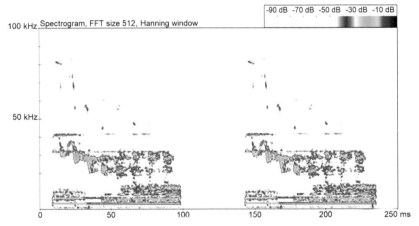

Figure 8.2.13   Adult greater horseshoe bat CF modified echolocation social calls and infant development calls (Maggie Andrews).

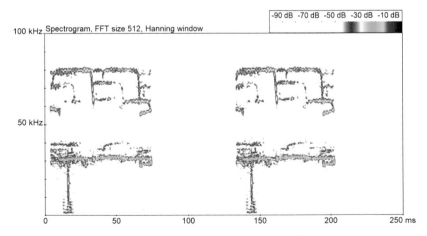

Figure 8.2.14   Infant greater horseshoe bat development calls, CF modified echolocation (Maggie Andrews).

Frequency-modulated modified echolocation calls

The adult FM modified echolocation calls are measured as a sequence of short calls (8.6±4.1 ms) (Figures 8.2.15 and 8.2.16). Measurements of the second harmonic of calls in a sequence of rising frequencies from 40–58 kHz to 60–78 kHz are followed by a series of echolocation calls at 80–84 kHz. The sequence can also be observed as a cascade of decreasing-frequency calls. Often a cascade of decreasing-frequency calls is followed by a sequence of increasing-frequency calls.

Infant FM modified echolocation calls are attempts at echolocation, but the FmaxE of the call is not sustained and the echolocation calls that follow at 60–78 kHz are below the adult frequency. The frequency range of the calls is similar to the adult FM modified echolocation calls but the calls are longer (17.3±1.7 ms).

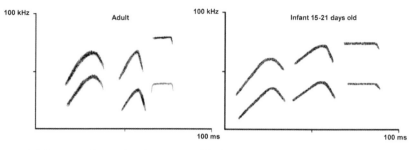

Figure 8.2.15   Adult greater horseshoe bat FM modified echolocation social calls and infant development calls.

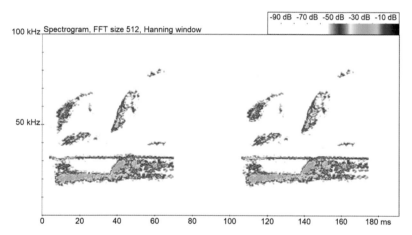

Figure 8.2.16   Adult greater horseshoe bat FM modified echolocation social calls in a nursery roost (Maggie Andrews).

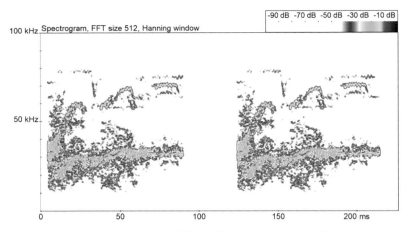

Figure 8.2.17   Infant greater horseshoe bat, FM modified echolocation calls in a nursery roost (Maggie Andrews).

Multiple-component FM calls

If the frequency rises then falls several times it is identified as a complex FM multiple-component call, and each variation in frequency is identified as a component (Figures 8.2.18 and 8.2.19). These complex FM calls have sufficient variation that they have enough information for individual or group identification.

If there are oscillations during the rising and falling components it is identified as a trill call (Figure 8.2.20). The presence of oscillations is sufficient to identify these calls but a sample can be measured. These advertisement calls occur mainly in hibernacula but also in and around nursery roosts.

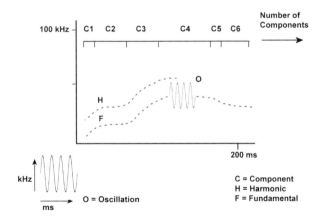

Figure 8.2.18   Measurement of greater horseshoe bat multiple-component FM social calls.

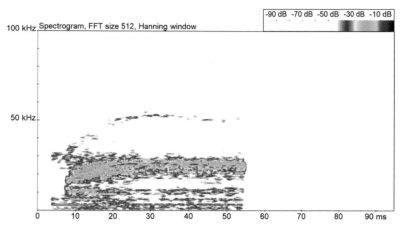

Figure 8.2.19   Adult greater horseshoe bat multiple-component FM social call, two-component call (Maggie Andrews).

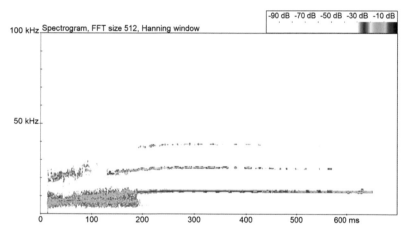

Figure 8.2.20   Adult greater horseshoe bat multiple-component FM social call, two-component trill call (Maggie Andrews).

Typically, the fundamental frequencies range from 10 to 29 kHz and the duration varies according to the number of components. The duration of complex FM calls with three components ranges from 20 to 163 ms (78.4±43.3 ms); this is significantly shorter than a call with five components, which ranges from 117 to 240 ms (167.3±48.4 ms) (Figures 8.2.21 and 8.2.22).

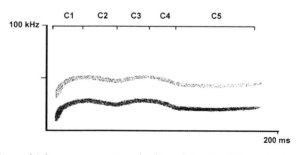

Figure 8.2.21   Complex multiple-component social call made by an adult greater horseshoe bat.

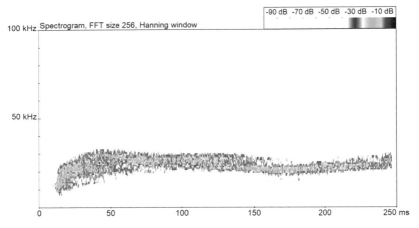

Figure 8.2.22 Complex multiple-component social call made by an adult greater horseshoe bat in a nursery roost (Maggie Andrews).

The relatively low frequencies of trill advertisement calls carry a longer distance in a cave hibernaculum than echolocation calls and are made at the roosting site away from the cave entrance, and trill calls occurred most frequently between 16.00 and 19.00 h (Andrews *et al.* 2006) (Figures 8.2.23 and 8.2.24).

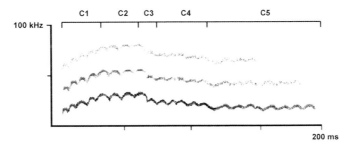

Figure 8.2.23 Multiple-component trill social call made by an adult greater horseshoe bat.

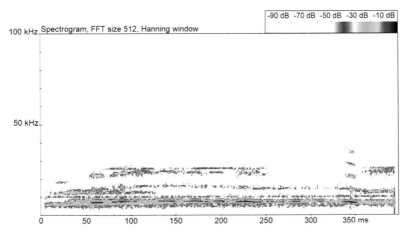

Figure 8.2.24 Multiple-component trill social call made by an adult greater horseshoe bat outside a nursery roost (Maggie Andrews).

Adult greater horseshoe bats make trill calls during social behaviour outside nursery roosts especially from May to July. The best time to record them is during emergence when the greater horseshoe bats fly around the exit hole before they disperse to forage (Andrews and Andrews 2016).

Distress calls
Distress calls are emitted by individuals in a state of physical duress, such as when held in the hand (Figures 8.2.25 and 8.2.26).

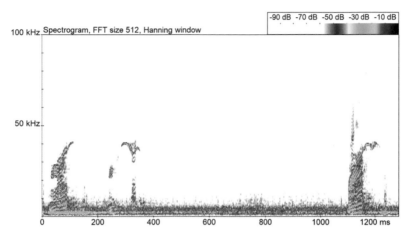

Figure 8.2.25   Distress call of a juvenile greater horseshoe bat held in the hand (Jon Russ).

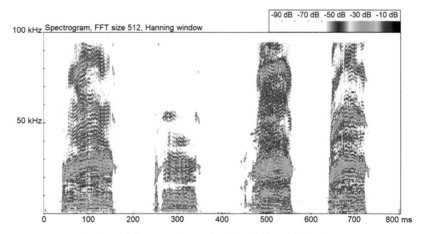

Figure 8.2.26   Distress call of an adult greater horseshoe bat held in the hand (Jon Russ).

SPECIES WITH SIMILAR OR OVERLAPPING ECHOLOCATION CALLS
None.

NOTES

Peter Andrews made the recordings of infant *R. ferrumequinum* ultrasonic calls while Tom McOwat monitored the infant bats in the hand. Peter Andrews also set up the modified time-expansion bat detectors, which enabled remote ultrasonic recordings to be made in the nursery roost and made a major contribution to the scientific analysis of *R. ferrumequinum* ultrasonic social calls.

Varying the intensity of the call using software is often needed to visualise the parameters, because some calls have a lot of noise associated with them and it is sometimes easier to recognise the call from the second harmonic.

Remote TE recording for 24 hours in October is recommended for identification of trill calls. Placement of a TE detector with external batteries and a battery-operated digital recorder between 10.00 and 11.00 h is advisable, when greater horseshoe bats are torpid. Periods when the bats are torpid can be established by estimation of variability in the number of echolocation calls per hour in a 24-hour period (Andrews *et al.* 2006).

Greater horseshoe bats do not disperse immediately and fly out and back in and then around the exit hole. It is advisable to have two bat workers with manual counters, one to count the bats out and the other to count bats flying back in.

8.3 Mediterranean horseshoe bat

Rhinolophus euryale (Blasius, 1853)

Eleni Papadatou, Panagiotis Georgiakakis and Artemis Kafkaletou-Diez

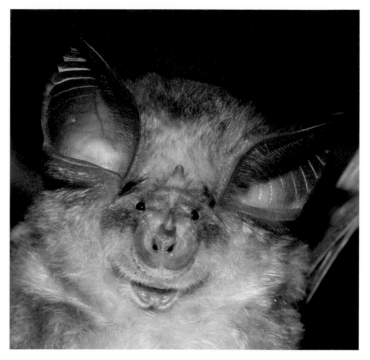

© Panagiotis Georgiakakis

DISTRIBUTION

The species occurs from northwest Africa throughout the European Mediterranean and western Asia Minor. It is absent from the Balearic Islands, Crete and Cyprus.

EMERGENCE

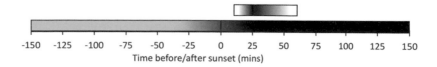

-150 -125 -100 -75 -50 -25 0 25 50 75 100 125 150

Time before/after sunset (mins)

FLIGHT AND FORAGING BEHAVIOUR

Very agile and manoeuvrable flight close to or within vegetation, such as tree canopies and even dense bushes. It may also hunt at the forest edge and close to the ground in open forest space. It catches its prey usually in flight, occasionally from a perch.

HABITAT

The species prefers deciduous and riparian forests for foraging, but it may also hunt in other habitats such as scrublands and olive groves. Open areas and conifer plantations are avoided.

In karst areas, summer, mating and winter roosts are usually found in caves. The species may also roost in disused mines outside karst areas or where suitable caves are limited. In the northern parts of its distribution, colonies may roost in roof spaces or other artificial spaces. Colonies in underground sites may be mixed with other species (e.g. *Rhinolophus*, *Myotis* and *Miniopterus*).

ECHOLOCATION

Measured parameters	Mean (range) FM-CF-FM call
Inter-pulse interval (ms)	83.8 (20.3–138.0)
Call duration (ms)	52.3 (15.3–91.6)
Frequency of maximum energy (kHz)	104.7 (102.0–106.4)
Start frequency (kHz)	92.5 (86.2–104.9)
End frequency (kHz)	89.5 (83.6–104.4)

Heterodyne
When tuned to around 104 kHz a continuous warbling can be heard. However, depending on the geographical location, as well as the sex and age of the animals, there may be an overlap in frequency range with Mehely's horseshoe bat and lesser horseshoe bat. The weaker fundamental or first harmonic can be heard at half this frequency, around 52 kHz.

Time expansion/full spectrum
The calls of Mediterranean horseshoe bats show the typical FM-CF-FM structure found in congeneric species, with a long CF component preceded and followed by two short FM components (Figures 8.3.1–8.3.3). As with all rhinolophids, most energy is contained in the second harmonic and the weaker fundamental has half its frequency. The CF component of calls emitted from bats in flight has an average FmaxE of around 104 kHz, though in western Europe it may be lower by approximately 2 kHz. Average peak resting frequency (i.e. frequency of calls from stationary individuals, such as roosting and hand-held bats) may vary between 104 and 106 kHz (lower in western Europe). The frequency may fluctuate depending on geographical location as well as the age of bats. Call frequency may further vary slightly as the individual adjusts its echolocation to compensate for the Doppler shift during flight. The FM components can be variable: if the call is weak, they may not even be visible, whereas in cluttered conditions their bandwidth may increase (Figure 8.3.2). Calls may be long, but sometimes groups of shorter calls with a higher repetition rate may be emitted by a bat; in these calls, often only the first call has an initial FM component (Figure 8.3.3).

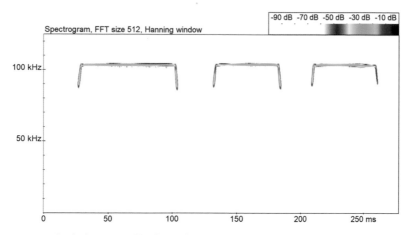

Figure 8.3.1   Typical echolocation calls of a Mediterranean horseshoe bat (Eleni Papadatou).

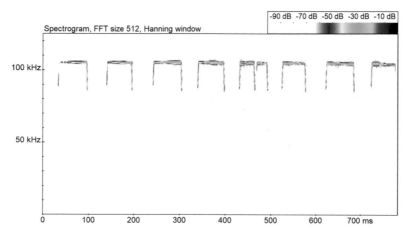

Figure 8.3.2   Increased bandwidth of FM component of the echolocation calls of a Mediterranean horseshoe bat (Eleni Papadatou).

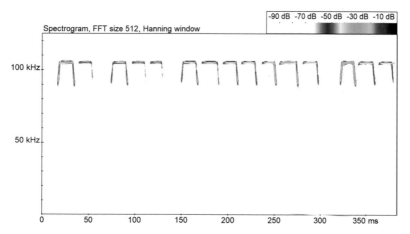

Figure 8.3.3   Decreased duration of echolocation calls of a Mediterranean horseshoe bat (Eleni Papadatou).

SOCIAL CALLS

Heterodyne
The fundamental frequencies of the social calls are in the range 16–37 kHz but it is not advisable to try to identify these calls using heterodyne detectors because the identification of each type of call requires broadband recordings and analysis with software.

Time expansion/full spectrum
A variety of social calls have been recorded for this species (Figures 8.3.4–8.3.7). The functions of these calls are not known.

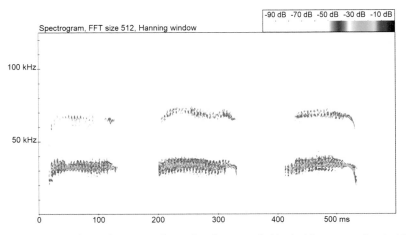

Figure 8.3.4   Social call of a Mediterranean horseshoe bat recorded in deciduous woodland with open grass areas (Phil Riddett).

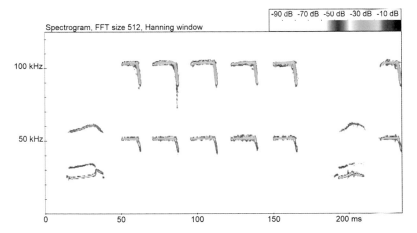

Figure 8.3.5   Social call of a Mediterranean horseshoe bat recorded at a cave entrance (Danilo Russo).

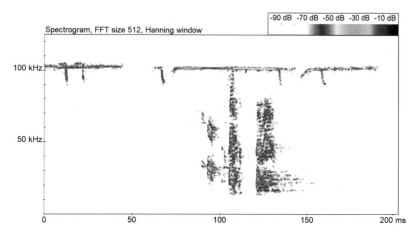

Figure 8.3.6   Social call of a Mediterranean horseshoe bat recorded at a cave entrance (Yves Bas).

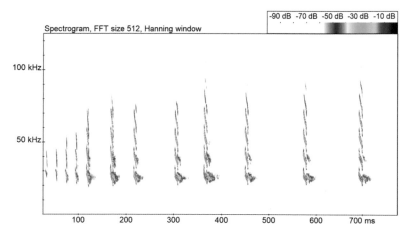

Figure 8.3.7   Social call of a Mediterranean horseshoe bat recorded at a cave entrance (Arjan Boonman).

SPECIES WITH SIMILAR OR OVERLAPPING ECHOLOCATION CALLS

Rhinolophus mehelyi, Rhinolophus hipposideros (see Chapter 7 species notes).

NOTES

FmaxE of the Mediterranean horseshoe bat may overlap, especially at the higher end of its frequency range, with that of Mehely's horseshoe bat and sometimes with that of the lesser horseshoe bat. Species identification based on calls with peak frequencies within the overlapping zone should be made with caution, and sometimes it may not be possible. This applies in particular when recorded bats have not been visually observed, in areas where the species occur in sympatry, or where their presence (or absence) is unknown. See Chapter 7.

The measured parameters are based on a sample of 47 individuals.

8.4 Mehely's horseshoe bat

Rhinolophus mehelyi (Matschie, 1901)

Mauro Mucedda and Ermanno Pidinchedda

© Mauro Mucedda

DISTRIBUTION

Mehely's horseshoe bat has a Mediterranean distribution, in north Africa (Morocco, Algeria, Tunisia, Libya and Egypt), southern Europe (southern Portugal, Spain, only one occurrence in France, Italy (only Sardinia, Sicily, Apulia)), the Balkans (Bulgaria, Bosnia and Herzegovina, Greece, Macedonia, Moldova, Montenegro, Romania, Serbia, Slovenia), Cyprus, Asia Minor, Anatolia, Transcaucasia, Iran and Afghanistan (Alcaldé *et al.* 2016).

EMERGENCE

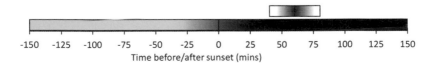

-150   -125   -100   -75   -50   -25   0   25   50   75   100   125   150
Time before/after sunset (mins)

FLIGHT AND FORAGING BEHAVIOUR

Flies slowly, with high manoeuvrability. Also glides. It flies low to the ground above grass and along the edges of bushes. Hunts by ambushing from a perch such as a branch.

HABITAT

Mehely's horseshoe bat is a typical troglophilous species; in Italy, it uses underground roosts in sites such as caves, mines and artificial galleries all year round. It forms mixed colonies with other troglophilous species, such as the Mediterranean horseshoe bat, Schreiber's bent-winged bat, Maghreb mouse-eared bat, the greater mouse-eared bat and the long-fingered bat. Hibernates with same species or with greater horseshoe bats (Mucedda *et al.* 2009).

In southwest Spain, Mehely's horseshoe bat forages in different Mediterranean wooded habitats, including loose oak woodland, olive groves/almond orchards, and also eucalyptus plantations. Tends not to forage in open habitats (Russo *et al.* 2005).

ECHOLOCATION

Measured parameters	Mean (range) FM-CF-FM call
Inter-pulse interval (ms)	6.5 (2.9–14.4)
Call duration (ms)	22.8 (14.9–34.1)
Frequency of maximum energy (kHz)	109.0 (104.1–113.3)
Start frequency (kHz)	95.4 (83.0–105.0)
End frequency (kHz)	89.6 (78.0–100.0)

Heterodyne
Tuning around 104–110 kHz a characteristic warbling is heard. If the bat is near the bat detector it is possible to hear the weaker first harmonic around 52–55 kHz.

Time expansion/full spectrum
Mehely's horseshoe bats emit FM-CF-FM echolocation calls with a long constant-frequency (CF) component, preceded by an ascending frequency-modulated (FM) component and followed by a descending FM component (Figure 8.4.1). The second harmonic contains the strongest intensity (or contains the maximum energy) in the call.

Resting and hand-held Mehely's horseshoe bats generally emit sequences of two, three or four pulses, more rarely five or six pulses. The first pulse of every sequence is generally a little shorter than the following pulses.

In Sardinia FmaxE of the echolocation calls of juveniles (on average 103.7 kHz) is significantly lower than that of adults (on average 107.7 kHz). FmaxE of Mehely's horseshoe bat at its lower frequency values, mainly for young bats, can overlap with the Mediterranean horseshoe bat, and at its higher frequency values can overlap with the lesser horseshoe bat (Russo *et al.* 2001).

Echolocation calls of Mehely's horseshoe bats show no sexual dimorphism (Russo *et al.* 2007b).

The mean FmaxE of Mehely's horseshoe bats in Sicily is higher (112.2 kHz) than in Sardinia (107.3 kHz).

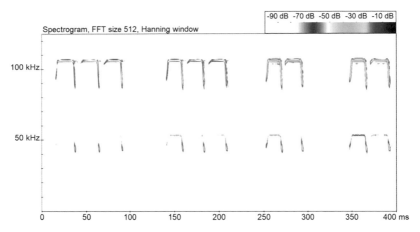

Figure 8.4.1   Echolocation calls of hand-held Mehely's horseshoe bats emitted in groups of 2 and 3 pulses (Mauro Mucedda).

SOCIAL CALLS

Heterodyne
Social calls are highly variable and are in the range 25–60 kHz.

Time expansion/full spectrum
Social calls emitted by Mehely's horseshoe bats appear as very complex and variable signals with many harmonics (Figures 8.4.2 and 8.4.3). The functions of these social calls are unknown.

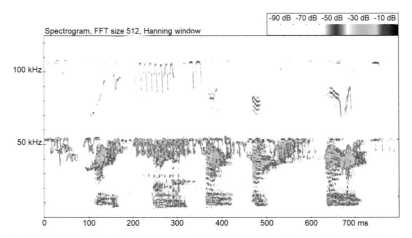

Figure 8.4.2   Social calls emitted by one or more Mehely's horseshoe bats inside a cave (Mauro Mucedda).

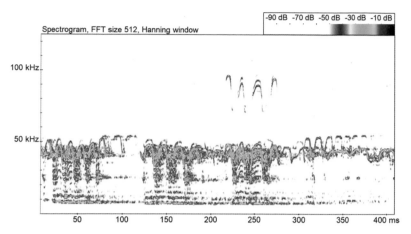

Figure 8.4.3   Different social calls emitted by one or more Mehely's horseshoe bats inside a cave (Mauro Mucedda).

SPECIES WITH SIMILAR OR OVERLAPPING ECHOLOCATION CALLS

Rhinolophus hipposideros, Rhinolophus euryale (see Chapter 7 species notes).

8.5 Blasius's horseshoe bat *Rhinolophus blasii* (Peters, 1866)

Panagiotis Georgiakakis

© Panagiotis Georgiakakis

DISTRIBUTION

A species of the eastern Mediterranean, including many islands, also occurring in northwest Africa (Morocco and Tunisia). It is present in the east Adriatic countries, the Balkans and southern Romania, Asia Minor and the Middle East. Probably extinct from Italy and Slovenia (Dietz and Kiefer 2016).

EMERGENCE

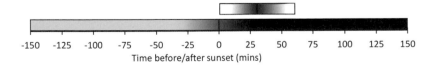

Time before/after sunset (mins)

FLIGHT AND FORAGING BEHAVIOUR

Agile flight, and can catch prey close to plants or other big objects but rarely from the ground (Dietz and Kiefer 2016, Siemers and Ivanova 2004). It usually flies at 0.5–5 m, constantly searching bushes for insects.

HABITAT

A lowland species rarely found higher than 1,000 m above sea level. It seems to prefer structured habitats with open wooded vegetation and scrubs, shrublands and oak forests. Wetlands are also regularly visited.

It roosts mainly in caves and other underground habitats (Dietz and Kiefer 2016). In Greece and Cyprus several medium-sized colonies are found in disused mines (Benda *et al.* 2007, Hanák *et al.* 2001; P. Georgiakakis, personal observation).

ECHOLOCATION

Measured parameters	Mean (range) FM-CF-FM call
Inter-pulse interval (ms)	77.8 (42.5–100.0)
Call duration (ms)	47.5 (29.0–65.8)
Frequency of maximum energy (kHz)	93.3 (92.0–94.9)
Start frequency (kHz)	85.3 (79.4–91.2)
End frequency (kHz)	88.9 (87.9–89.9)

Heterodyne

A continuous warbling sound is heard when the detector is tuned to around 93 kHz. The echolocation calls of horseshoe bats are distinct from those of all other species of bat found in Europe.

Time expansion/full spectrum

Like all rhinolophids, Blasius's horseshoe bat emits FM-CF-FM echolocation calls, i.e. calls with a constant-frequency main component, with brief frequency-modulated components at the start and end. In very weak calls, these FM sweeps are barely visible in a spectrogram. Blasius's horseshoe bat calls are unmistakable, due to their unique frequencies: frequency of maximum energy (FmaxE) is about 96 kHz in resting animals and *c.*93 kHz in free flight. Inter-pulse interval and call duration are highly variable, especially in resting animals, since they occasionally emit short calls in groups of 2–5 (Figures 8.5.1 and 8.5.2). In free flight, the FmaxE may fluctuate (Figures 8.5.3 and 8.5.4).

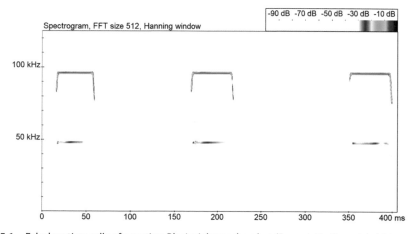

Figure 8.5.1   Echolocation calls of a resting Blasius's horseshoe bat (Panagiotis Georgiakakis).

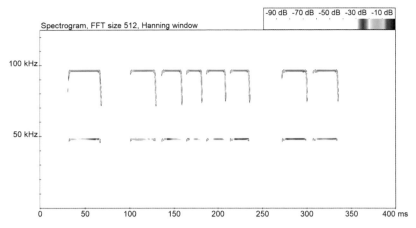

Figure 8.5.2 Echolocation calls of a resting Blasius's horseshoe bat, emitted in groups (Panagiotis Georgiakakis).

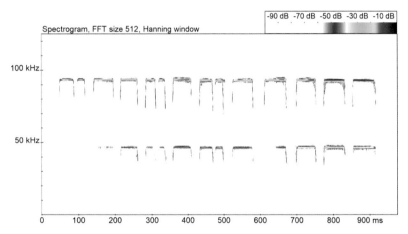

Figure 8.5.3 Echolocation calls of a Blasius's horseshoe bat flying above a stream (Panagiotis Georgiakakis).

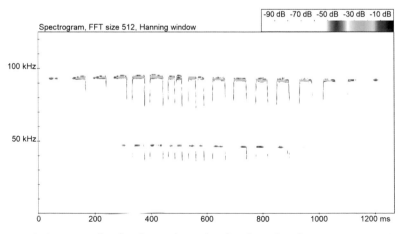

Figure 8.5.4 Echolocation calls of a Blasius's horseshoe bat flying besides riparian vegetation (Panagiotis Georgiakakis).

SOCIAL CALLS

Heterodyne
No information available

Time expansion/full spectrum
No information available.

SPECIES WITH SIMILAR OR OVERLAPPING ECHOLOCATION CALLS

None.

8.6 Daubenton's bat *Myotis daubentonii* (Kuhl, 1819)

Jon Russ

© René Janssen

DISTRIBUTION

Common and widespread throughout Europe up to about 63°N. Often limited by hilly and mountainous regions in the Mediterranean region.

EMERGENCE

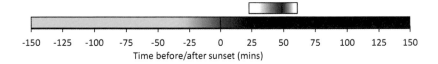

Time before/after sunset (mins)

FLIGHT AND FORAGING BEHAVIOUR

Fast and agile with a fast wingbeat – sometimes even whirring. Often hunts just 5–25 cm above the water, like a hovercraft. During flight over water, the turns are very long. Natterer's bat in the same situation is much more agile, often 'flipping over' to change direction, has a more 'hectic' flight style and generally forages higher above the water. Daubenton's bat trawls insects from the water surface. Can often be seen to leave small circular ripples where it has caught an insect. During mid-summer to autumn, when larger prey items are available, occasionally observed in very fast straight flight close to the surface of the water (Van De Sijpe 2011).

HABITAT

Ponds, lakes, rivers and also small streams. Feeds over calm water and avoids areas of turbulence or riffles and also areas containing duckweed. Occasionally forages along riparian treelines. Although usually associated with water, Daubenton's bat can also be found in other habitats such as woodland paths and treelines. In woodland, forages in the middle of the path, compared to whiskered bats, for example, which tend to 'hug' the vegetation edge without straying into the open. Generally, this is true whenever Daubenton's bat is found near cluttered habitats. If foraging along a treeline it rarely flies within 2 m of the vegetation and when over water will avoid the overhanging trees and bushes near the edge of the water.

Summer roosts in crevices in buildings and bridges, holes in trees and roost boxes. Hibernation recorded in underground sites such as caves, mines and cellars but also crevices in bridges and under hanging tiles. Also found in culverts, particularly during the autumnal swarming period.

ECHOLOCATION

Measured parameters	Mean (range) FM call
Inter-pulse interval (ms)	75.5 (27.5–186.0)
Call duration (ms)	3.2 (1.4–5.8)
Frequency of maximum energy (kHz)	47.0 (41.8–56.5)
Start frequency (kHz)	81.1 (50.3–109.7)
End frequency (kHz)	29.4 (22.4–38.6)

Heterodyne

Daubenton's bat echolocation is heard as a rapid series of clicks on the detector. Pulse repetition rate is very fast and regular compared with the irregular sound of the pipistrelle. The rhythm is often likened to the sound of a cat purring. Although its pulse repetition rate is more regular than that of other *Myotis* bats, much caution is needed in attempting to tell *Myotis* bats apart based on heterodyne calls. The calls of pipistrelles flying in clutter can sound similar to those of the Daubenton's bat. However, it can be distinguished from pipistrelles, for example, by tuning down below the frequencies of pipistrelle calls to 35 kHz, where dry 'clicks' can be heard.

Time expansion/full spectrum

Produces frequency-modulated sweeps starting at about 85 kHz and ending at about 25 kHz, and often a 'heel' or bend at around 40 kHz is present (Figure 8.6.1). Generally, produces an S-shaped call. In most situations, the bat seems to produce calls of this shape, and it is only when an individual is flying in a very cluttered environment that the calls straighten out (Figure 8.6.2). The average pulse duration is around 3.2 ms. Frequency containing maximum energy (FmaxE) is generally difficult to determine as peak energy is distributed through a wide range of frequencies. Often, because the bat is foraging close to the water, the call contains many 'missing' frequencies as a result of interference between the call reflected from the water surface and the directly recorded call (Figure 8.6.3). Feeding buzzes are produced during insect capture (Figure 8.6.4), and these can sometimes be quite long (Figure 8.6.5).

Daubenton's bats occasionally use quite long echolocation signals from mid-summer to autumn (Van De Sijpe 2011). During this time, many large insects are present, as opposed to spring and early summer, when only small non-biting midges floating or flying very low above the water surface are available, and it is thought that this results in a change in the bat's acoustic hunting behaviour. During fast straight cruising flight when the bat is flying close to the water surface, short periods of silence alternate with very strong signals. The longest durations are around 8–14 ms following a period of silence (Figure 8.6.6) and then gradually the call duration decreases to the 'normal' range.

Although the peak frequencies are variable in all three species, the pond bat generally uses lower peak frequencies (median 41 kHz for durations of 1–4 ms and median 35 kHz for durations >14 ms) than Daubenton's bat (median 49 kHz for durations of 1–4 ms and median 37 kHz for durations of 8–14 ms) and long-fingered bat (median 45 kHz for durations of 1–4 ms) (Van De Sijpe 2011). End frequencies are significantly higher in the long-fingered bat (median 32 kHz, for durations of 1–4 ms) than in the other two species (pond bat median 25 kHz, Daubenton's bat median 26 kHz, for durations of 1–4 ms), quasi-constant-frequency parts in the middle of the signal are only found in the pond bat, and long-fingered bats sometimes used pulse series with alternating end frequencies (Van De Sijpe 2011).

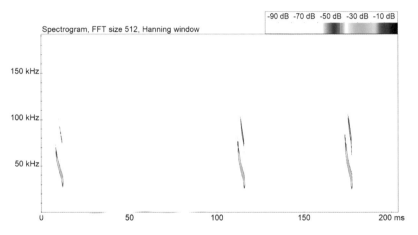

Figure 8.6.1   Echolocation calls of a Daubenton's bat flying over a pond (Jon Russ).

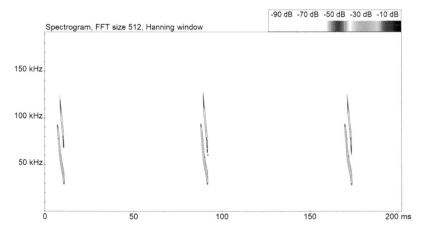

Figure 8.6.2   Echolocation calls of hand-released Daubenton's bat (Jon Russ).

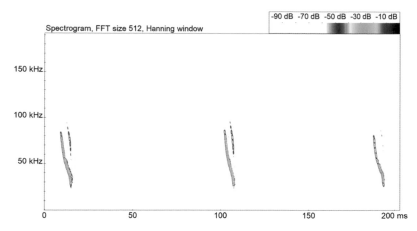

Figure 8.6.3   Echolocation calls of a Daubenton's bat flying close to the water surface, showing the classic interference pattern observed in this species (Jon Russ).

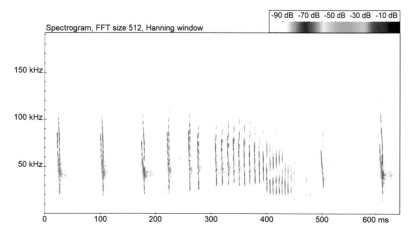

Figure 8.6.4   Feeding buzz produced by a Daubenton's bat during prey capture (Jon Russ).

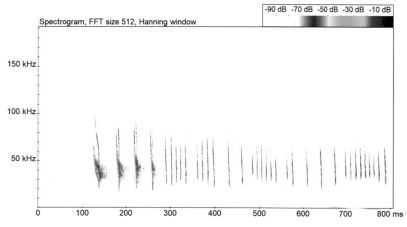

Figure 8.6.5   Unusually long Daubenton's bat feeding buzz (Marc Van De Sijpe).

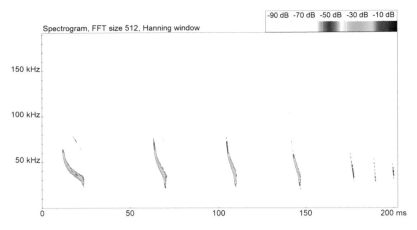

Figure 8.6.6   Long Daubenton's bat echolocation pulses observed during mid-summer to autumn (Marc Van De Sijpe).

SOCIAL CALLS

Heterodyne
The most commonly heard call is occasionally heard as a loud extra call slipped into the echolocation call sequence. Harsh-sounding calls are also frequently heard at swarming sites during the autumnal mating period.

Time expansion/full spectrum
Daubenton's bat occasionally produces calls that appear bent at the top like an umbrella handle (Figures 8.6.7 and 8.6.8), and these are frequently recorded at swarming sites in the autumn, often varying in maximum frequency (Figure 8.6.9) and start frequency (Figure 8.6.10), although they can also be recorded at foraging sites. The species also produces a wide variety of other calls at swarming sites (e.g. Figures 8.6.11–8.6.15).

Distress calls, recorded from bats held in the hand, consist of a series of frequency-modulated sweeps of very short duration with an FmaxE of about 30 kHz (Figure 8.6.16).

At roost sites, a wide range of loud social calls are heard (e.g. Figures 8.6.17 and 8.6.18).

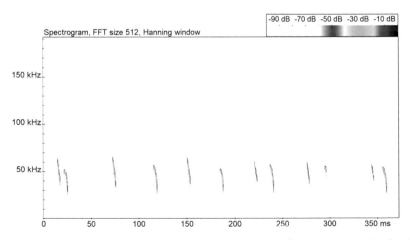

Figure 8.6.7   Calls recorded from Daubenton's bat flying at a height of 2 m over a river (Jon Russ).

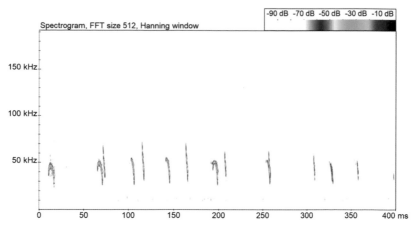

Figure 8.6.8   Calls recorded from Daubenton's bat flying very close to the water surface (Jon Russ).

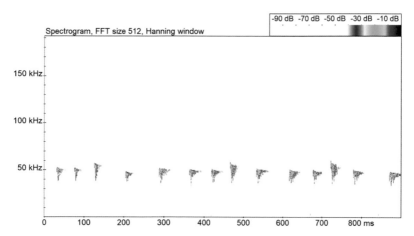

Figure 8.6.9   Social calls from Daubenton's bat recorded at the entrance to a cave (Jon Russ).

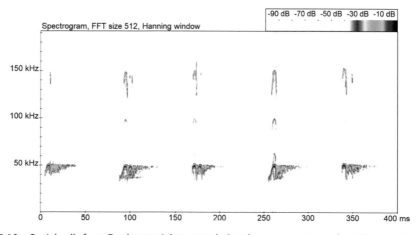

Figure 8.6.10   Social calls from Daubenton's bat recorded at the entrance to a culvert (Jon Russ).

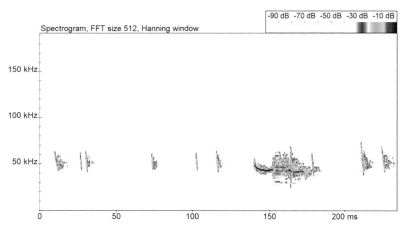

Figure 8.6.11   Social calls from Daubenton's bat recorded at the entrance to a culvert (Jon Russ).

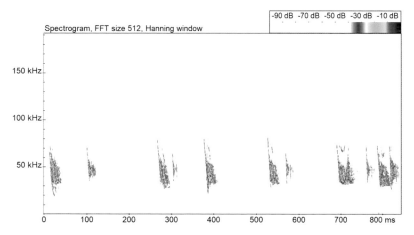

Figure 8.6.12   Social calls from Daubenton's bat recorded at the entrance to a culvert (Jon Russ).

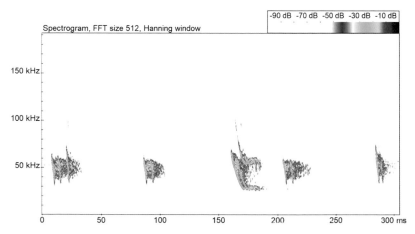

Figure 8.6.13   Social calls from Daubenton's bat recorded at the entrance to a culvert (Jon Russ).

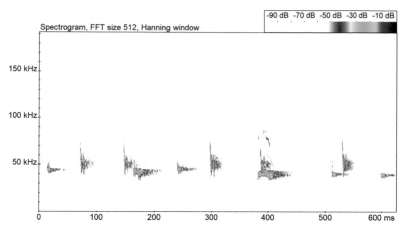

Figure 8.6.14   Social calls from Daubenton's bat recorded at the entrance to a culvert (Jon Russ).

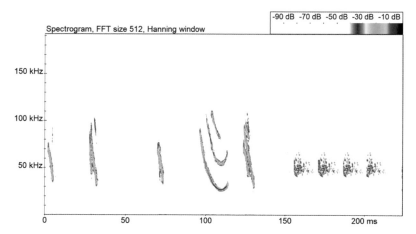

Figure 8.6.15   Example of a social call recorded from Daubenton's bats swarming in a cave (Arjan Boonman).

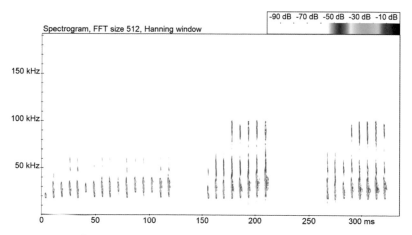

Figure 8.6.16   Distress calls recorded from Daubenton's bat held in the hand (Jon Russ).

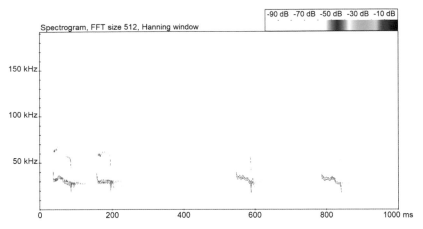

Figure 8.6.17   Calls recorded inside a Daubenton's bat maternity roost (Phil Riddett).

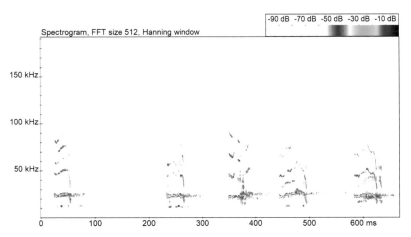

Figure 8.6.18   A further example of calls recorded inside a Daubenton's bat maternity roost (Phil Riddett).

SPECIES WITH SIMILAR OR OVERLAPPING ECHOLOCATION CALLS

Myotis cappaccini, Myotis dasycneme (see Chapter 7 species notes).

8.7 Pond bat *Myotis dasycneme* (Boie, 1825)

Carola van den Tempel, Marc Van De Sijpe and Arjan Boonman

© René Janssen

DISTRIBUTION

The pond bat is distributed over northern parts of central Europe starting in the Netherlands in the west into Siberia in the east. The northern border reaches from Denmark and southern Sweden to Estonia and across Russia as far east as the Central Siberian Plateau, China and Kazakhstan. The southern border is less clear. It is known to occur in northern France, northern Germany, the Czech Republic, Hungary and Ukraine, but only in some places in Romania.

EMERGENCE

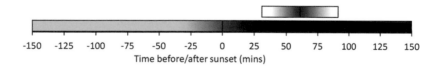

Time before/after sunset (mins)

-150 -125 -100 -75 -50 -25 0 25 50 75 100 125 150

FLIGHT AND FORAGING BEHAVIOUR

On emerging from the roost, the pond bat usually keeps flying low, following the contours of houses on the way to the nearest waterway. On rivers and canals used for commuting, pond bats tend to stay away from the banks (unlike Daubenton's bat) and fly about 26 cm above the water surface in a straight line, reaching speeds of up to 9.7 m·s⁻¹ during commuting (Britton *et al.* 1997), which is faster than most other European bats. Over the foraging areas, the flight behaviour is similar to Daubenton's bat but with the bat making more turns and occasionally shooting high up in the air, probably to chase a moth.

The bat's commuting behaviour and echolocation remain unchanged when entering extremely dense fog in the early morning, which is a common phenomenon over canals on windless nights in late summer. Pond bats often forage over large lakes, but they have also been observed flying over the sea. Because of the relatively long wings and higher mass, the pond bat is less manoeuvrable than Daubenton's bat, yet its trawling behaviour appears identical. The water needs to be clear of vegetation because most prey is taken with the feet (hawking) or tail membrane (trawling) from the water. The main prey are chironomids, including pupae, followed by beetles and Trichoptera (caddisflies) (Britton *et al.* 1997). A study on a population in northern Germany found very similar results, but with a larger percentage of Trichoptera and a small percentage of moths (Sommer and Sommer 1997). The strong seasonality of Trichoptera may play a role in its occurrence in the bat's diet.

HABITAT

The pond bat usually forages over large waterways, such as canals, slow-flowing rivers and lakes. It is hardly seen above small isolated water bodies. It is occasionally found foraging above meadows.

Pond bats use many different kinds of roosts in different seasons. During summer, females and their young can be found in houses, in cavity walls, behind cladding, under roofs and in church attics or towers. A group of females can contain from 40 to 750 individuals. To fly from the roost to water for feeding, the bats follow contours of elements in the landscape. Rows of trees can, therefore, be very important for the species. During the mating season, males have their own territory, which can be found in bat boxes or tree holes, although pond bats are also found at swarming sites. In winter, they hibernate in underground sites such as limestone quarries, caves and bunkers. The species is migratory. Banded individuals in the Netherlands have frequently turned up 250 km south (as the crow flies, but the bat is likely to have travelled well over 300 km). One banded individual was retrieved in the Eifel, in Germany, 312 km from its release point. Other Dutch populations have been shown to have much shorter migration routes. The species prefers to hang fairly exposed as opposed to hiding in deep crevices.

ECHOLOCATION

Measured parameters	Mean (range)	
	FM-qCF call	qCF call
Inter-pulse interval (ms)	71.5 (15.5–110.5)	131.0 (111.2–217.3)
Call duration (ms)	(4.0–8.0)	16.0 (8.0–20.0)
Frequency of maximum energy (kHz)	39.0 (37.0–41.0)	35.0 (33.0–38.0)
Start frequency (kHz)	85.0 (65.0–90.0)	46.0 (40.0–56.0)
End frequency (kHz)	26.0 (25.0–35.0)	28 (24.0–30.0)

Heterodyne
While flying over the water the pond bat usually has long qCF signals. These are heard as 'slaps' at around the frequency of 36 kHz. These slaps have a large interval and are usually very loud.

During swarming or in a more closed habitat pond bats, like all *Myotis* species, have steep FM sounds. The mean frequency is usually around 39 kHz with a high repetition rate.

Time expansion/full spectrum

In typical cluttered environments, pond bats use short (2–4 ms) FM pulses that never start as high as Daubenton's bat in clutter (95–100 kHz), starting at a maximum of 85 kHz (Figure 8.7.1). However, in very cluttered habitats, such as when swarming at roost entrances, the maximum frequency can become quite high, occasionally creeping into the Daubenton's bat range (Figure 8.7.2). During commuting, flight pulses may still start at this high frequency, but pulse duration tends to be longer, with the lengthening occurring around 35 kHz (Figure 8.7.3). Inter-pulse intervals in this situation are typically 90 ms. When hunting over land, calls of medium duration can be produced of around 14 ms. The more open the environment, the more pulses are emitted, with much lower starting frequencies. Typical echolocation calls above open water are qCF signals that start at 46 kHz and have a long-duration part (8–20 ms) at around 35 kHz from knee to heel (Figure 8.7.4). The final FM tail is short and drops steeply. This type of call is easily distinguishable from Daubenton's bat, in case both are foraging above the same surface.

Feeding buzzes show the typical decrease in frequency and inter-pulse interval for most vespertilionid bats (Figure 8.7.5) and can sometimes be produced very quickly after the open-environment calls (Figure 8.7.6). In some cases, pond bats seem to fly 'silently', only occasionally emitting a very loud 20 ms pulse. Some recordings suggest that in reality the bat is emitting very quiet short FM calls preceding these loud qCF emissions, which usually remain undetected from the far riverbank. This may occur when stealthily approaching hearing moths are present. Occasionally the downward FM portion of the call is not so pronounced (Figure 8.7.7).

Although the peak frequencies are variable in all three species, the pond bat generally uses lower peak frequencies (median 41 kHz for durations of 1–4 ms and median 35 kHz for durations > 14 ms) than Daubenton's bat (median 49 kHz for durations of 1–4 ms and median 37 kHz for durations of 8–14 ms) and long-fingered bat (median 45 kHz for durations of 1–4 ms) (Van De Sijpe 2011). End frequencies are significantly higher in the long-fingered bat (median 32 kHz, for durations of 1–4 ms) than in the other two species (pond bat median 25 kHz, Daubenton's bat median 26 kHz, for durations of 1–4 ms), quasi-constant-frequency parts in the middle of the signal are only found in the pond bat, and long-fingered bats sometimes used pulse series with alternating end frequencies (Van De Sijpe 2011).

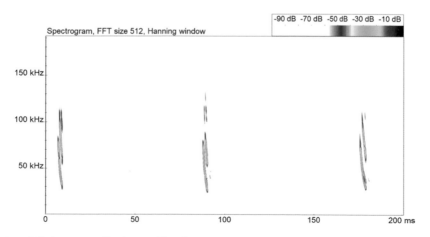

Figure 8.7.1   Echolocation calls of a pond bat flying over a graveyard near to a roost (Marc Van De Sijpe).

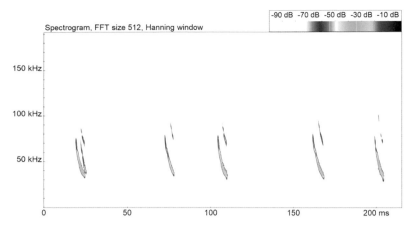

Figure 8.7.2   Echolocation calls of a pond bat flying in a very cluttered environment next to a roost entrance during swarming (Marc Van De Sijpe).

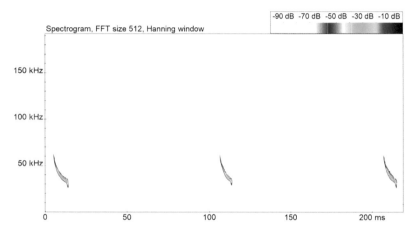

Figure 8.7.3   Echolocation calls of a pond bat commuting over a woodland road (Marc Van De Sijpe).

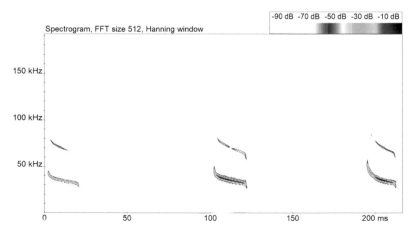

Figure 8.7.4   Echolocation calls of a pond bat hunting over a 30 m canal flanked with reeds and poplars (Marc Van De Sijpe).

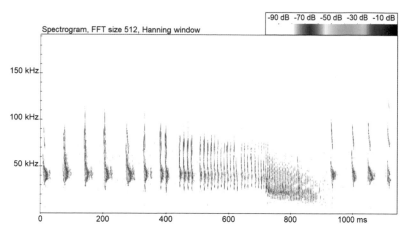

Figure 8.7.5　Typical feeding buzz of a pond bat foraging over a canal (Marc Van De Sijpe).

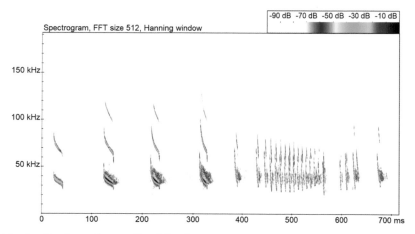

Figure 8.7.6　Feeding buzz of a pond bat following an open-environment call (Marc Van De Sijpe).

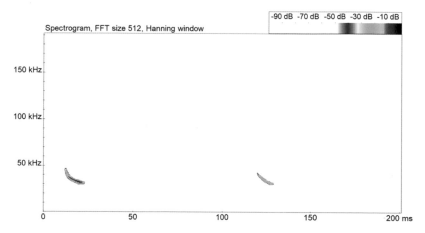

Figure 8.7.7　Echolocation calls of a pond bat commuting over pasture, showing a reduced FM tail (Marc Van De Sijpe).

SOCIAL CALLS

Heterodyne
Pond bats do not seem to have typical mating calls like pipistrelles or nyctaloids. Social calls of pond bats leaving the maternity roost can be heard within the range 16–20 kHz and can usually be heard without a detector. Social calls from the hand (distress calls) are short pulses which are very loud and repeated quickly.

Time expansion/full spectrum
When leaving a maternity roost many different calls are produced. A long qCF call from 23 to 10 kHz during 60 ms is emitted (Figure 8.7.8) but more frequently upward hooked calls from 40 ms are produced and followed by an FM call of only 10 ms (Figure 8.7.9). Calls with a qCF structure can also be produced which are upward hooked with a peak at 16 kHz and with peaks also at 26 kHz (Figure 8.7.10).

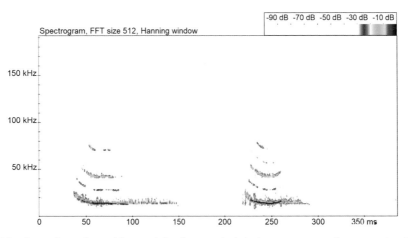

Figure 8.7.8   Recording of pond bat social calls recorded during emergence from a maternity colony in mid-June when young are present (Carola van den Tempel).

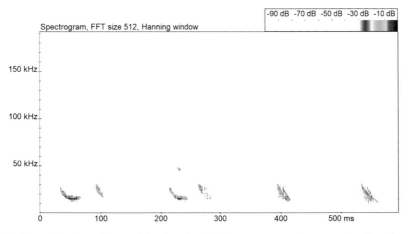

Figure 8.7.9   Example of pond bat social call emitted during emergence from a maternity colony (Carola van den Tempel).

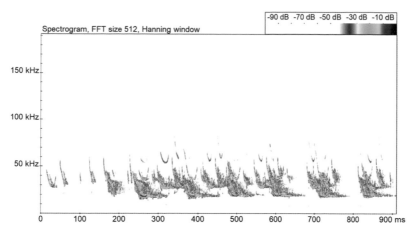

Figure 8.7.10   A further example of pond bat social call during emergence from a maternity colony (Carola van den Tempel).

In the hand, pond bats produce very steep 'distress calls' which start at around 63 kHz and end at around 15 kHz. The pulses are steep (2 ms) and have an interval of about 75 ms (Figure 8.7.11)

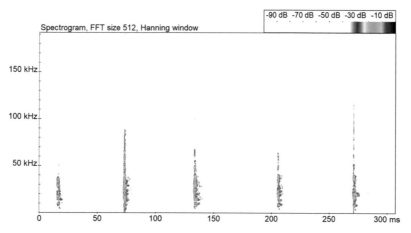

Figure 8.7.11   Distress calls of a pond bat held in the hand during bat-box checks in August (Carola van den Tempel).

Typical mating calls are unknown. However, during the mating season calls similar to Daubenton's bat mating calls are produced close to maternity roosts. These downward hooked calls are within the range of pond bat parameters (start frequency 49 kHz, end frequency 33 kHz) and last 8 ms (Figure 8.7.12). This type of call is frequently recorded when bats are foraging (e.g. Figures 8.7.13 and 8.7.14).

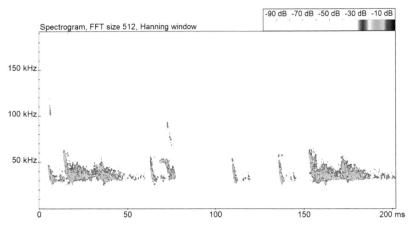

Figure 8.7.12   Social calls of pond bats recorded during swarming at a maternity roost containing 100+ bats during August (Carola van den Tempel).

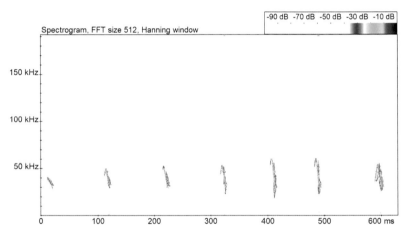

Figure 8.7.13   Pond bat social calls emitted during flight low over the water surface (Marc Van De Sijpe).

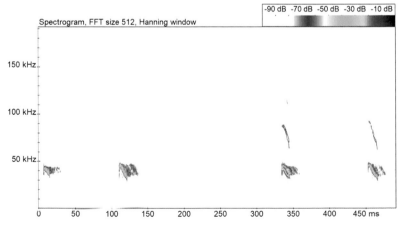

Figure 8.7.14   A further example of pond bat social calls emitted during flight low over the water surface (Marc Van De Sijpe).

A variety of other social calls are emitted. For example, a call which starts at around 38 kHz and ends at around 27 kHz, exceptionally lasting for about 31 ms (Figure 8.7.15) and a call which has a duration of 6–10 ms with an interval of 90 ms that starts at around 40 kHz and ends at around 21 kHz, which is sometimes slightly shaped like a backwards 'S' (Figure 8.7.16).

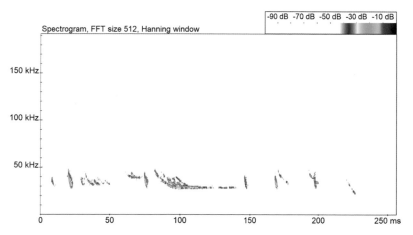

Figure 8.7.15   Social call of a pond bat recorded near the roost with a flashlight playing over or near the entrance. Possible distress call (Carola van den Tempel).

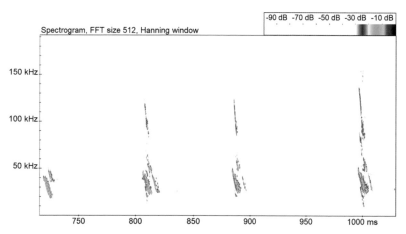

Figure 8.7.16   Social call of a pond bat recorded near a maternity roost with a flashlight playing over or near the entrance. Possible distress call (Carola van den Tempel).

During swarming at the maternity roost, an arch-shaped call is occasionally produced with a duration of around 31 ms which starts and ends at about 37 kHz with a maximum frequency of around 45 kHz (Figure 8.7.17).

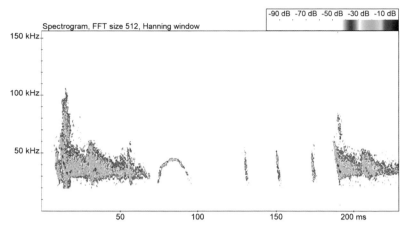

Figure 8.7.17   Social calls of a pond bat recorded during swarming at a roost containing 100+ bats during August (Carola van den Tempel).

Social calls have been recorded from bats perched on a tree (Figure 8.7.18).

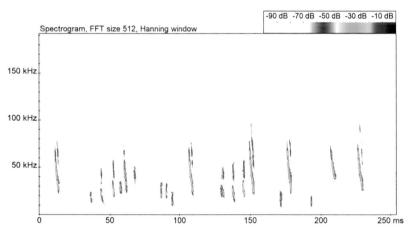

Figure 8.7.18   Social calls of a pond bat, one of a group of three resting on a tree (Marc Van De Sijpe).

Social calls have also been recorded at a cave entrance during the autumn at known swarming and hibernation sites (Figure 8.7.19).

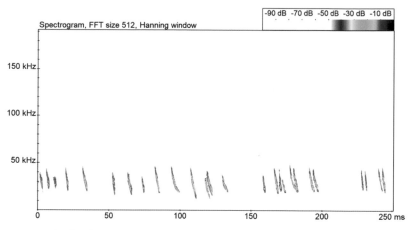

Figure 8.7.19   Social calls of pond bats at a cave entrance known to be a swarming and hibernation site (Jonathan Demaret).

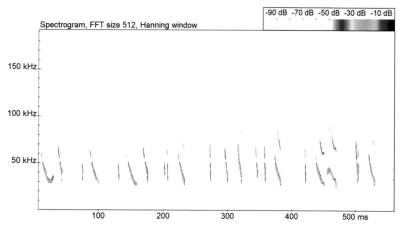

Figure 8.7.20   Social calls of pond bats at a cave entrance known to be a swarming and hibernation site (Jonathan Demaret).

SPECIES WITH SIMILAR OR OVERLAPPING ECHOLOCATION CALLS

Myotis daubentonii (see Chapter 7 species notes).

8.8 Long-fingered bat

Myotis capaccinii (Bonaparte, 1837)

Daniela Hamidović, Eleni Papadatou and Marc Van De Sijpe

© Dragan Fixa Pelić

DISTRIBUTION

The species has a circum-Mediterranean distribution that ranges from western to south-eastern Europe, all the way to southeast Asia and northwest Africa. In Europe, it is present in Spain, France, Italy, Slovenia, Croatia, Bosnia and Herzegovina, Serbia, Montenegro, Albania, Macedonia, Bulgaria, Romania, Turkey and Greece, while in Switzerland it is regionally extinct. Its range partially overlaps with that of Daubenton's bat (Guillen 1999, Paunović 2016).

EMERGENCE

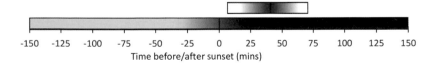

Time before/after sunset (mins)

FLIGHT AND FORAGING BEHAVIOUR

The long-fingered bat is specialised for aquatic habitats. It trawls for prey over the calm and smooth surface of water bodies such as rivers and lakes using its feet and uropatagium (Kalko 1990, Spitzenberger and von Helversen 2001, Almenar *et al.* 2006, 2008). Its flight is slow and manoeuvrable, often down to a few centimetres from the water surface, sometimes touching it with its wingtips (Daniela Hamidović, personal observation).

HABITAT

Forages over water bodies such as rivers and lakes. It may also hunt over lagoons and even the sea, especially where other suitable water surfaces may not be available (Dietz and Kiefer 2016; Eleni Papadatou personal observation). It rarely hawks for prey near the canopy and the vegetation edge (Kalko 1990; Daniela Hamidović, personal observation).

The species roosts almost exclusively in underground sites, primarily caves, although it may roost in disused mines, cellars or waterway tunnels (UNEP/EUROBATS 2016). Unlike the pond bat or Daubenton's bat, there is only scarce evidence of overground roost use (Farina *et al.* 1999). It is therefore typically found in limestone areas with large water bodies such as rivers and lakes with rich riparian vegetation, although it may also be found in other areas with suitable underground sites (Dietz and Kiefer 2016).

ECHOLOCATION

Measured parameters	Mean (range) FM call
Inter-pulse interval (ms)	59.19 (13.4–174.0)
Call duration (ms)	3.9 (2.1–6.9)
Frequency of maximum energy (kHz)	52.8 (41.8–67.2)
Start frequency (kHz)	88.0 (64.6–105.0)
End frequency (kHz)	37.8 (28.0–45.0)

Heterodyne

When flying over the water, the long-fingered bat produces continuous steep and short FM signals that are heard as continuous dry 'ticks' which are loudest when tuning the bat detector between 45 and 50 kHz. It is not possible to discriminate the long-fingered bat from Daubenton's bat with certainty when the two species occur together, though the latter species may emit calls with greatest amplitude between 40 and 45 kHz.

Time expansion/full spectrum

In clutter the long-fingered bat can emit quite a broadband call (Figure 8.8.1) which occasionally can go as high as 105 kHz, sweeping down to around 28 kHz (Figure 8.8.2). As the bat moves into a slightly less cluttered habitat the maximum frequency begins to drop, as does the minimum frequency (Figure 8.8.3). In all habitat types, the species typically emits a short narrowband sigmoidal FM signal that sweeps down from about 88 kHz to about 38 kHz (Figure 8.8.4). Mean call duration is about 4 ms and inter-pulse interval is approximately 60 ms. The bandwidth decreases in more open habitats (Figure 8.8.5). In many signals, there is often a strong interference pattern present when bats fly close to the water (see Figures 8.8.2 and 8.8.3). A second harmonic is often present when bats emerge from a roost in large numbers, while calls from bats flying low over the water usually do not have a second harmonic (Hamidović 2005).

The feeding buzz usually consists of two components, 'buzz I' and 'buzz II', but buzz II is short and may be missing when the bat is hunting fish (Hamidović 2005, Aihartza *et al.* 2008, Aizpurua *et al.* 2014).

Although the peak frequencies are variable in all three species, the pond bat generally uses lower peak frequencies (median 41 kHz for durations of 1–4 ms and median 35 kHz for durations >14 ms) than Daubenton's bat (median 49 kHz for durations of 1–4 ms and median 37 kHz for durations of 8–14 ms) and long-fingered bat (median 45 kHz for durations of 1–4 ms) (Van De Sijpe 2011). End frequencies are significantly higher in the long-fingered bat (median 32 kHz, for durations of 1–4 ms) than in the other two species (pond bat median 25 kHz, Daubenton's bat median 26 kHz, for durations of 1–4 ms), quasi-constant-frequency parts in the middle of the signal are only found in the pond bat, and long-fingered bats sometimes used pulse series with alternating end frequencies (Van De Sijpe 2011).

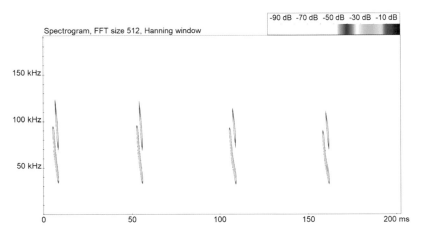

Figure 8.8.1   Echolocation calls of a long-fingered bat released from the hand (Daniela Hamidović).

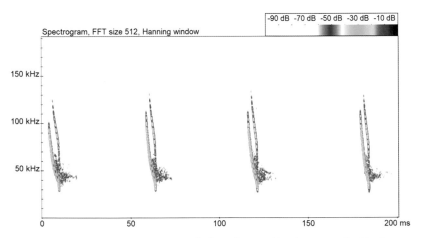

Figure 8.8.2   Very broadband echolocation calls of a long-fingered bat (Marc Van De Sijpe).

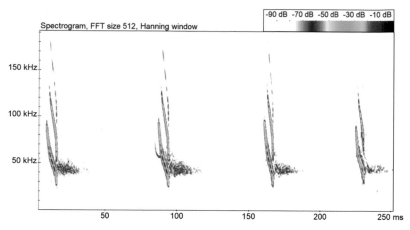

Figure 8.8.3   Echolocation calls of a long-fingered bat in a slightly cluttered habitat flying over a pond (Marc Van De Sijpe).

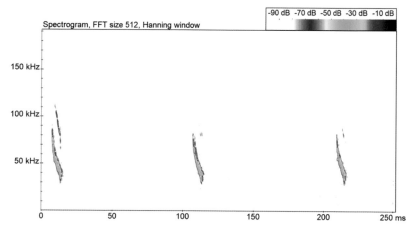

Figure 8.8.4   Typical echolocation calls of a long-fingered bat flying over a pond (Marc Van De Sijpe).

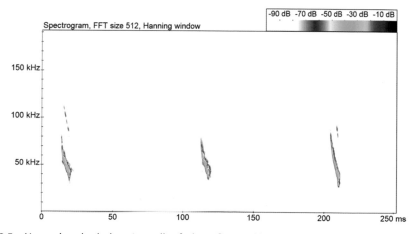

Figure 8.8.5   Narrowband echolocation calls of a long-fingered bat (Marc Van De Sijpe).

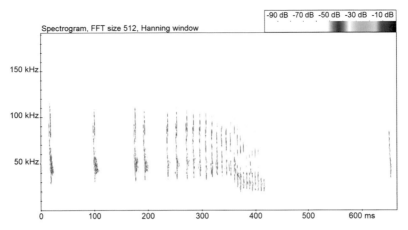

Figure 8.8.6   The feeding buzz produced by a long-fingered bat (Daniela Hamidović).

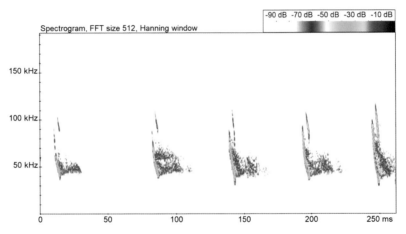

Figure 8.8.7   The distinctive alternating end frequency of a long-fingered bat (Marc Van De Sijpe).

SOCIAL CALLS

A variety of social calls have been recorded at cave entrances and from bats foraging over water, the functions of which are unknown (Figures 8.8.8–8.8.13).

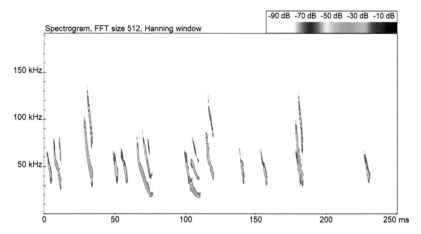

Figure 8.8.8   Social calls of a long-fingered bat emitted while foraging over water (Marc Van De Sijpe).

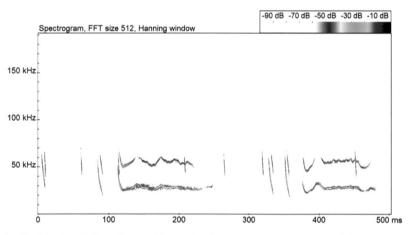

Figure 8.8.9   Social calls of a long-fingered bat emitted at a cave entrance in an oak forest (Yves Bas).

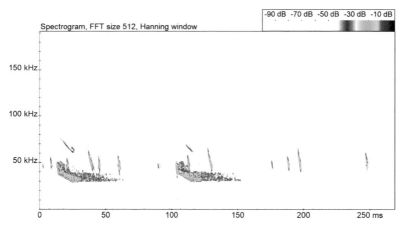

Figure 8.8.10   A further example of social calls of a long-fingered bat emitted at a cave entrance in an oak forest (Yves Bas).

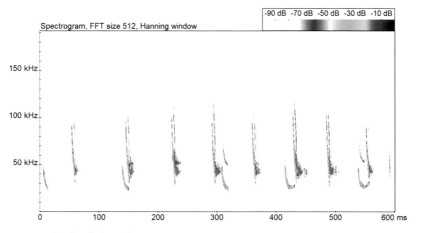

Figure 8.8.11   Social calls of a long-fingered bat emitted over a calm pool in a river (Marc Van De Sijpe).

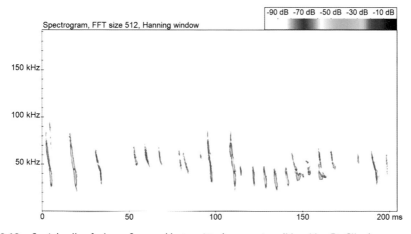

Figure 8.8.12   Social calls of a long-fingered bat emitted over a river (Marc Van De Sijpe).

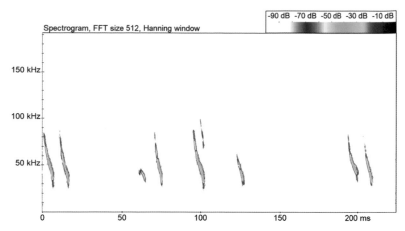

Figure 8.8.13   Social call of a long-fingered bat emitted over a river (Marc Van De Sijpe).

SPECIES WITH SIMILAR OR OVERLAPPING ECHOLOCATION CALLS

Myotis daubentonii (see Chapter 7 species notes).

NOTES

Measured parameters are from samples from Croatia (emerging from the roost, commuting through clutter, hand-released and flying over a river; Hamidović 2005; number of calls 383) and Greece (emerging from the roost, hand-released, swarming and free flight; number of calls 56).

8.9 Brandt's bat *Myotis brandtii* (Eversham, 1845)

Alex Lefevre and Marc Van De Sijpe

© René Janssen

DISTRIBUTION

A Eurasian bat species with a predominantly central and northern presence. Its distribution is from southern Scotland and England (with a single record from Ireland) to mainly northern and central Europe. In France, the species is absent in the west with an increasing abundance to the east. Reaches as far as Bulgaria, Romania and Ukraine to the east. Some occasional recordings have been made in the north of Greece and central Italy. In many European regions, the occurrence of Brandt's bat is not well known, since it seems difficult to distinguish this species from whiskered bat.

EMERGENCE

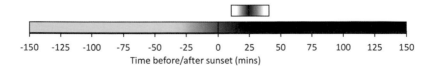

-150   -125   -100   -75   -50   -25   0   25   50   75   100   125   150

Time before/after sunset (mins)

FLIGHT AND FORAGING BEHAVIOUR

Occurs mainly in woodland, marshes and other damp areas. In forests, they have a very agile flight. The species seems to be more linked to forests than whiskered bats. In northern Europe, Brandt's bat even hunts in mountain forests up to the treeline. Flies at different heights (1–10 m) from close to the ground vegetation up to the canopy. Foraging occurs along edges of woodland and forest roads in a straight line. In the forest, they can be seen hunting for extended times by circling in open spaces between the herbaceous and scrub layers of the understorey and the base of the canopy. They also exploit open spaces higher in the canopy. From the ground, the silhouettes can be seen crossing small clearings in the upper canopy. With its somewhat larger wings, this species seems to be more manoeuvrable than whiskered bat.

HABITAT

Brandt's bat has a strong preference for broadleaf damp forests (mainly beech and oak forests) and wetlands, marshes, moorland. Even mountain pine forests up to an altitude of over 1,500 m are important. In western Europe, the species also occurs in parks, gardens, treelines and hedges. Open habitats are avoided.

Roosts in tree holes, cracks in trunks and behind exfoliating bark in summer as well as in bat boxes. In buildings often in the roof void along the ridge or under the ridge cavity. Hibernation sites are located in caves and mines.

ECHOLOCATION

Measured parameters	Mean (range) FM call
Inter-pulse interval (ms)	88.0 (56.7–190.0)
Call duration (ms)	3.5 (1.5–7.0)
Frequency of maximum energy (kHz)	42.0 (37.0–50.0)
Start frequency (kHz)	91.6 (59.0–121.9)
End frequency (kHz)	34.0 (23.0–41.8)

Heterodyne
The species can be heard at a maximum distance of 5–7 m. The rhythm is slower (compared to whiskered bat) and regular.

Time expansion/full spectrum
Brandt's bats use FM calls of large bandwidth that resemble those of whiskered bats, Daubenton's bat and other *Myotis* species (Figures 8.9.1 and 8.9.2). Close to vegetation, in clutter, the calls have a short duration (1–2 ms) (Figures 8.9.3 and 8.9.4), and occasionally have starting hooks when in extreme clutter (Figure 8.9.5). The starting frequencies can go up to 140 kHz, the end frequencies are usually at 30–40 kHz. A bit further from the vegetation, for example in open woodland, Brandt's bats use slightly longer calls with a duration up to 5 ms and a slightly lower end frequency. These calls are curved with a knee at 55–60 kHz and a heel at 40 kHz. According to Barataud (2015), the species can be clearly distinguished from whiskered bats (see notes).

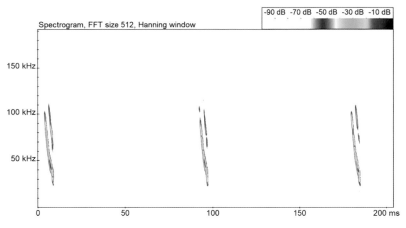

Figure 8.9.1   Echolocation calls of Brandt's bat foraging along a woodland path (Marc Van De Sijpe).

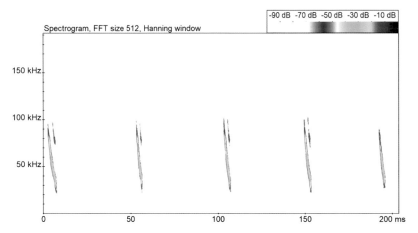

Figure 8.9.2   Echolocation calls of Brandt's bat foraging along woodland edge (Marc Van De Sijpe).

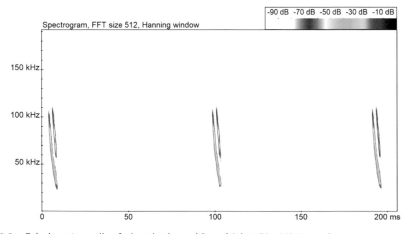

Figure 8.9.3   Echolocation calls of a hand-released Brandt's bat (Karri Kuitunen).

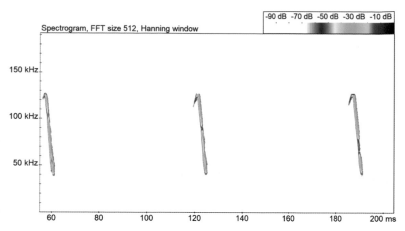

Figure 8.9.4   Echolocation calls of a Brandt's bat at a forest roost (Risto Lindstedt).

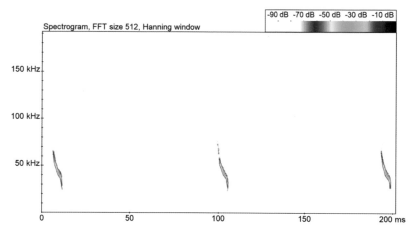

Figure 8.9.5   Echolocation calls of a Brandt's bat flying in the open (Marc Van De Sijpe).

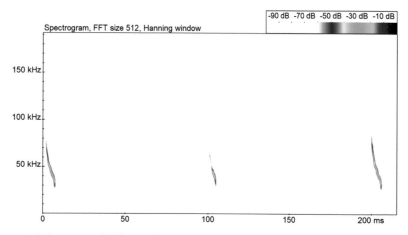

Figure 8.9.6   Echolocation calls of a Brandt's bat flying over a field, showing a 'warbling' in the echolocation call (Simon Dutilleul).

SOCIAL CALLS

Heterodyne
Social calls are rarely observed during flight, even near their roosts or at autumn swarming sites. When handled, Brandt's bat is relatively quiet compared to Natterer's or whiskered bat.

Time expansion/full spectrum
Distress calls, emitted by bats under physical duress, consist of a series of downward-sweeping components and are often heard when a bat is held in the hand (Figures 8.9.7 and 8.9.8).

Social calls are occasionally recorded in the vicinity of roosts and usually consist of a series of V-shaped calls and FM sweeps (Figures 8.9.9–8.9.11).

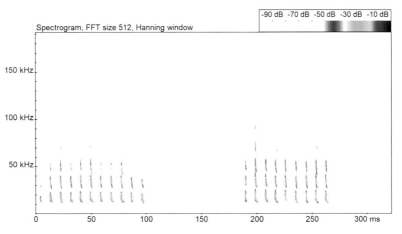

Figure 8.9.7   Distress calls of a Brandt's bat held in the hand (Risto Lindstedt).

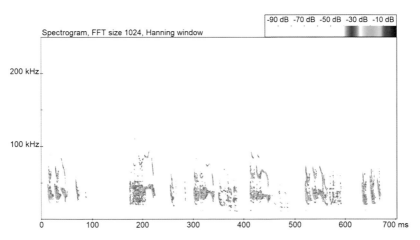

Figure 8.9.8   Social calls of a Brandt's bat held in the hand (Harry Lehto).

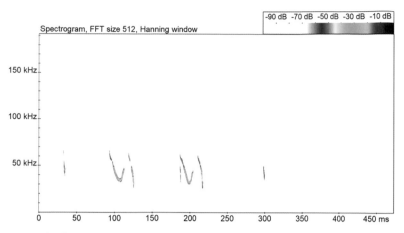

Figure 8.9.9   Social calls of a Brandt's bat recorded near a roost (Rich Flight).

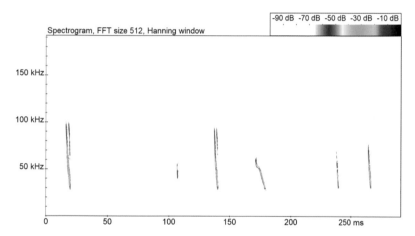

Figure 8.9.10   Social calls of a Brandt's bat recorded near a roost (Rich Flight).

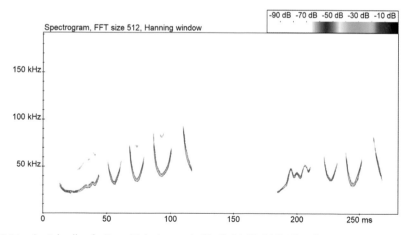

Figure 8.9.11   Social calls of a Brandt's bat recorded in flight (Kari Miettinen).

SPECIES WITH SIMILAR OR OVERLAPPING ECHOLOCATION CALLS

Myotis mystacinus, Myotis daubentonii, Myotis alcathoe (see Chapter 7 species notes).

NOTES

Barataud (2015) uses a purely auditory analysis of the steep FM signal played in time expansion. Observers need to listen to the energy distribution (explosive start or final whack or absence of peak) of the signals. This needs good recordings, but also the necessary skills and a lot of specific training.

8.10 Whiskered bat *Myotis mystacinus* (Kuhl, 1817)

Alex Lefevre and Marc Van De Sijpe

© René Janssen

DISTRIBUTION

The name 'whiskered' bat has been applied for a long time to a very large and difficult-to-determine group. The whiskered bat *Myotis mystacinus* is widely distributed across Europe, from the northern part of the Iberian peninsula to the southern part of Scandinavia (above 64°N). The eastern distribution border cannot be assessed, owing to confusion with David's myotis *M. davidii*, but there are confirmed records from the Caucasus. There are no records from the extreme southern part of Italy. The species has also been recorded at the Turkish west coast and in Israel.

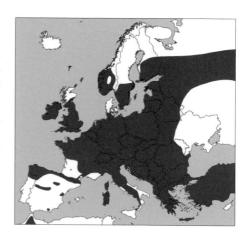

EMERGENCE

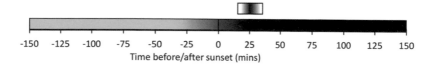

-150	-125	-100	-75	-50	-25	0	25	50	75	100	125	150

Time before/after sunset (mins)

FLIGHT AND FORAGING BEHAVIOUR

This small bat has a very manoeuvrable and rapid flight, alternated with brief gliding periods. Gliding can also frequently be observed around the canopy of forests. Prefers foraging at heights between 1 and 6 m, in wide circuits along large calm water bodies such as ponds and lakes. Forages along vegetation edges, including hedges, woodland paths, gallery forests and treelines, but also above gardens, parks and villages. Is less tied to forests and water than Brandt's bat. The diet consists of flying insects including Diptera, Arachnida, Hymenoptera and Lepidoptera. Normally they forage on flying insects, but on some occasions prey can be collected directly from foliage.

HABITAT

The whiskered bat is a species of open and semi-open but structured small-scale landscapes with isolated patches of forests connected by hedges and woodland edges. It is also frequently observed in villages and their surroundings (gardens, orchards and meadows) containing different types of water bodies. Seems to be less often observed in forests than Brandt's bat.

Roosts in cavities in buildings, such as attic spaces, behind cladding and in cracks in walls during the summer. Also behind peeling bark. In winter hibernates in caves and mines, rarely in rock crevices.

ECHOLOCATION

Measured parameters	Mean (range) FM call
Inter-pulse interval (ms)	113.0 (66.7–251.5)
Call duration (ms)	4.2 (3.0–6.4)
Frequency of maximum energy (kHz)	47.5 (39.2–68.5)
Start frequency (kHz)	88.1 (65.0–120.0)
End frequency (kHz)	32.4 (25.6–43.3)

Heterodyne
The sound is the typical 'dry clicks' of *Myotis* species but quieter. Unlike most other *Myotis* species such as Daubenton's bat, the whiskered bat can only be heard at a maximum distance of 5–7 m. The rhythm is fast and regular.

Time expansion/full spectrum
The most common echolocation calls of the whiskered bat are brief broadband FM calls with a duration of 2–4 ms (Figures 8.10.1 and 8.10.2). The starting frequency can reach 130–140 kHz, the end frequency is at 30 kHz on most occasions. The inter-pulse interval varies between 80 and 200 ms during the search phase with no harmonics. When the bats fly further away from obstacles the duration can increase a little and exceed 4 ms (Figures 8.10.3 and 8.10.4). These calls are bent, with a knee at 60 kHz and a heel at 40 kHz. In dense clutter conditions, the duration reduces further to about 1 ms and the end frequency increases up to 40 kHz, as shown for example in Figure 8.10.5. During the feeding buzz, harmonics can be observed and calls start at 65–85 kHz and drop to around 25 kHz. The separation of steep FM signals from those of Brandt's bat is impossible, according to most authors. A thorough analysis of hundreds of signals has shown that under comparable conditions those of whiskered bats are higher with shorter duration and more calls per second (11–14 calls/second).

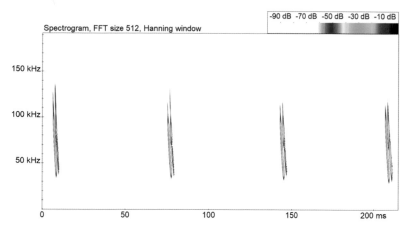

Figure 8.10.1   Echolocation calls of a juvenile male whiskered bat flying over a woodland path after hand release (Marc Van De Sijpe).

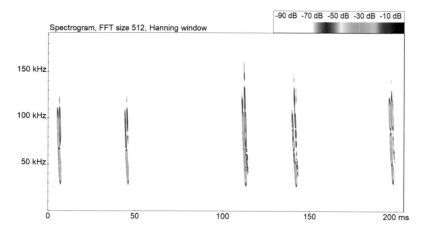

Figure 8.10.2   Echolocation calls of an adult female whiskered bat flying over a pasture after hand release (Marc Van De Sijpe).

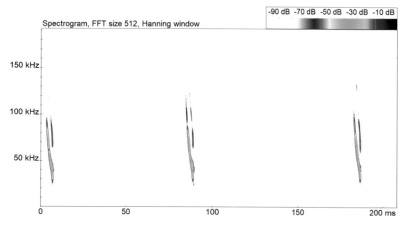

Figure 8.10.3   Echolocation calls of 4 ms duration of an adult male whiskered bat flying over a pasture after hand release (Marc Van De Sijpe).

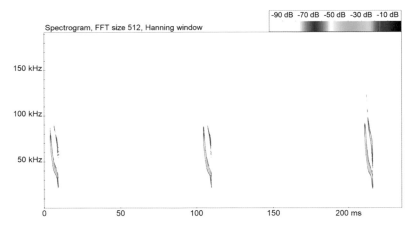

Figure 8.10.4   Echolocation calls of a whiskered bat hunting along a hedgerow (Marc Van De Sijpe).

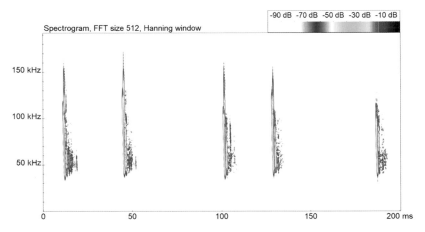

Figure 8.10.5   Echolocation calls of a whiskered bat flying around in a church attic, the maternity roost (Marc Van De Sijpe).

SOCIAL CALLS

Heterodyne
Social calls are rarely heard during flight, but they can be recorded close to the roost. The duration is around 12 ms. When held in the hand or in a holding bag, whiskered bat is typically more vocal than Daubenton's bat and Brandt's bat.

Time expansion/full spectrum
Distress calls of whiskered bats consist of a series of short FM calls of low bandwidth and low frequency quickly repeated one after another, resembling the distress calls of other *Myotis* species (Figure 8.10.6).

Various other social calls can be recorded at the entrance of maternity roosts during emergence or swarming in the early morning. These include V-shapes with a duration of almost 20 ms and the lowest frequency around 27 kHz as well as broadband FM calls that end with a small qCF part (Figures 8.10.7–8.10.9). Series of short FM calls like distress calls can also be heard during the morning swarming (Figure 8.10.10) and from the entrance of the roost just before take-off (Figure 8.10.11).

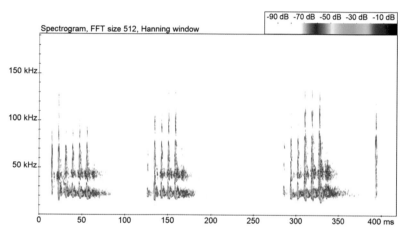

Figure 8.10.6 Distress calls of an adult male whiskered bat in the hand, just before release (Marc Van De Sijpe).

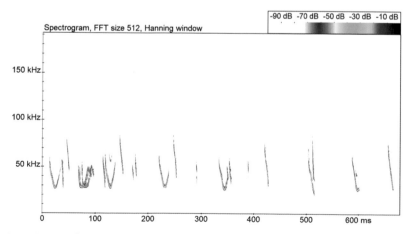

Figure 8.10.7 Variety of social calls of whiskered bat when swarming around the entrance to a maternity roost in the attic of a castle, in the early morning (Marc Van De Sijpe).

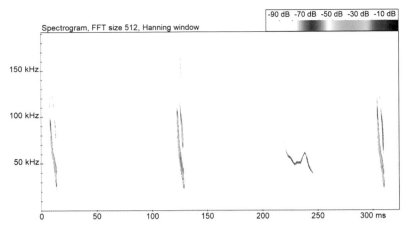

Figure 8.10.8   Social call amidst echolocation calls of swarming whiskered bats at the entrance of the maternity roost (Marc Van De Sijpe).

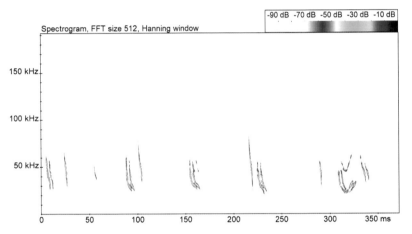

Figure 8.10.9   Social calls of whiskered bats during morning swarming (Marc Van De Sijpe).

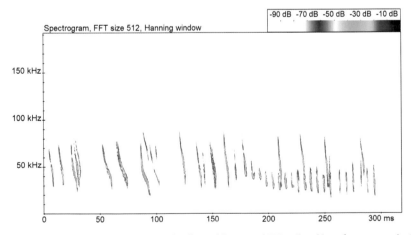

Figure 8.10.10   Various whiskered bat social calls and bursts of FM calls of low frequency during morning swarming (Marc Van De Sijpe).

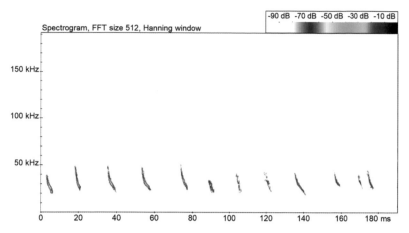

Figure 8.10.11   Low-frequency social calls emitted by whiskered bats at the entrance to a maternity roost in the attic of a castle, just before emergence after dusk (Marc Van De Sijpe).

SPECIES WITH SIMILAR OR OVERLAPPING ECHOLOCATION CALLS

Myotis brandtii, Myotis daubentonii, Myotis alcathoe (see Chapter 7 species notes).

8.11 David's myotis *Myotis davidii* (Peters, 1869)

Jon Russ and Panagiotis Georgiakakis

© Panagiotis Georgiakakis

DISTRIBUTION

David's myotis occurs throughout the Balkan countries (including Crete and the other Greek islands) through eastern Romania and southern Ukraine up to the Caucasus. There have been single records from Turkey. Further data are required to determine its eastern and southern distribution beyond this. Several old records attributed to whiskered bats *Myotis mystacinus* belong to this species.

EMERGENCE

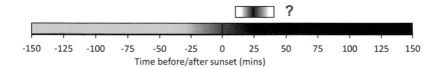

Time before/after sunset (mins)

FLIGHT AND FORAGING BEHAVIOUR

There is little information on flight and foraging behaviour, although it is likely to be similar to that of the whiskered bat, having a rapid and highly manoeuvrable flight, foraging along vegetation edges and woodland paths, in the crowns of trees and above ponds.

HABITAT

Largely associated with steppe habitats or steppe-like open land and floodplains. In the Caucasus also in mountain steppes up to more than 2,000 m.

Nursery roosts reported in construction joints in bridges and in rock crevices. Individuals roost in a wide variety of crevices.

ECHOLOCATION

Measured parameters (based on limited data)	Mean (range) FM call
Inter-pulse interval (ms)	113.3 (82.1–151.3)
Call duration (ms)	3.7 (2.7–4.3)
Frequency of maximum energy (kHz)	47.9 (43.2–56.4)
Start frequency (kHz)	91.6 (79.1–106.9)
End frequency (kHz)	40.8 (39.00–42.4)

Heterodyne

No data. Presumed to be similar to the whiskered bat (see section 8.10).

Time expansion/full spectrum

There are very few data on the echolocation of David's myotis. However, the echolocation seems to be very similar to that of the whiskered bat. In cluttered habitats, such as when the bat is flying in a cave entrance, calls start at around 110 kHz and sweep down to around 23 kHz with a duration of approximately 3.5 ms (Figure 8.11.1). Calls can have a slight knee at around 42 kHz in such situations but can also be very straight (e.g. Figure 8.11.2).

Typically calls in edge habitats such as along woodland edge and wide woodland paths start at around 95 kHz and sweep down to around 35 kHz (Figure 8.11.3), with calls of longer duration in slightly more open habitats (Figure 8.11.4).

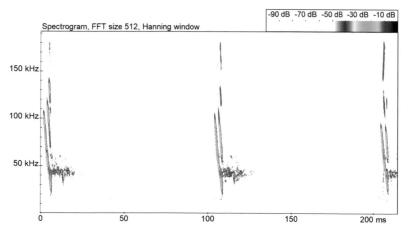

Figure 8.11.1   Echolocation calls of David's myotis recorded in a cave entrance (Panagiotis Georgiakakis).

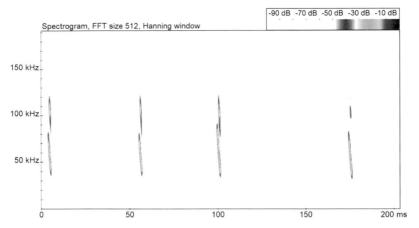

Figure 8.11.2   Echolocation calls of David's myotis recorded in a cave entrance (Panagiotis Georgiakakis).

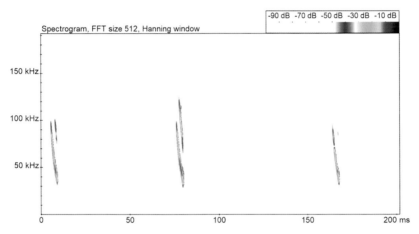

Figure 8.11.3   Echolocation calls of David's myotis recorded on a woodland path (Panagiotis Georgiakakis).

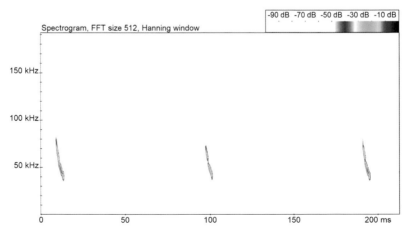

Figure 8.11.4   Echolocation calls of David's myotis recorded on a woodland path (Panagiotis Georgiakakis).

SOCIAL CALLS

Heterodyne
No data.

Time expansion/full spectrum
Social calls have been recorded from bats calling in the hand when under physical duress. The calls comprise a series of quite narrowband components with strong harmonics that go up and down in frequency (Figure 8.11.5) and can be harsh, relatively broadband and of long duration (Figure 8.11.6).

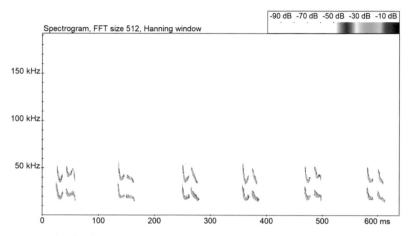

Figure 8.11.5   Social calls of David's myotis recorded in the hand (Panagiotis Georgiakakis).

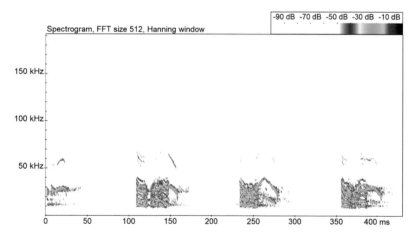

Figure 8.11.6   Social calls of David's myotis recorded in the hand (Panagiotis Georgiakakis).

SPECIES WITH SIMILAR OR OVERLAPPING ECHOLOCATION CALLS

Little information. *Myotis mystacinus, Myotis brandtii* (see Chapter 7 species notes).

NOTES

The presence of *Myotis aurascens* in Europe was proposed by Benda and Tsytsulina (2000) in a morphological revision of the *M. mystacinus* group. According to this study, many published records of *M. mystacinus* belonged to *M. aurascens*. The most recent studies (Benda *et al.* 2016, Dundarova *et al.* 2017) suggest the inclusion of *M. aurascens* and *M. nipalensis* in *M. davidii*. This view has been accepted by EUROBATS but not (yet) by Fauna Europaea or IUCN.

8.12 Alcathoe whiskered bat

Myotis alcathoe (von Helversen and Heller, 2001)

Alex Lefevre, Marc Van De Sijpe and Julia Hafner

© Daniel Whitby

DISTRIBUTION

A Eurasian species with a dispersed but mainly central European presence. Its distribution ranges from northern Spain to southern Sweden with occasional records from England, Italy, Hungary and northern Greece. Although there are abundant records from some areas, such as France and Hungary (probably due to more focused bat research), the species appears to be rare in most of its range. The species is also present in the Caucasus region, reaching as far as Bulgaria, Romania, Moldova, Ukraine and Azerbaijan. In many European regions, the occurrence of Alcathoe whiskered bat is not well known and is patchily distributed. One of the reasons is that this species was only recently discovered (2001). Identification is based mainly on genetic characterisation, due to the difficulty of distinguishing this species visually from other 'whiskered' bat species.

EMERGENCE

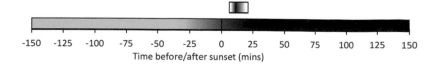

Time before/after sunset (mins)

FLIGHT AND FORAGING BEHAVIOUR

In forests, they have a very agile flight. Foraging occurs in dense vegetation and along forest roads in a straight line. Hunting individuals have been observed in forests up to an altitude of 800 m.

Radio-tracked individuals were shown to fly at a height of 3–10 m, from the scrub layer upwards, but spent most of their time in the canopy of oak forests. Observations made in France showed that the species adapted its foraging behaviour to the canopy as soon as other *Myotis* bat species arrived, indicating that this species starts hunting early after sunset to avoid interspecies competition and uses ecological niches such as the canopy more than most other congeners. Alcathoe whiskered bat is considered together with Natterer's bat and Geoffroy's bat as an edge-space aerial hawking forager.

HABITAT

This smallest European *Myotis* bat occurs mainly in riparian vegetation and valley forests. It seems to be the most specialised bat species of native primaeval broadleaved forests in Europe. Like Brandt's bat, Alcathoe whiskered bat has a strong preference for broadleaf damp forests. Typical habitat for this species is small brooks, creeks or ponds surrounded by old-growth deciduous woodland, especially old oak–ash–hornbeam forests. The species also occurs in forests surrounded by treelines and many hedgerows. A closed canopy is preferred, especially for forest roads. Open habitats such as large forest cuttings or meadows are avoided.

ECHOLOCATION

Measured parameters	Mean (range) FM call
Inter-pulse interval (ms)	85.0 (50.0–95.0)
Call duration (ms)	2.6 (1.5–5.0)
Frequency of maximum energy (kHz)	52.4 (50.0–65.0)
Start frequency (kHz)	100.0 (91.2–130.0)
End frequency (kHz)	42.9 (35.0–50.0)

Heterodyne
The species can be recorded with a heterodyne bat detector at a maximum distance of 8–15 m. The high end frequency seems to be characteristic. Unlike many other non-*Myotis* species, Alcathoe whiskered bat does not have a particular call frequency, and the call sweeps from 125 kHz to 43 kHz very rapidly. The best frequency is usually at 52 kHz (Hafner *et al.* 2015). Try tuning a bat detector to about 50 kHz and listen for tonal differences. The calls are audible with no clear change in sound quality. The repetition rate is a fast rhythm with between 11 and 20 calls per second.

Time expansion/full spectrum

The echolocation of this species is so adapted to detect fine objects that it is better able to detect mist nets than most other *Myotis* species. The end frequency is higher than Geoffroy's bat (<40 kHz) and as short as Bechstein's bat (Figure 8.12.1). Close to vegetation, the calls have a short duration (1–2 ms). The starting frequencies can go up to 130 kHz. Geoffroy's bat can use much higher starting frequencies, up to 160–180 kHz, but not all detectors can pick up these high frequencies. The end frequencies of Alcathoe whiskered bats are usually above 43 kHz and are the highest found in European *Myotis* species. Occasionally the echolocation calls can have a slightly lower end frequency, down to 36 kHz (durations up to 3 ms) or 33 kHz (durations 3–5 ms) (Barataud 2015). The typical curved *Myotis* calls have a heel at 45–55 kHz and a knee at 65 kHz (Figure 8.12.2). The frequency of the heel usually decreases with increasing duration: 55 kHz (3 ms), 50 kHz (3–5 ms) and 45 kHz (5 ms) (Pfeiffer *et al.* 2015). Occasionally the frequency containing maximum energy (FmaxE) can be above 50 kHz (Figure 8.12.3).

The Alcathoe whiskered bat can be separated from its congeners on echolocation calls mainly by using the start and end frequencies. Bat-detector recordings from this species can easily be overlooked as a pipistrelle or Geoffroy's bat flying in a closed environment except for the higher FmaxE. The evaluation of the call shapes in the spectrogram can aid in distinguishing between Alcathoe whiskered bat (sigmoid structure with a heel at 65 kHz and knee at 45–55 kHz), pipistrelle in cluttered situations (a beginning of qCF at the end of the call is often visible even in very short-duration calls) and Geoffroy's bat (call ending is linear, not sigmoid). Some other *Myotis* species (Bechstein's bat, whiskered bat, Brandt's bat) can have end frequencies above 40 kHz when they are in extreme clutter conditions, but usually only for a short time. As soon as they enter a less cluttered environment, the end frequencies of these species readily drop below 40 kHz.

Feeding buzzes are typical of the *Myotis* species (Figure 8.12.4).

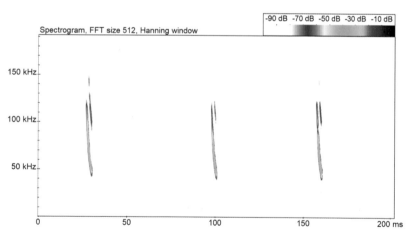

Figure 8.12.1   Typical echolocation calls of an Alcathoe whiskered bat foraging over a woodland ride (Marc Van De Sijpe).

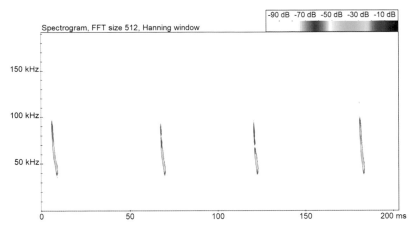

Figure 8.12.2   Echolocation calls of Alcathoe whiskered bat flying over a river (Marc Van De Sijpe).

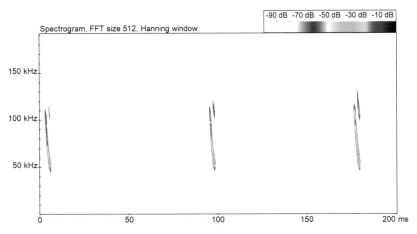

Figure 8.12.3   Echolocation calls of Alcathoe whiskered bat with FmaxE above 50 kHz (Marc Van De Sijpe).

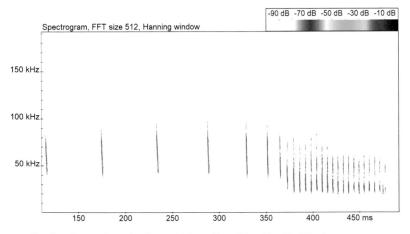

Figure 8.12.4   Feeding buzz of an Alcathoe whiskered bat (Marc Van De Sijpe).

SOCIAL CALLS

Heterodyne
Published data on the social calls of this species are very limited. They are rarely observed during flight, or even in the vicinity of their roosts or at autumn swarming sites.

Time expansion/full spectrum
Distress calls are broadband and harsh (Figure 8.12.5).

Calls from maternity roosts and at dawn swarming generally consist of frequency-modulated sweeps with a hooked tail (Figures 8.12.6–8.12.8) (Hafner *et al.* 2015).

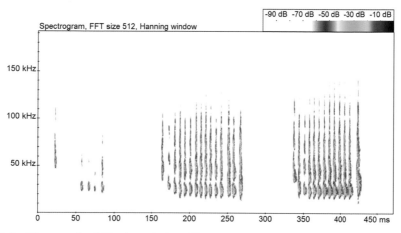

Figure 8.12.5   Distress calls of Alcathoe whiskered bat held in the hand (Phil Riddett).

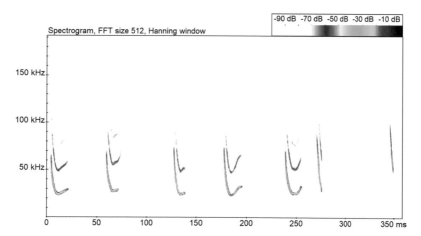

Figure 8.12.6   Social calls of an Alcathoe whiskered bat recorded at evening emergence (Julia Hafner).

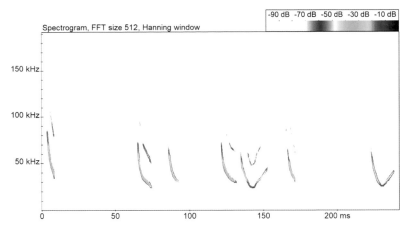

Figure 8.12.7   A further example of social calls of an Alcathoe whiskered bat recorded at evening emergence (Julia Hafner).

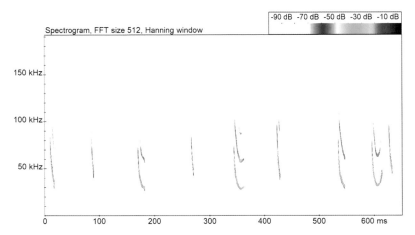

Figure 8.12.8   Social calls of an Alcathoe whiskered bat recorded at dawn swarming (Julia Hafner).

SPECIES WITH SIMILAR OR OVERLAPPING ECHOLOCATION CALLS

Myotis emarginatus (see Chapter 7 species notes).

8.13 Geoffroy's bat *Myotis emarginatus* (Geoffroy, 1806)

Horst Schauer-Weisshahn and Claude Steck

© René Janssen

DISTRIBUTION

As a thermophilic bat species with a main geographical distribution in the Mediterranean, Geoffroy's bat has its northern distribution boundary in the Netherlands, Germany and southern Poland (Dietz and Kiefer 2014).

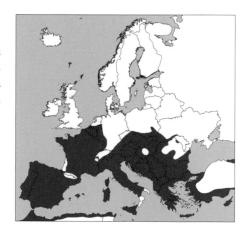

EMERGENCE

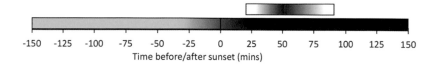

Time before/after sunset (mins)

FLIGHT AND FORAGING BEHAVIOUR

Highly manoeuvrable flight pattern close to vegetation, gleaning prey from surfaces (foliage gleaning) and spiders probably from cobwebs. Flight height ranging from about ground level up to treetops. In cowsheds, they glean prey from the ceiling in a pendulum-like flight pattern. Commuting flights close to vegetation or along streams. Range of nursery colonies mostly a few kilometres (<8 km), but distances up to 16 km between roost and foraging habitat recorded (Steck and Brinkmann 2015). When foraging, flight speed relatively slow (slower than Natterer's bat).

HABITAT

In central Europe, during parturition and lactation, Geoffroy's bats frequently hunt in livestock sheds (e.g. Steck and Brinkmann 2006, Dekker *et al.* 2013), as observed in other parts of Europe as well. Apart from this, Geoffroy's bat is found in a wide variety of habitats. Foraging in old-growth forests, thinned forest stands, riparian forests, dry forests, forest edges, olive groves, orchards, hedgerows and scrubland (e.g. Flaquer *et al.* 2008, Zahn *et al.* 2010, Goiti *et al.* 2011, Steck and Brinkmann 2015). The most important factor is high cover of woody plants combined with high structural diversity.

In the northern part of the distribution area, nursery roosts occur exclusively in buildings; in the Mediterranean also in caves (e.g. Spada 2009, Dietz and Kiefer 2014). Roosts of solitary individuals or small groups beneath roof overhangs/eaves, rock coves or tree cavities (e.g. Krull *et al.* 1991, Steck and Brinkmann 2015). Hibernation roosts in underground sites such as caves, tunnels, galleries (e.g. Goiti *et al.* 2011).

ECHOLOCATION

Measured parameters	Mean (range)	
	FM linear	**FM curvilinear**
Inter-pulse interval (ms)	53.1 (16.9–125.5)	70.1 (32.9–149.5)
Call duration (ms)	2.1 (1.1–3.6)	3.5 (1.4–6.1)
Frequency of maximum energy (kHz)	64.5 (48.8–106.4)	52.5 (43.9–92.8)
Start frequency (kHz)	129.5 (94.7–169.9)	113.2 (61.5–145.5)
End frequency (kHz)	39.6 (29.3–55.7)	35.1 (27.3–49.8)

Heterodyne

Since Geoffroy's bat hunts in a variety of habitats, its call structure is quite variable. The variability of the usually steep frequency-modulated calls expresses itself less in form than in duration and bandwidth. Still, there seem to be two major call types that vary slightly in their form. These are steep linear modulated calls closer to clutter and slightly curvilinear modulated calls in more open habitat. The difference between these two call types is quite difficult to hear using a heterodyne detector.

Geoffroy's bat is commonly recognised acoustically by its steep linear downward modulated FM signal (a dry-sounding 'tick') of short duration and short inter-pulse intervals with an end frequency above 40 kHz. The energy is often distributed fairly equally over the entire call, which makes it very difficult to discern a certain frequency containing maximum energy (FmaxE). The slightly curvilinear modulated call can be more easily confused with calls of other *Myotis* species due to the lower end frequency.

As the bat approaches cluttered habitat the signals get very faint. The pulse intervals change very rapidly with now and then a call missing due to the very agile flight behaviour. This makes it sound a bit like running a finger along the teeth of a comb with increasing and decreasing speed and one or two teeth missing.

In listening to the calls there is no distinct difference between search-phase calls and approach calls, except that the inter-pulse intervals get shorter. The final buzz with its distinct downward shift in frequencies is quite similar to those of other *Myotis* species.

The calls of Geoffroy's bat can most easily be confused with those of the Alcathoe whiskered bat, which usually an have end frequency just slightly above Geoffroy's bat. But, like Geoffroy's bat, the Alcathoe bat can reach end frequencies below 40 kHz. Furthermore, depending on flight situation and distance from the detector, calls of other *Myotis* species may also show end frequencies just below 40 kHz (e.g. Natterer's bat, Bechstein's bat, whiskered bat and Brandt's bat, all of which are also capable of producing short, steep and linear frequency-modulated calls). Although the Natterer's bat can be distinguished from the Geoffroy's bat if an end frequency below 25 kHz is observed, the likelihood of confusion between Geoffroy's bat and other *Myotis* species based on acoustics remains high. Visual observations of flight behaviour and physical attributes (e.g. underside fur colour) may give additional clues for species identification.

Time expansion/full spectrum

Computer analysis of the two major call types of Geoffroy's bat shows a distinctive difference in the values of the measured call parameters.

The steep linear modulated call is used in commuting flight close to clutter, when swarming, in search flight close to clutter, and in the approach phase before the final buzz (Figure 8.13.1). The bandwidth of this call type lies between a mean start frequency of 130 kHz down to a mean end frequency of about 40 kHz but can reach a maximum of 170 kHz to a minimum of 29 kHz. Call duration varies between 1.1 and 3.6 ms with a mean of 2.1 ms. Depending on flight situation the inter-pulse interval can be very constant or highly variable, probably due to undetected calls, and shows a mean duration of 53 ms with a range of 17–126 ms. The energy of this call type shows a bell-shaped distribution with no distinct peak and a mean value of 65 kHz but a high variability in a range between 49 and 106 kHz.

In comparison, the curvilinear modulated call type which is used as search call in more open habitat shows a smaller, slightly downward-shifted bandwidth between a mean start frequency of 113 kHz and a mean end frequency of 35 kHz (Figures 8.13.2 and 8.13.3). The maximum frequency of these calls can reach up to about 146 kHz and may sweep down to a minimum of 27 kHz. Call duration increases to a mean of 3.5 ms with a variation between 1.4 and 6.1 ms. The inter-pulse interval also increases to a mean duration of 70 kHz with a range of 33–150 ms but shows the same variability as the linear call type. The energy of the curvilinear modulated call type also shows a bell-shaped distribution with a mean value of 53 kHz and a downward-shifted variability in a range between 44 and 93 kHz. Sequences of this call type often show a regular alternation between relatively loud and quiet calls. This may be caused by 'scanning behaviour' where the bat is aiming its sonar beam in different directions and is therefore more or less orientated to the microphone (see Seibert *et al.* 2013).

Schumm *et al.* (1991) found for calls emitted in open space that Geoffroy's bat uses very long calls of up to 7.2 ms with a shallow modulated 'tail'-component (CF) around 51 kHz. The calls have a lower start frequency around 95 kHz and sweep down to about 43 kHz. They also observed hovering by the Geoffroy's bat, with echolocation calls resembling final buzz calls.

The calls of the final buzz, comprising 2–10 pulses, are very faint (and therefore seldom recorded), very short (0.2–0.6 ms) and have a start frequency of about 45 kHz sweeping down to an end frequency of 35–20 kHz (Schumm *et al.* 1991) (Figure 8.13.4).

While computer analysis opens more possibilities to discern characteristic echolocation calls of different *Myotis* species, there is still a huge overlap in call parameter values and call structures depending on flight situation and hunting behaviour. Calls of Alcathoe whiskered bat might be distinguished from the steep linear FM calls of Geoffroy's bat with end frequencies above 40 kHz by a more prominent curvilinear modulation, lower start frequency and usually a lower call repetition rate (e.g. Bartsch 2012). Also, the end frequency seems to vary less in call sequences of Alcathoe whiskered bat (C. Dietz, personal communication). The curvilinear modulated call of Geoffroy's bat can be more easily confused with the calls of other *Myotis* species (see Chapter 7). Hence, utmost care has to be taken before pronouncing on the presence or absence of Geoffroy's bat based on acoustic data alone.

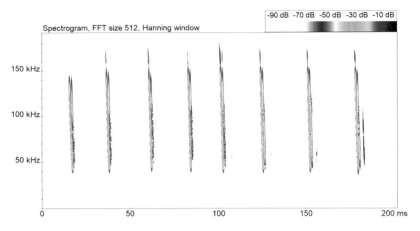

Figure 8.13.1   Steep linear FM echolocation calls of Geoffroy's bat in closed habitat (Horst Schauer-Weisshahn and Claude Steck).

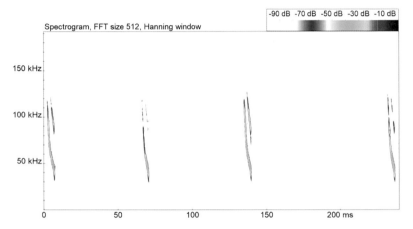

Figure 8.13.2   Slightly curvilinear modulated echolocation calls of Geoffroy's bat in open habitat (Horst Schauer-Weisshahn and Claude Steck).

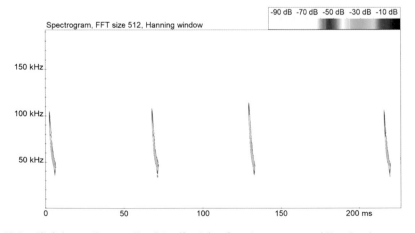

Figure 8.13.3   Slightly curvilinear calls of Geoffroy's bat foraging over a pond (Jon Russ).

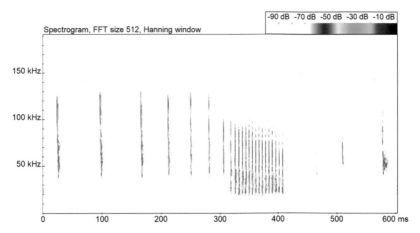

Figure 8.13.4   Feeding buzz of Geoffroy's bat feeding over a pond (Jon Russ).

SOCIAL CALLS

Heterodyne
No data.

Time expansion
Social calls of the Geoffroy's bat are very rarely recorded. Distress calls are produced when the bat is under physical duress (Figure 8.13.5). These are similar to distress calls of other European bat species.

Wimmer and Kugelschafter (2017) report two similar social calls recorded at the entrance of an abandoned mine in southwestern Germany, which is used as a hibernaculum by many Geoffroy's bats. The calls could be assigned to Geoffroy's bat from photographs taken simultaneously with the recordings. The recorded social call shown in Wimmer and Kugelschafter (2017) is a multipart trill (Figure 8.13.6). They compare it to the social calls of the whiskered bat as described in Russ (2012). It also shows similarities to the recording of the distress calls of a Geoffroy's bat held in the hand in Middleton *et al.* (2014). Due to the small number of recordings, Wimmer and Kugelschafter (2017) offer no measurements of the recorded calls.

Recordings from a maternity roost of Geoffroy's bat in southwestern Germany show a variety of social calls (Figures 8.13.7 and 8.13.8). It was not possible to discern whether these calls were from adult or juvenile bats.

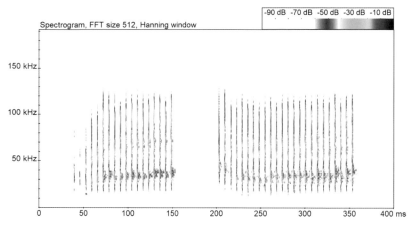

Figure 8.13.5   Distress calls of Geoffroy's bat held in the hand (Phil Riddett).

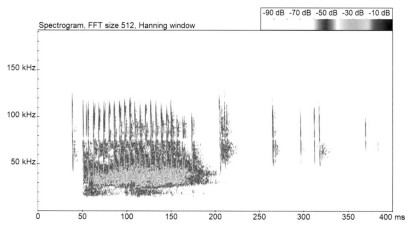

Figure 8.13.6   Social call of a Geoffroy's bat recorded at the entrance of an abandoned mine (Bernadette Wimmer and Karl Kugelschafter).

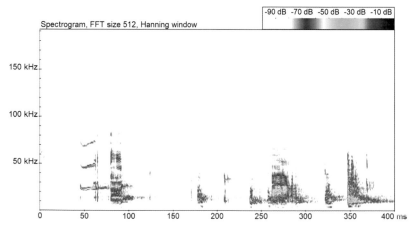

Figure 8.13.7   Social call of a Geoffroy's bat recorded in a maternity roost (Horst Schauer-Weisshahn and Claude Steck).

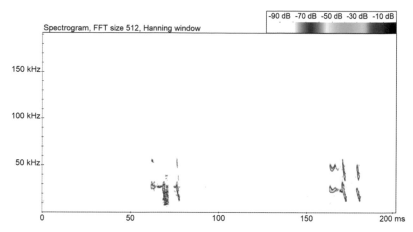

Figure 8.13.8   Social call of a Geoffroy's bat recorded in a maternity roost (Horst Schauer-Weisshahn and Claude Steck).

SPECIES WITH SIMILAR OR OVERLAPPING ECHOLOCATION CALLS

Myotis nattereri, Myotis crypticus, Myotis escalerai, Myotis alcathoe (see Chapter 7 species notes).

NOTES

The measured values of the usually very faint calls of Geoffroy's bat show a high variability that may be due to a variety of reasons. Distance and orientation of the bat to the microphone, air moisture and temperature, microphone type and therefore its frequency response, analysis software and settings such as threshold value, manual or automatic measurements etc. may influence the values of the measured call parameters. As a consequence, it will often be very difficult if not impossible to distinguish Geoffroy's bat from other *Myotis* species just by the calls. This applies especially to recordings made during passive monitoring, where additional information such as flight behaviour is not available.

8.14 Natterer's bat *Myotis nattereri* (Kuhl, 1817)

Marc Van De Sijpe and Alex Lefevre

© René Janssen

DISTRIBUTION

Widely distributed across Europe, from the Mediterranean region including most islands in the south, up to northern Scotland and the southern parts of Scandinavia in the north. The populations reported in the Iberian peninsula and in some parts of southwestern France may represent the cryptic sibling species, Iberian Natterer's bat *Myotis escalerai*.

EMERGENCE

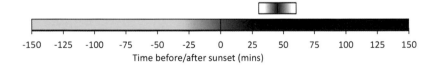

Time before/after sunset (mins)

FLIGHT AND FORAGING BEHAVIOUR

Flight is slow and highly manoeuvrable, including sudden flipping while changing direction, sometimes hovering, or turning in a confined space. Forages by gleaning from surfaces and aerial hawking, low over the ground and water surface and higher up in the canopy of trees. This bat preys on silent and immobile prey within confined spaces. Unlike other clutter-dwelling species like Bechstein's bat, long-eared bats and greater mouse-eared bat, Natterer's bat does not primarily rely on prey-generated sounds when searching for prey within the foliage or on the ground. They rely rather on active echolocation or sensing the surface with the fringe of hairs on the tail membrane (Siemers and Swift 2006). Experiments show that this bat can detect prey items a few centimetres from clutter by using loud, short echolocation calls with extreme bandwidth (Siemers and Schnitzler 2000). The diet consists of a wide variety of arthropods, including web-building spiders, lepidopteran larvae, harvestmen and diurnal flies. The latter are caught by gleaning, by aerial hawking very close to obstacles (e.g. spiders hanging in their web), or by random rakes (Swift and Racey 2002). Low over water the flight style of Natterer's bat differs from Daubenton's bat. Natterer's bat is more manoeuvrable, continuously changing angle, whereas Daubenton's bat keeps flying in a horizontal plane (Limpens and Feenstra 1997).

HABITAT

A typical woodland bat, occurring in all kinds of forest habitats, broadleaved and coniferous, as well as in more anthropogenic habitats that resemble woodland, such as orchards or parkland. Although usually absent in vast open landscapes, Natterer's bat can be found in open habitat near woodland, such as meadows or pastures, where it hunts low over the grassy vegetation or around single trees. Also hunts over water. Just like Geoffroy's bats, Natterer's bats visit cattlesheds, probably attracted by the wealth of diurnal flies which can be caught from ceilings and walls (Dietz *et al.* 2007).

ECHOLOCATION

Measured parameters	Mean (range) FM call
Inter-pulse interval (ms)	80.1 (25.0–189.0)
Call duration (ms)	4.7 (1.9–7.1)
Frequency of maximum energy (kHz)	46.9 (36.0–67.0)
Start frequency (kHz)	107.0 (72.0–180.0)
End frequency (kHz)	23.0 (10.0–40.0)

Heterodyne
Natterer's bat can be heard in heterodyne as a rattling sound, a series of dry ticks, thereby resembling other *Myotis* species. This species has some of the fastest acoustic signals of all European bats, up to 40 signals per second. Experienced users of heterodyne detectors can identify this bat under good observation conditions based on the heterodyne sounds, the rhythm and simultaneous visual observation of the flight behaviour. The sounds of this species often resemble those of stubble burning. The pulse repetition rate is variable, often including bursts of sound at extremely fast rates when the bat is exploring the space very close to vegetation. In transit flight in the open Natterer's bats can produce a series of dry clicks in a slow and very regular rhythm, causing confusion with the greater mouse-eared bat. Caution is needed when using solely heterodyne to determine Natterer's bats, which holds for most *Myotis* species.

Time expansion/full spectrum
The calls most often encountered usually have a very short duration (2–4 ms) and an extreme bandwidth (Figure 8.14.1). Along with Geoffroy's bat, this species uses the highest starting frequencies known in Europe, up to 180 kHz (Skiba 2003). Unfortunately, these high frequencies are of little help in the field, since most microphones are unable to detect them, and even very sensitive microphones will only pick up these extreme frequencies at the start of the calls when the bats are flying very close to the microphone. The end frequency is

variable, generally lower than in other *Myotis* species. When Natterer's bats are exploring the undergrowth or swarming around roosts, one can observe a series of signals with extremely low end frequencies, down to 10 kHz. The audible-frequency part of these calls can be heard even without a detector. This audible sound is a faint rattling, very like the sound of a Natterer's bat one can hear using a heterodyne detector. When viewed in a spectrogram, the short calls of large bandwidth often show a convex curvature, a characteristic feature which is also usually present when high end frequencies are present (Figure 8.14.3). It is most pronounced in the high-frequency start of the signal (above 120 kHz), but sometimes obvious at lower frequencies too. Calls of longer duration emitted in open spaces have a concave curvature, as in many other *Myotis* species (Figure 8.14.4). Frequency containing maximum energy (FmaxE) is generally difficult to determine due to the wide range of frequencies and is less important in terms of identification than bandwidth, starting frequency, end frequency and pulse shape. The combination of the latter features often allows reliable identification of time-expanded recordings.

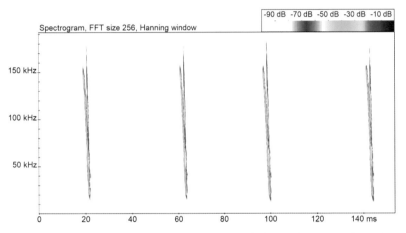

Figure 8.14.1   Echolocation calls of a Natterer's bat exploring the base of a tree in a forest. Duration is short, bandwidth and starting frequency are very high, end frequency is very low, and the pulses show a slight convex curvature (Marc Van De Sijpe and Alex Lefevre).

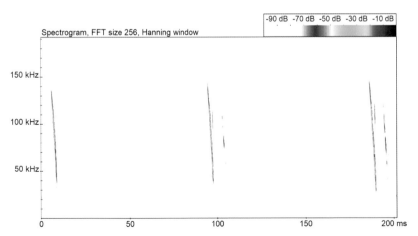

Figure 8.14.2   Echolocation calls of Natterer's bats shortly after they emerged from the maternity roost in a bat box. Extreme starting frequencies (up to 175 kHz) and very low end frequency (13 kHz), resulting in extreme bandwidths (exceeding 160 kHz) (Marc Van De Sijpe and Alex Lefevre).

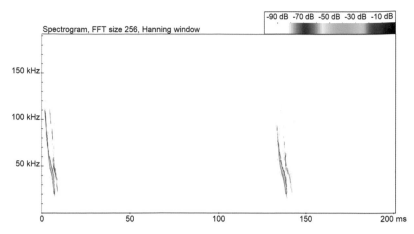

Figure 8.14.3   Natterer's bat using short echolocation calls with high end frequencies (40 kHz). Note that the convex curvature remains visible (Marc Van De Sijpe and Alex Lefevre).

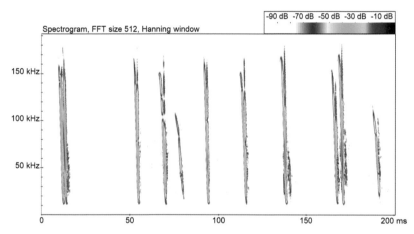

Figure 8.14.4   The same Natterer's bat crossing a more open space in the forest. Concave curvature, duration exceeds 5 ms, and less extreme start/end frequencies (Marc Van De Sijpe and Alex Lefevre).

SOCIAL CALLS

Heterodyne
Because of the longer duration and the qCF or U-shaped parts, the social calls of Natterer's bats can be perceived as wet or smacking sounds in heterodyne.

Time expansion/full spectrum
As in other *Myotis* species, social calls are seldom heard in the hunting habitat but more frequently near the maternity roost when there are juveniles in the roost (June to July). The most common social calls of this species are U shapes with legs of very high bandwidth (Pfalzer and Kusch 2003) (Figures 8.14.5 and 8.14.6). These calls can be encountered in the immediate vicinity of the roost during morning swarming, at evening emergence, or from bats in the roost. The U shapes can sometimes be repeated shortly one after another or even merge to create an undulating wave (Figure 8.14.7). Close to the roost sites other types of social calls can be recorded too, including sweeps with a qCF part at the end or in the middle (Figure 8.14.9) as well as a wide variety of low-frequency squawks, probably mother–young

interactions (Figures 8.14.10 and 8.14.11). Natterer's bats swarming in front of large underground sites in late summer can use extremely broadband FM calls with an inverse U shape at the high-frequency start (Figure 8.14.12). Distress calls of Natterer's bats in the hand just before release are bursts of quickly repeated FM calls of usually small bandwidth and low frequency (Figure 8.14.13), sometimes with a U shape at the end (Figure 8.14.14).

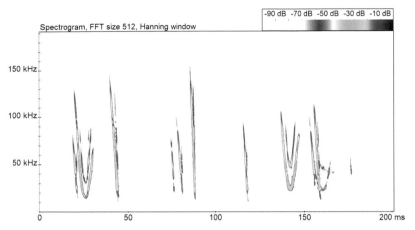

Figure 8.14.5   Social calls – U shapes with legs of broad bandwidth. Recorded near a maternity roost of Natterer's bat in a bat box, during evening emergence (Marc Van De Sijpe and Alex Lefevre).

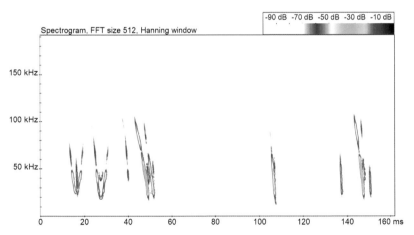

Figure 8.14.6   Natterer's bat social calls – U shapes merging into an undulating call (Marc Van De Sijpe and Alex Lefevre).

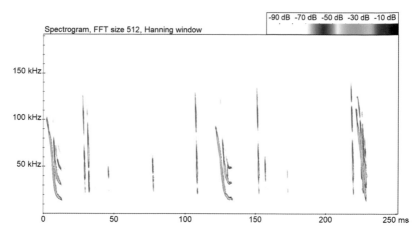

Figure 8.14.7   Social calls – FM-qCF shapes. Recorded near a maternity roost of Natterer's bat in a bat box, during evening emergence (Marc Van De Sijpe and Alex Lefevre).

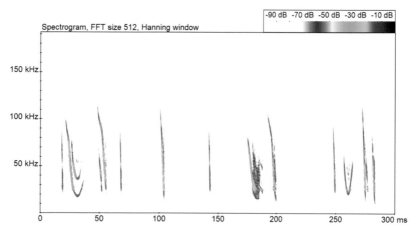

Figure 8.14.8   Natterer's bat – FM calls with a bend in the upper-frequency part (Marc Van De Sijpe and Alex Lefevre).

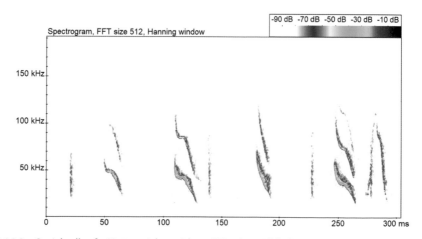

Figure 8.14.9   Social calls of a Natterer's bat with a qCF in the middle (Marc Van De Sijpe and Alex Lefevre).

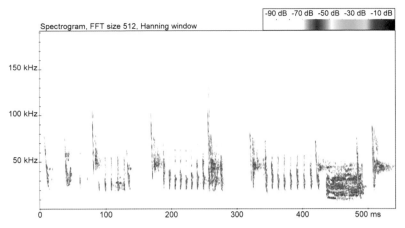

Figure 8.14.10   Low-frequency social calls of Natterer's bats in the maternity roost, a bat box (Marc Van De Sijpe and Alex Lefevre).

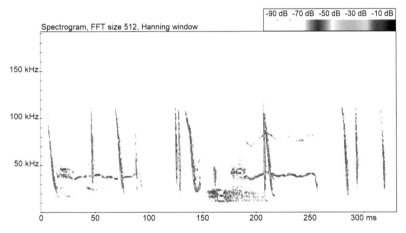

Figure 8.14.11   Social calls of Natterer's bats in the maternity roost. Long undulating calls (Marc Van De Sijpe and Alex Lefevre).

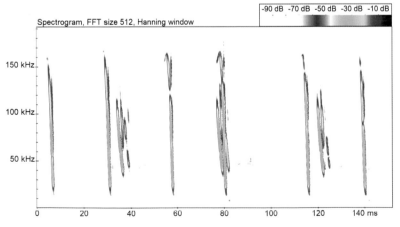

Figure 8.14.12   Natterer's bat, special calls with inverse U shape at the high-frequency start during swarming behaviour at the entrance to an underground site (Marc Van De Sijpe and Alex Lefevre).

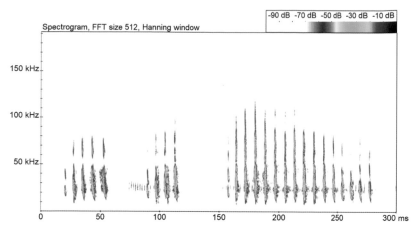

Figure 8.14.13   Distress calls of a female Natterer's bat in the hand (Marc Van De Sijpe and Alex Lefevre).

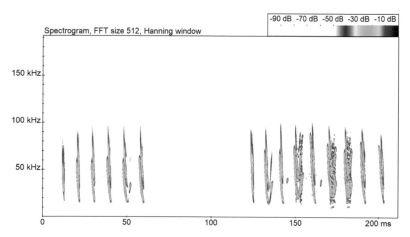

Figure 8.14.14   Distress calls of a male Natterer's bat in the hand (Marc Van De Sijpe and Alex Lefevre).

SPECIES WITH SIMILAR OR OVERLAPPING ECHOLOCATION CALLS

Myotis emarginatus, Myotis crypticus, Myotis escalerai, Myotis alcathoe (see Chapter 7 species notes).

NOTES

Echolocation calls appear to be very similar in structure to those of *Myotis crypticus* and *M. escalerai*. However, these species appear to occupy quite distinct areas within Europe, with little overlap.

8.15 Cryptic myotis

Myotis crypticus (Ruedi, Ibáñez, Salicini, Juste and Puechmaille, 2019)

Jon Russ

© Angel Ruiz Elizalde

DISTRIBUTION

The cryptic myotis, discovered in 2019, is mostly distributed across European countries bordering the Mediterranean Sea, from the mountain areas of provinces in central and northern Spain, southern France west to Austria, north to Switzerland, and south to most of the Italian peninsula. Also, to the north and west of the Alps, for example in western Switzerland or Rhône-Alpes (France) (Salicini *et al.* 2011, Puechmaille *et al.* 2012, Juste *et al.* 2019).

EMERGENCE

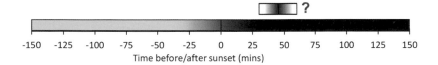

-150   -125   -100   -75   -50   -25   0   25   50   75   100   125   150

Time before/after sunset (mins)

FLIGHT AND FORAGING BEHAVIOUR

No information available.

HABITAT

Found in a wide range of altitudes from sea level to 1,000 m. It feeds in forest and grassland habitats. Roosts in tree hollows and man-made structures.

ECHOLOCATION

Measured parameters (based on limited data)	Mean (range) FM call
Inter-pulse interval (ms)	82.0 (40.0–180.0)
Call duration (ms)	4.3 (1.8–7.7)
Frequency of maximum energy (kHz)	45.9 (37.0–64.0)
Start frequency (kHz)	101.0 (72.0–149.0)
End frequency (kHz)	20.0 (11.0–36.0)

Heterodyne
No data. Presumed to be similar to Natterer's bat (see section 8.14).

Time expansion/full spectrum
Very few data. In clutter, a situation the bats are often encountered in, call bandwidth is very high and calls are of very short duration (2–4 ms) (Figures 8.15.1 and 8.15.2). Start frequencies can be as high as 150 kHz, sweeping down to as low as 11 kHz. Very broadband calls start with a small hook (see Figures 8.15.1 and 8.15.2). Often show a convex curvature with short-duration calls. When the bat is flying in a slightly more open habitat the bandwidth decreases (e.g. Figures 8.15.3 and 8.15.4).

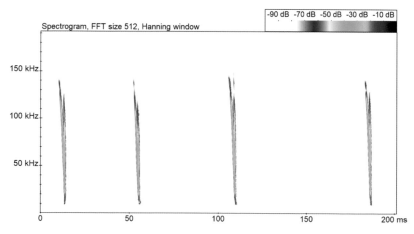

Figure 8.15.1   Echolocation calls of a cryptic myotis in clutter during hand release (Sébastien Puechmaille).

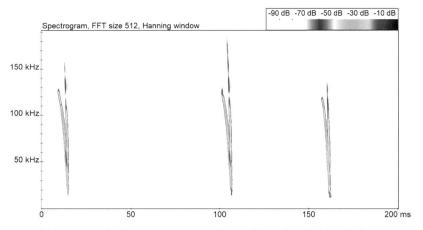

Figure 8.15.2   Echolocation calls of a cryptic myotis in clutter during hand release (Sébastien Puechmaille).

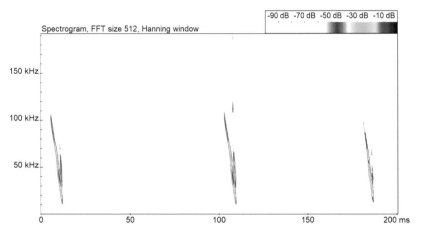

Figure 8.15.3   Echolocation calls of a cryptic myotis a few seconds after hand release (Sébastien Puechmaille).

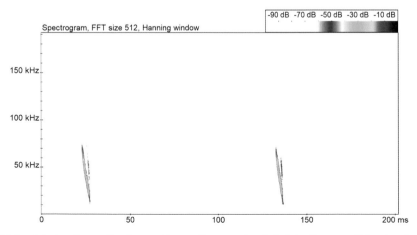

Figure 8.15.4   Echolocation calls of a cryptic myotis emerging fully into the open following hand release (Sébastien Puechmaille).

SOCIAL CALLS

Heterodyne
Distress calls are harsh, and loudest at around 25 kHz.

Time expansion/full spectrum
Distress calls, emitted when a bat is held in the hand, for example, consist of a series of FM sweeps produced in rapid succession (Figure 8.15.5).

Figure 8.15.5   Distress calls of a cryptic myotis (Sébastien Puechmaille).

SPECIES WITH SIMILAR OR OVERLAPPING ECHOLOCATION CALLS

Myotis nattereri, Myotis emarginatus, Myotis escalerai, Myotis alcathoe (see Chapter 7 species notes).

NOTES

Echolocation calls appear to be very similar in structure to those of *Myotis nattereri* and *Myotis escalerai*. However, these species appear to occupy quite distinct areas within Europe, with little overlap.

8.16 Iberian Natterer's bat *Myotis escalerai* (Cabrera, 1904)

Jon Russ, Alex Lefevre and Marc Van De Sijpe

© Jens Rydell

DISTRIBUTION

The Iberian Natterer's bat occurs throughout the Iberian peninsula north to the French Pyrenees, as well as on the Balearic Islands.

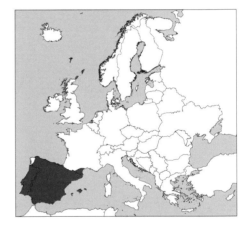

EMERGENCE

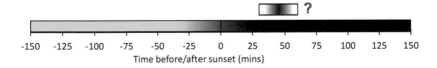

Time before/after sunset (mins)

FLIGHT AND FORAGING BEHAVIOUR

Flight is much like that of Natterer's bat, being relatively slow and highly manoeuvrable, sometimes hovering, or turning in a confined space. Probably forages by gleaning from surfaces and aerial hawking, low over the ground and water surfaces and higher up in the canopy of trees.

HABITAT

In Spain can be found both in arid coastal areas and in the wettest and coldest areas of the interior. Includes beech and mountain pine forests but more frequent in holm oak woodland.

Roosts in large numbers in caves to form nursery colonies. Old mines and castle under-crofts. Also in gaps in churches, cattle byres and barns.

ECHOLOCATION

Measured parameters (based on limited data)	Mean (range) FM call
Inter-pulse interval (ms)	80.0 (40.0–180.0)
Call duration (ms)	4.3 (1.9–7.9)
Frequency of maximum energy (kHz)	45.0 (37.0–65.0)
Start frequency (kHz)	100.0 (75.0–155.0)
End frequency (kHz)	27.0 (10.0–39.0)

Heterodyne
No data. Presumed to be similar to Natterer's bat (see section 8.14).

Time expansion/full spectrum
In clutter, a situation the bats are often encountered in, call bandwidth is very high and calls are of very short duration (2–4 ms) (Figures 8.16.1 and 8.16.2). Start frequencies can be as high as 155 kHz, sweeping down to as low as 10 kHz. Very broadband calls start with a small hook (see Figure 8.16.1). When viewed in a spectrogram, the short calls of large bandwidth often show a convex curvature, a characteristic feature. When the bat is flying in a slightly more open habitat the bandwidth decreases (e.g. Figures 8.16.3 and 8.16.4). The inter-pulse interval and the duration increase.

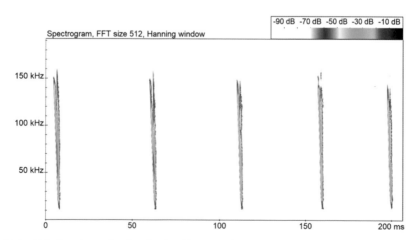

Figure 8.16.1   Echolocation calls of an Iberian Natterer's bat recorded in a cave entrance (Alex Lefevre).

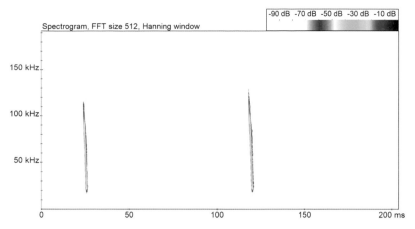

Figure 8.16.2   Echolocation calls of an Iberian Natterer's bat exiting a cave (Daniel Fernández Alonso).

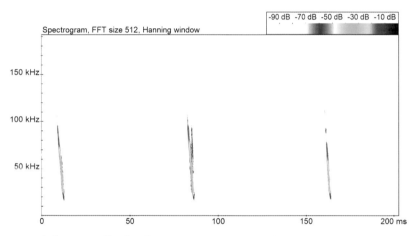

Figure 8.16.3   Echolocation calls of an Iberian Natterer's bat outside a cave entrance (Marc Van De Sijpe).

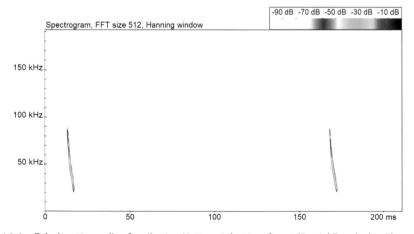

Figure 8.16.4   Echolocation calls of an Iberian Natterer's bat in a forest (Daniel Fernández Alonso).

SOCIAL CALLS

Heterodyne
No data.

Time expansion/full spectrum
Calls have been recorded at cave entrances which consist of slight variations on the echolocation call, usually being kinked or hooked (e.g. Figures 8.16.5 and 8.16.6). The function of these calls is unknown.

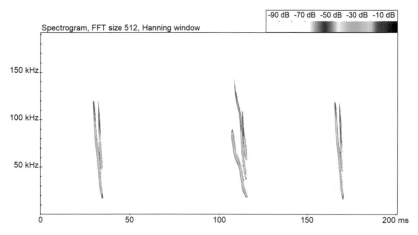

Figure 8.16.5   Possible social call of an Iberian Natterer's bat within an echolocation call sequence recorded at a cave entrance (Marc Van De Sijpe).

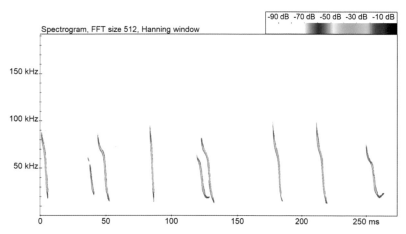

Figure 8.16.6   A further example of possible Iberian Natterer's bat social calls recorded at a cave entrance (Marc Van De Sijpe).

SPECIES WITH SIMILAR OR OVERLAPPING ECHOLOCATION CALLS

Myotis nattereri, Myotis crypticus, Myotis emarginatus, Myotis alcathoe (see Chapter 7 species notes).

NOTES

Echolocation calls appear to be very similar in structure to those of *Myotis nattereri* and *Myotis crypticus*. However, these species appear to occupy quite distinct areas within Europe, with little overlap.

8.17 Bechstein's bat *Myotis bechsteinii* (Kuhl, 1818)

Stephanie Murphy, Clara Gonzalez Hernandez and Jon Russ

© René Janssen

DISTRIBUTION

Bechstein's bat has a broad distribution throughout Europe, from the Iberian peninsula to the Caucasus mountains in the east, and as far as southern Scandinavia in the north. However, it is regarded as a rare species throughout its distribution area.

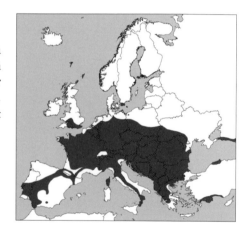

EMERGENCE

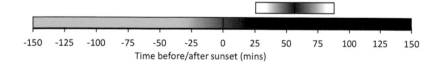

-150   -125   -100   -75   -50   -25   0   25   50   75   100   125   150

Time before/after sunset (mins)

FLIGHT AND FORAGING BEHAVIOUR

Bechstein's bats are characterised by their long ears and broad wings. They have a slow fluttering flight that is agile and adapted to foraging in dense vegetation. Prey is caught and consumed during flight from high in the canopy to low on the ground by gleaning. Occasionally observed fly-catching from a perch. Metabarcoding dietary studies in the UK have found that the diversity of taxa consumed predominantly comprised Diptera (26 taxa) and Lepidoptera (26 taxa).

HABITAT

Bechstein's bat is predominantly associated with ancient broadleaf woodlands (Greenaway and Hill 2004), and previous studies in the UK have shown a strong association with oak and ash woodland with a well-developed understorey (Hill and Greenaway 2006). In Europe, deciduous beech and oak wood comprise the majority of habitat, but they have also been recorded in parkland, orchard and open canopy in woodland. Metabarcoding dietary studies in the UK have found that taxa consumed were at least in part tree-associated (74%), with fewer being associated with open habitats (18%) and wetlands (8%).

Roosts are found in tree holes, cavities in trunks and often bat boxes. Very few roosts known in buildings. Hibernates in winter in underground sites and tree holes.

ECHOLOCATION

Measured parameters	Mean (range) FM call
Inter-pulse interval (ms)	89.7 (32.7–178)
Call duration (ms)	3.3 (1.5–6.9)
Frequency of maximum energy (kHz)	48.5 (36–69.6)
Start frequency (kHz)	97.0 (60.0–136.0)
End frequency (kHz)	35.0 (24.9–42.3)

Heterodyne

Like other *Myotis* species, Bechstein's bats can be heard in heterodyne as a sequence of dry clicks with a slow repetition rate (inter-pulse intervals >90 ms). In some cases, a characteristic dry explosive end can be identified in these calls. This explosive end to the call is characteristic of Bechstein's bat, Brandt's bat, Geoffrey's bat, lesser mouse-eared bat, greater mouse-eared bat, whiskered bat and Daubenton's bat (Barataud 2015).

Time expansion/full spectrum

Most Bechstein's bat calls consist of calls with short pulse durations (median of 3.13 ms), an average bandwidth of 90 kHz and a weak second harmonic present in good-quality calls (Figure 8.17.1). In addition, calls are characterised by high start frequencies (up to 135 kHz) and end frequencies ranging between 25 and 42 kHz. The start frequency of the calls is susceptible to atmospheric attenuation and its value is very variable depending on the position and distance of the microphone in relation to the echolocating bat, and therefore caution is recommended when utilising this parameter for species identification. In cluttered situations (which are encountered quite frequently) the calls can be very broadband (Figure 8.17.2). Conversely, as the bat moves more into the open the bandwidth decreases, the duration decreases and the heel in the call becomes more pronounced (Figures 8.17.3–8.17.6, moving away from clutter).

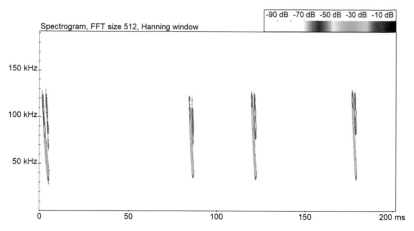

Figure 8.17.1   Echolocation calls of Bechstein's bat recorded near the roost (Alex Lefevre).

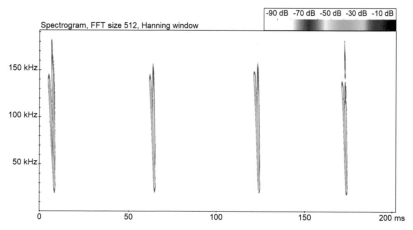

Figure 8.17.2   Echolocation calls of Bechstein's bat recorded in a forest (Alex Lefevre).

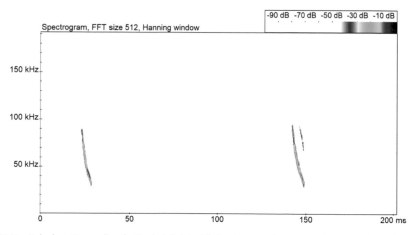

Figure 8.17.3   Echolocation calls of a Bechstein's bat flying in a semi-open environment (Stephanie Murphy).

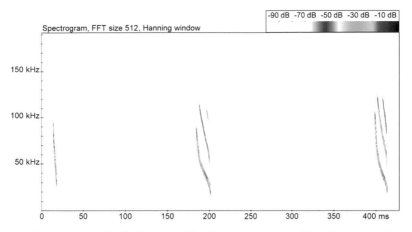

Figure 8.17.4   Echolocation calls of a Bechstein's bat flying in an open habitat (Alex Lefevre).

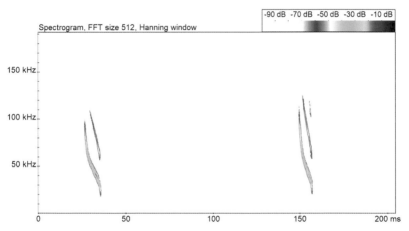

Figure 8.17.5   Further example of the echolocation calls of a Bechstein's bat flying in an open habitat (Alex Lefevre).

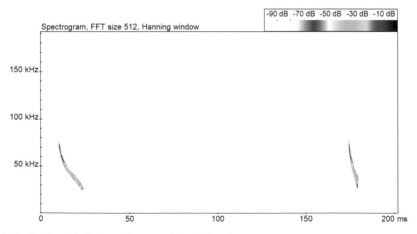

Figure 8.17.6   Bechstein's flying in the open (Alex Lefevre).

SOCIAL CALLS

Heterodyne
When listened to in heterodyne, Bechstein's social calls can be perceived as rapid crackling clicks.

Time expansion/full spectrum
Bechstein's social calls are usually heard near maternity roosts when juveniles are present. These calls are frequency-modulated. Typical calls have an FM structure with a slight concave curvature (Pfalzer and Kusch 2003), and harmonics are usually present in good-quality recordings (e.g. Figures 8.17.7 and 8.17.8). These social calls can be encountered during tandem flights, in the proximity of the roost by bats inside the roost as well as from flying bats, and have also been found to be used as a contact signal between individuals (Pfalzer and Kusch 2003). The calls have a typical duration of 18–20 ms with a start frequency as high as 115 kHz and an average end frequency of 26 kHz.

A wide variety of social calls are produced during dawn swarming around roost sites, with bats calling from the roost and in flight (e.g. Figures 8.17.9–8.17.11) and also from the roost following swarming (Figure 8.17.12). Note that the call shown in Figure 8.17.12 is also present in Figures 8.17.9 and 8.17.11.

While under physical duress individuals emit a series of short-duration FM pulses in quick succession (Figure 8.17.13). The frequency containing maximum energy (FmaxE) is at about 20 kHz.

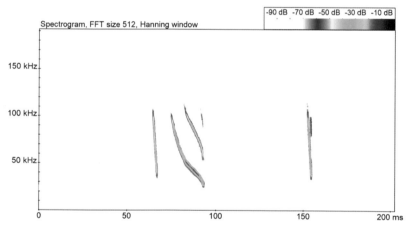

Figure 8.17.7   Social call and echolocation calls of a Bechstein's bat recorded near a maternity colony (Alex Lefevre).

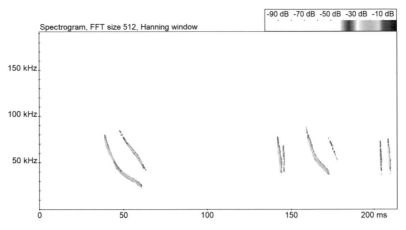

Figure 8.17.8   Social calls of a Bechstein's bat (with faint echolocation calls) (Iain Hysom).

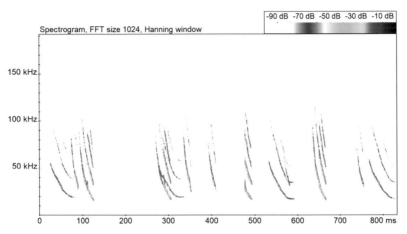

Figure 8.17.9   Social calls of Bechstein's bats recorded during swarming activity around a bat box (Marc Van De Sijpe).

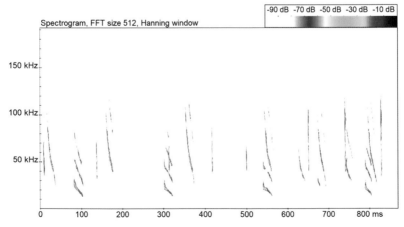

Figure 8.17.10   Social calls of Bechstein's bats recorded during swarming activity around a bat box (Marc Van De Sijpe).

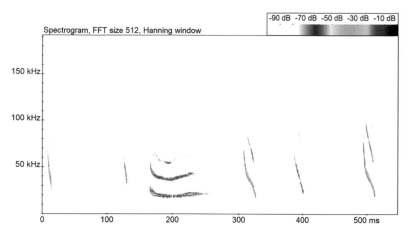

Figure 8.17.11   Social calls of Bechstein's bats recorded during swarming activity around a bat box (Marc Van De Sijpe).

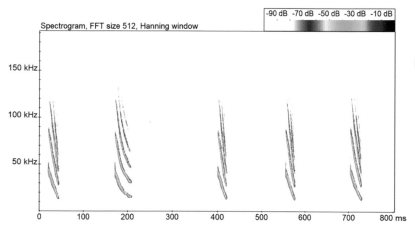

Figure 8.17.12   Social calls of Bechstein's bats recorded calling from a bat box following dawn swarming (Marc Van De Sijpe).

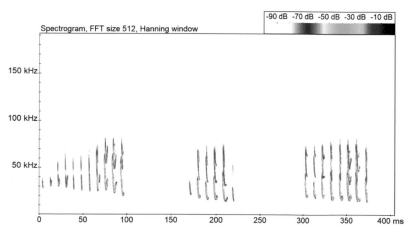

Figure 8.17.13   Bechstein's bat distress calls (Jon Russ).

SPECIES WITH SIMILAR OR OVERLAPPING ECHOLOCATION CALLS

Myotis emarginatus, Myotis daubentonii, Myotis myotis, Myotis blythii (see Chapter 7 species notes).

8.18 Greater mouse-eared bat
Myotis myotis (Borkhausen, 1797)

Ana Rainho and Joanna Furmankiewicz

© Ana Rainho

DISTRIBUTION

The greater mouse-eared bat occurs in Europe south of the North and Baltic Seas; from the Iberian peninsula east to the western regions of Poland, Ukraine, Turkey and the Levant (Arlettaz *et al.* 1997b, Castella *et al.* 2000). Some individuals are occasionally observed in southern UK and Sweden (Coroiu *et al.* 2016). It does not occur in some Mediterranean islands such as Corsica and Sardinia where the Maghreb mouse-eared bat *Myotis punicus* is present (Castella *et al.* 2000).

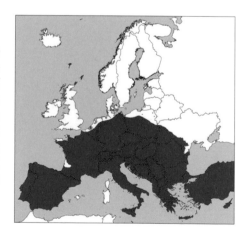

EMERGENCE

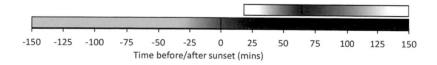

-150   -125   -100   -75   -50   -25   0   25   50   75   100   125   150
Time before/after sunset (mins)

FLIGHT AND FORAGING BEHAVIOUR

Fast commuting flights range from 5.6 to 11.1 m·s^{-1}, generally flying directly to foraging areas (Rainho and Palmeirim 1999). The greater mouse-eared bat can forage by aerial hawking, capturing flying prey on the wing, but more often captures prey from the ground (Arlettaz 1996). While foraging, a bat flies continuously and at moderate speed close to the ground (30–70 cm) and when attempting to make a capture it drops over its prey with wings wide open, supposedly to increase capture success (Arlettaz 1996, Rainho et al. 2010).

HABITAT

The greater mouse-eared bat often captures prey directly from the ground, selecting foraging areas with freshly cut pastures or cereal crops, forest debris or bare ground (Arlettaz 1999). Greater mouse-eared bats feed on soil surface-dwelling arthropods, with carabid beetles as the preferred prey in central Europe (Arlettaz 1996, Siemers and Güttinger 2006) and crickets also showing up as a major prey in the southern parts of its range (Ramos Pereira et al. 2002). Bats can be found foraging in a variety of habitats where the undergrowth is reduced and their preferred prey is available such as freshly mown meadows, short-grazed pastures, forests, woodland orchards, among others located in the vicinity of their roosts (Audet 1990, Zahn et al. 2006, Rainho et al. 2010, Rudolph et al. 2009).

Typically, an underground dweller during most of the year in the southern part of its range, maternity colonies of the greater mouse-eared bat are established in large buildings in central Europe (Rodrigues et al. 2003).

The greater and lesser mouse-eared bats are closely related sibling species, but despite the similarity in morphology and choice of roosting places, the two species use different foraging habitats (Arlettaz 1999), and therefore prey on different arthropods (Arlettaz et al. 1997a). The lesser mouse-eared bat does not commonly use forested habitats and preys mainly on bush-crickets, which it often picks from grass stalks while in flight (Arlettaz 1999, Siemers and Güttinger 2006).

ECHOLOCATION

Measured parameters	Mean (range) FM call
Inter-pulse interval (ms)	112.4 (42.9–189.3)
Call duration (ms)	5.7 (1.8–10.5)
Frequency of maximum energy (kHz)	37.8 (27.1–57.2)
Start frequency (kHz)	62.9 (49.1–105.1)
End frequency (kHz)	28.3 (23.2–38.7)

Heterodyne

The calls of the greater mouse-eared bat can be heard as relatively strong dry clicks with a relatively low repetition rate. These are best heard just below 30 kHz in open environments and around 35 kHz (to 37 kHz) in more cluttered areas or near the vegetation.

The calls produced by this species are very similar to the calls of the lesser mouse-eared bat, and it is not possible to differentiate these two species using the heterodyne system.

Time expansion/full spectrum

Greater mouse-eared bats produce short (1.8–10.5 ms), broadband frequency-modulated calls that sweep down from around 105 kHz to 25 kHz in cluttered habitats (Figure 8.18.1), occasionally with starting hooks (Figure 8.18.2), gradually decreasing in bandwidth as the bat moves into more open habitats (Figures 8.18.3–8.18.5). Mean call end frequency is around 28 kHz, pulse duration is about 5.7 ms, and the inter-pulse interval is 112 ms. The end frequency is rather similar in all habitat situations, averaging 27 kHz in open habitats and 30 kHz in cluttered areas. Duration ranges from 1.8 to 6.5 ms in pulses recorded in cluttered environments, and from 4.2 to 10.5 ms in sequences recorded in open areas.

The greater mouse-eared bat relies on listening for prey-generated sounds (e.g. rustling sounds or mating calls) to locate food on the ground (Arlettaz *et al.* 2001). While hunting by passive listening and before landing, the greater mouse-eared bat generally emits a short loud buzz, but reduces call amplitude prior to touching the ground (Russo *et al.* 2007a). This 'whispering echolocation' mode helps bats to process prey-generated sounds while preventing prey sensitive to their echolocation calls eluding capture (Arlettaz *et al.* 2001, Russo *et al.* 2007a). Typical feeding buzzes are emitted by this species during aerial hawking, showing the usual pattern of increased pulse repetition rate (Figure 8.18.6).

Both navigation and approach calls emitted by the greater mouse-eared bat are similar to the calls produced by its sibling the lesser mouse-eared bat. Some significant differences are nevertheless found in bats hand-released in an open environment, with greater mouse-eared bat calls showing a lower frequency of maximum energy (FmaxE) and a longer inter-pulse interval than those of the lesser mouse-eared bat, which seems not to produce the loud buzz frequently emitted by the greater mouse-eared bat prior to attempting prey capture (Russo *et al.* 2007a). Even recognising these differences, it is generally considered that it is not possible to separate the two species through their echolocation calls alone.

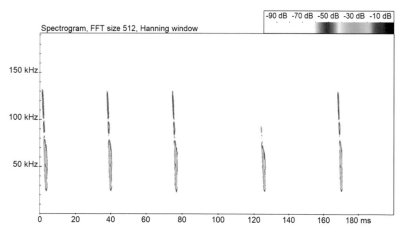

Figure 8.18.1   Echolocation calls of a greater mouse-eared bat in a cluttered forest environment (Ana Rainho).

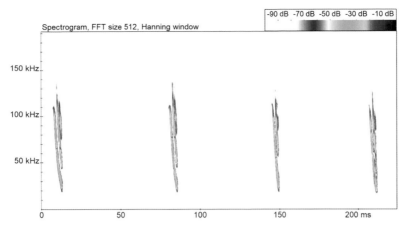

Figure 8.18.2   Echolocation calls of a greater mouse-eared bat in a cave, with starting hooks to its calls (Marc Van De Sijpe).

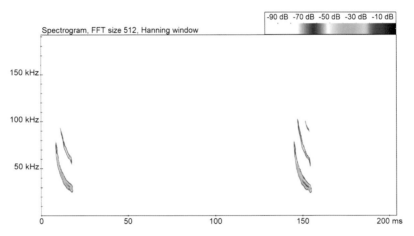

Figure 8.18.3   Echolocation calls of a greater mouse-eared bat a semi-open environment (Marc Van De Sijpe).

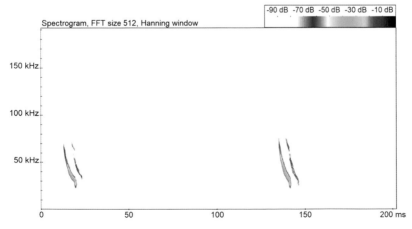

Figure 8.18.4   Echolocation calls of a greater mouse-eared bat flying in the open over grassland (Marc Van De Sijpe).

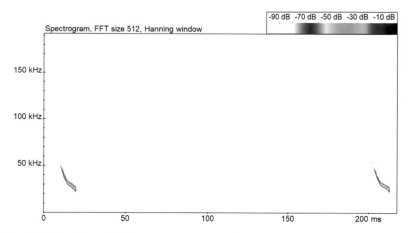

Figure 8.18.5   Echolocation calls of a greater mouse-eared bat flying in the open (Marc Van De Sijpe).

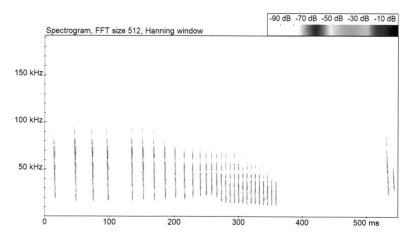

Figure 8.18.6   Feeding buzz of a greater mouse-eared bat (Marc Van De Sijpe).

SOCIAL CALLS

Heterodyne
A variety of harsh-sounding social calls can be heard, which are usually of low frequency.

Time expansion/full spectrum
Many social calls are documented in the literature. Walter and Schnitzler (2019) describe echolocation-like, low-frequency sweeps and long, broadband squawks during agonistic interactions of males. Pfalzer and Kusch (2003) describe single or pairs of short curved pulses of modulated frequency ('cheeps'), while Zahn and Dippel (1997) describe five different social calls (chirping, W-note, scolding, screaming and cheeping) produced by the greater mouse-eared bat during the mating season and copulation. Those calls were confirmed in underground mating sites in the Czech Republic, and new syllables were described: chevron-like (named V), M-note FM (the inverted W-note, i.e. short sinusoidal FM with an upward start), arched FM (A), downward FM (I), upward FM (K), downward FM with short CF part at the end (L) (J. Furmankiewicz and T. Postawa, unpublished). Moreover, it appeared that the mating repertoire of the greater mouse-eared bat is much more diversified and contains many combinations of syllables and phrases sung in a non-hierarchical

manner. So far, about 200 syllable combinations have been identified, but a much higher value might be expected (e.g. Figures 8.18.7 and 8.18.8). The songs were recorded during the mating period from groups of 2–3 individuals (usually a male with females), perching on the ceiling of the underground roost. This roost was also occupied by a cluster of females at the same time and solitary males, and also groups of 2–3 females and 2–3 males. Those songs are similar to those of the lesser mouse-eared bat, but they seem to have higher spectral parameters (i.e. FmaxE and mean frequency) and the vocal repertoire is more complex (J. Furmankiewicz and T. Postawa, unpublished). Further study is required.

During the mating season, males seem to defend their mating territory using a series of repeated trills of FM sweeps (Figures 8.18.9 and 8.18.10). Other calls are commonly produced inside the roost, some of which are long multi-harmonic signals of constant frequency and broadband while others show a complex structure combining multi-harmonic elements of sinusoidally modulated frequencies.

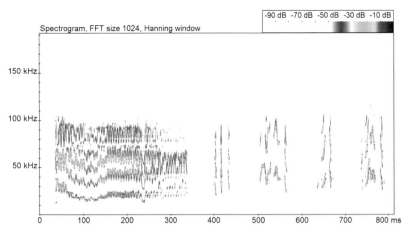

Figure 8.18.7   Mating vocalisations produced by a male greater mouse-eared bat during the mating season. The songs do not appear in their recorded sequence (Joanna Furmankiewicz).

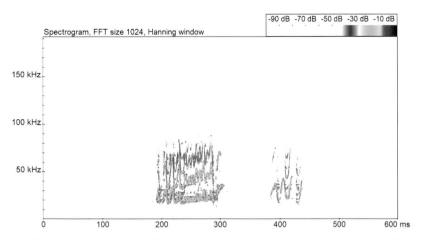

Figure 8.18.8   Further example of social calls produced by a male greater mouse-eared bat during the mating season (Phil Riddett).

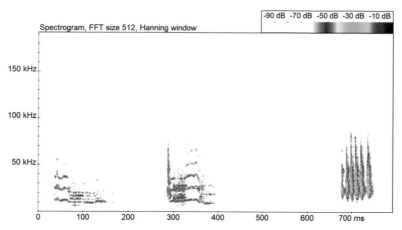

Figure 8.18.9   Social calls recorded from a greater mouse-eared bat roosting by a pond in a wood (Phil Riddett).

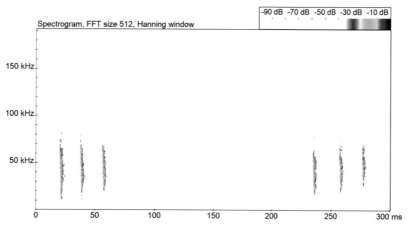

Figure 8.18.10   Social calls produced by a male greater mouse-eared bat in its territory during the mating season (Phil Riddett).

SPECIES WITH SIMILAR OR OVERLAPPING ECHOLOCATION CALLS

Myotis emarginatus, Myotis daubentonii, Myotis blythii, Myotis nattereri (see Chapter 7 species notes).

8.19 Lesser mouse-eared bat *Myotis blythii* (Tomes, 1857)

Ana Rainho and Joanna Furmankiewicz

© Péter Estók

DISTRIBUTION

The lesser mouse-eared bat occurs throughout Mediterranean Europe from the Iberian peninsula north to central France, Switzerland, the Czech Republic, Slovakia and Ukraine and eastwards to the Balkan peninsula and parts of Turkey (Dietz *et al.* 2009, Juste and Paunović 2016a). It does not occur in some Mediterranean islands such as Corsica, Sardinia and Malta where the Maghreb mouse-eared bat *Myotis punicus* is present (Dietz *et al.* 2009).

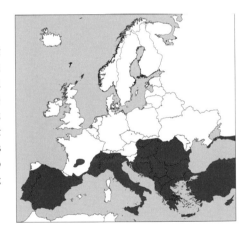

EMERGENCE

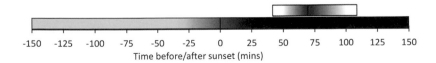

-150   -125   -100   -75   -50   -25   0   25   50   75   100   125   150

Time before/after sunset (mins)

FLIGHT AND FORAGING BEHAVIOUR

The lesser mouse-eared bat forages with a relatively slow continuous search flight at a low height, usually below 2 m from the ground, with obvious prospecting paths over the grassy vegetation (Arlettaz 1996, Güttinger *et al.* 1998). The bats have an agile flight that enables them to glean their preferred prey – bush-crickets – from the blades of the grass (Arlettaz *et al.* 1997a, Siemers *et al.* 2011, Schmieder *et al.* 2014). They usually hover and pick the prey from the vegetation, but less frequently they capture prey by landing on the ground or by aerial hawking (Arlettaz 1996, Güttinger *et al.* 1998).

HABITAT

The diet of the lesser mouse-eared bat is often dominated by bush-crickets, insects that are particularly abundant in grassy habitats (Arlettaz 1996, Siemers *et al.* 2011). As a consequence, the lesser mouse-eared bat typically hunts in steppe-like, long-grass meadows (Arlettaz 1999, Güttinger *et al.* 1998). This species does not commonly use forested habitats but it may forage at forest edges. It avoids urban areas, open fields and other habitats where prey abundance is very low (Arlettaz 1999).

The lesser mouse-eared bat is typically an underground dweller in the southern parts of its range. Underground sites are also used by this species during winter in central Europe, but the maternity roosts are established in the attics of large buildings (Dietz *et al.* 2009).

Its sibling species the greater mouse-eared bat is closely related, but, despite the similarity in morphology and choice of roosting places, the two species use different foraging habitats (Arlettaz 1999) and, as a consequence, prey on different arthropods (Arlettaz *et al.* 1997a, Siemers *et al.* 2011). the greater mouse-eared bat feeds on ground arthropods, foraging in a variety of habitats where the undergrowth is sparse.

ECHOLOCATION

Measured parameters	Mean (range) FM call
Inter-pulse interval (ms)	95.8 (32.3–237.6)
Call duration (ms)	4.6 (1.8–9.7)
Frequency of maximum energy (kHz)	41.5 (29.4–55.6)
Start frequency (kHz)	69.8 (43.4–102.8)
End frequency (kHz)	29.8 (21.1–39.4)

Heterodyne
The calls of the lesser mouse-eared bat can be heard as fairly strong dry clicks with a regular repetition rate. These are best heard just around 30 kHz in open environments and around 37 kHz in more cluttered areas or near the vegetation.

The echolocation calls of the lesser mouse-eared and greater mouse-eared bats are very similar (Russo and Jones 2002, Russo *et al.* 2007a), and it is not possible to differentiate the two species using the heterodyne.

Time expansion/full spectrum
The lesser mouse-eared bat uses short (1.8–9.7 ms), broadband frequency-modulated navigation calls that sweep down from 100 to 32 kHz in cluttered habitats (Figure 8.19.1) and often show a 'hook' in very cluttered habitats (Figure 8.19.2). In open habitats, calls sweep down from around 69 to 27 kHz (Figure 8.19.3). Mean call end frequency is 30 kHz,

pulse duration is about 4.6 ms, and the inter-pulse interval is 96 ms (Figure 8.19.4). Duration ranges from 1.8 to 4.6 ms in pulses recorded in cluttered environments and from 4.1 to 9.7 ms in sequences recorded in open areas. The transition from clutter to semi-open habitats can be seen in Figure 8.19.5.

While foraging, the lesser mouse-eared seems to rely on passive listening, eavesdropping on the songs of its preferred prey (Jones *et al.* 2011). Before attempting capture, the lesser mouse-eared bat reduces call amplitude (Russo *et al.* 2007a), entering a 'whispering echolocation' mode that helps bats to process prey-generated sounds while preventing prey sensitive to their echolocation calls eluding capture (Arlettaz *et al.* 2001, Russo *et al.* 2007a).

Both navigation and approach calls emitted by the lesser mouse-eared bat are similar to the calls produced by its sibling, the greater mouse-eared bat. Some significant differences have nevertheless been found in bats hand-released in an open environment, with lesser mouse-eared calls showing a higher frequency containing maximum energy (FmaxE) and a shorter inter-pulse interval than the calls of the greater mouse-eared (Russo and Jones 2002). During the approach phase, the lesser mouse-eared seems not to produce the loud buzz frequently emitted by the greater mouse-eared prior to attempting prey capture (Russo *et al.* 2007a). Even recognising these differences, it is generally considered that separation of the two species is not possible through their echolocation calls alone.

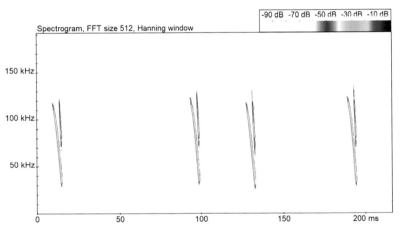

Figure 8.19.1   Echolocation calls of a lesser mouse-eared bat at the entrance of an underground roost (Ana Rainho).

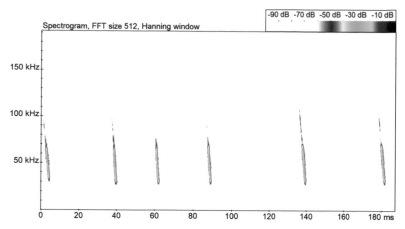

Figure 8.19.2   Echolocation calls of hand-released lesser mouse-eared bats in a cave entrance (Jon Russ).

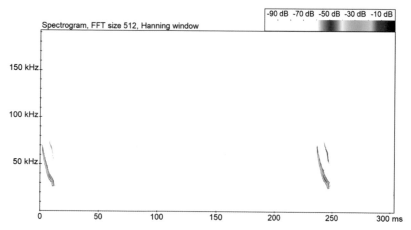

Figure 8.19.3   Echolocation calls of a lesser mouse-eared bat flying over pasture (Ana Rainho).

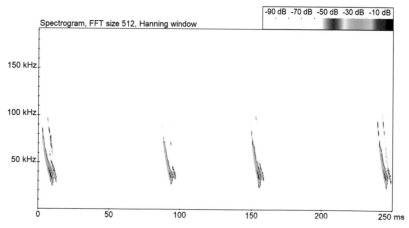

Figure 8.19.4   Echolocation calls of a lesser mouse-eared flying over a farmland area with sparse small olive trees (Ana Rainho).

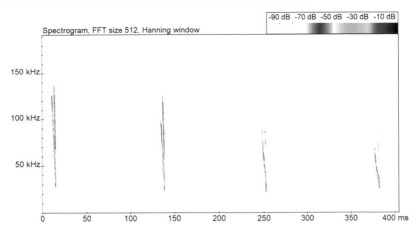

Figure 8.19.5   Echolocation calls of hand-released lesser mouse-eared bats emerging from a cave (Jon Russ).

SOCIAL CALLS

Heterodyne
Distress calls sound like those typically produced by a bat under physical duress: harsh and loud, usually loudest at around 25 kHz. Other social calls can be heard in the vicinity of roosts.

Time expansion/full spectrum
Distress calls, emitted from bats under physical duress, generally consist of a series of downward sweeps rapidly repeated (Figure 8.19.6)

The study of social vocalisations in underground mating sites of this species reveals that the mating repertoire of the lesser mouse-eared is very similar to that of the greater mouse-eared (J. Furmankiewicz and T. Postawa, unpublished). Both species use the same syllables in their mating songs and combine them (syllables into phrases and phrases into songs) in a non-hierarchical manner to emit one of the most diversified mating repertoires among bats. However, vocalisations of the lesser mouse-eared shows higher complexity and different proportions of particular syllables and phrases. Males in underground sites (usually abandoned mines) perch on the ceiling, alone or accompanied by 1–3 individuals (usually females). Some sites are also inhabited by a colony of females. In this case, many males (about 100–200 individuals) perch in the vicinity of the colony, but without body contact, so probably every individual occupies its own small territory on the ceiling. Males sing their songs perching on the ceiling; however, some solitary individuals remain silent. Bats in pairs emit chirps, screams and scoldings (for details of these calls see paragraph below and description of the greater mouse-eared bat, section 8.18). No social vocalisations in flight have been observed.

The songs are broadband frequency and multi-harmonic, with at least 2–3 harmonics and a maximum of 7–10 (Figures 8.19.7–8.19.10). Depending on the song type, the frequency bandwidth ranges between from about 4 kHz up to 124 kHz, and the duration from about 5 ms up to 900 ms. Several syllables were identified in the mating vocalisations of the lesser mouse-eared. The most frequently recorded syllables were chevron-like (named V) and downward FM (I), much less frequently noted were short sinusoidal W-note FM with a downward start, short sinusoidal M-note FM (the inverted W-shaped, i.e. with an upward start), upward FM (K), downward FM with short CF part at the end (L) and multiple sinusoidal FM (C, named chirping by Zahn and Dippel 1997), and rarely arch-shaped FM (A) and series of constant pulses switching to harmonic noise (S, scolding after Zahn and Dippel 1997, Figure 8.19.11). So far, about 700 phrase combination and about 1,400 syllable combinations have been identified in five underground mating sites of this species in western Ukraine, but much more is expected. The individual song contains from one up to 16 syllables. The function of some syllables phrases (e.g. S and a burst of I) requires further study.

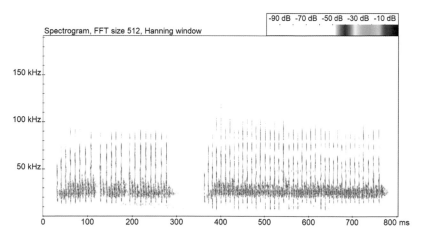

Figure 8.19.6   Distress calls of a hand-held lesser mouse-eared bat (Jon Russ).

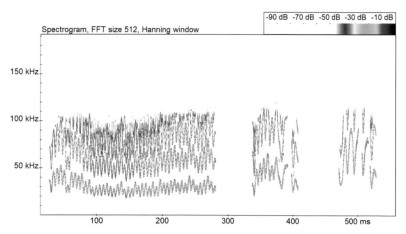

Figure 8.19.7   Example of randomly chosen song parts produced by lesser mouse-eared bats, inserted into a single file (Joanna Furmankiewicz).

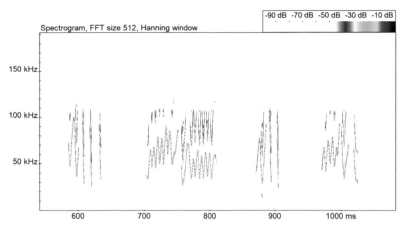

Figure 8.19.8   Example of randomly chosen song parts produced by lesser mouse-eared bats, inserted into a single file (Joanna Furmankiewicz).

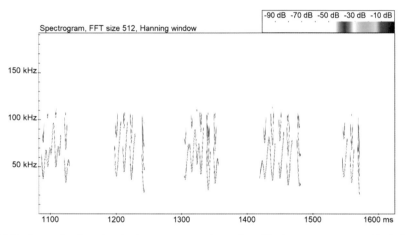

Figure 8.19.9   Example of randomly chosen song parts produced by lesser mouse-eared bats, inserted into a single file (Joanna Furmankiewicz).

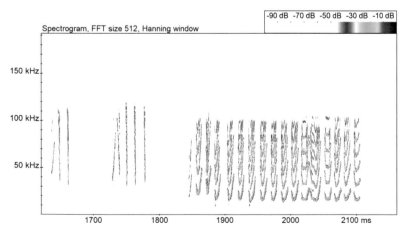

Figure 8.19.10   Example of randomly chosen song parts produced by lesser mouse-eared bats, inserted into a single file (Joanna Furmankiewicz).

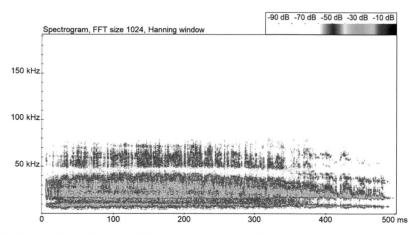

Figure 8.19.11   Scolding call produced by a lesser mouse-eared bat at a mating site (Joanna Furmankiewicz).

SPECIES WITH SIMILAR OR OVERLAPPING ECHOLOCATION CALLS

Myotis emarginatus, Myotis daubentonii, Myotis myotis, Myotis nattereri (see Chapter 7 species notes).

8.20 Maghreb mouse-eared bat

Myotis punicus (Felten, 1977)

Mauro Mucedda, Ermanno Pidinchedda and Gaetano Fichera

© Mauro Mucedda

DISTRIBUTION

The Maghreb mouse-eared bat has a Mediterranean distribution, occurring in northwest Africa (Morocco, Algeria, Tunisia, western Libya) and the islands of Malta, Corsica (France), Sardinia and recently Sicily (Italy) (Juste and Paunović 2016b).

EMERGENCE

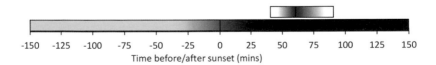

-150 -125 -100 -75 -50 -25 0 25 50 75 100 125 150

Time before/after sunset (mins)

FLIGHT AND FORAGING BEHAVIOUR

The Maghreb mouse-eared bat flies low, zigzagging backwards and forwards at a height of 10–50 cm above the ground. It can hover, and it captures prey on the ground from grass. During the night, the bat alternates phases of hunting and resting in temporary shelters or perches (abandoned houses, caves, trees) (Carrier 2009, Chenaval 2010, Poupart 2011).

HABITAT

The Maghreb mouse-eared bat in Sardinia occurs mainly in areas of holm and cork oak woodland, low scrubland and commercial forestry (Bosso *et al.* 2016). The foraging habitats more frequently used by the Maghreb mouse-eared bat in Corsica are open areas such as pasture (grazing land), grassland, but also low Mediterranean scrubland and agricultural environments (Beuneux 2004, Carrier 2009, Chenaval 2010, Poupart 2011).

The Maghreb mouse-eared bat is a troglophilous species that roosts in natural caves, mines and other man-made subterranean structures, more rarely in bridges and buildings (Beuneux 2004, Baron and Vella 2010, Mucedda *et al.* 1995). They form mixed nursery colonies with some other cave-dwelling bats such as Schreiber's bent-winged bat, long-fingered bat, Mediterranean horseshoe bat and Mehely's horseshoe bat (Beuneux 2004, Mucedda *et al.* 1995).

ECHOLOCATION

Measured parameters	Mean (range) FM call
Inter-pulse interval (ms)	111.0 (35.0–263.0)
Call duration (ms)	7.1 (3.7–10.7)
Frequency of maximum energy (kHz)	41.2 (30.1–74.5)
Start frequency (kHz)	81.9 (45.5–124.0)
End frequency (kHz)	23.1 (18.0–27.1)

Heterodyne
The Maghreb mouse-eared bat emits ticks, without an obvious peak in amplitude.

Time expansion/full spectrum
The Maghreb mouse-eared bat emits steep frequency-modulated (FM) signals (sweeps), starting at about 82 kHz and ending at about 23 kHz (Figure 8.20.1). The frequency containing maximum energy (FmaxE), which is not easy to determine, is about 41 kHz and the average pulse duration is about 7 ms. Energy is concentrated in the second half of the signal. In open space, the signals frequently have a final sigmoid twist, with a heel followed by a short FM descending tail.

In cluttered environments, calls are straighter, of shorter duration, have a higher repetition rate (a lower inter-pulse interval), with a higher FmaxE, and often a greater bandwidth (and therefore it could occasionally be confused with Natterer's bat) (Figure 8.20.2).

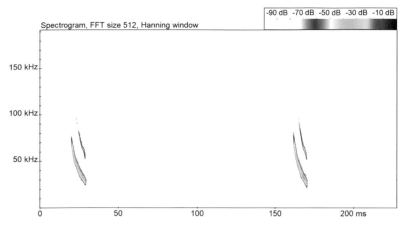

Figure 8.20.1   Echolocation calls of the Maghreb mouse-eared bat recorded in open space after emerging from the roost (Mauro Mucedda).

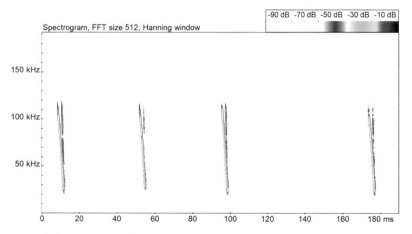

Figure 8.20.2   Echolocation calls of the Maghreb mouse-eared bat recorded emerging from an artificial gallery, in cluttered environment due to vegetation in the entrance (Mauro Mucedda).

SOCIAL CALLS

Heterodyne
Social calls are occasionally heard, consisting of loud, short rasps.

Time expansion/full spectrum
Social calls produced by Maghreb mouse-eared bats may consist of repeated short sequences of FM pulses (Figure 8.20.3).

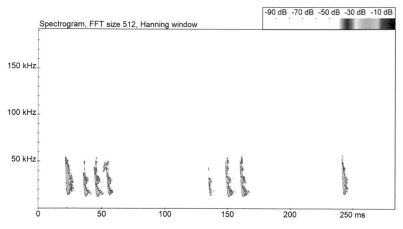

Figure 8.20.3   Social calls produced by a single Maghreb mouse-eared bat (probably male) roosting in a hole in the ceiling of a cave (Mauro Mucedda).

Distress calls produced by the Maghreb mouse-eared bat consist of a repeated long sequence of short-duration FM pulses, with a variable number of components (Figure 8.20.4).

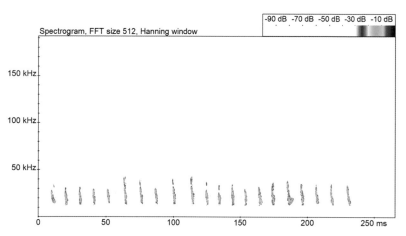

Figure 8.20.4   Distress calls produced by a hand-held Maghreb mouse-eared bat in the entrance of a cave (Mauro Mucedda).

SPECIES WITH SIMILAR OR OVERLAPPING ECHOLOCATION CALLS

Myotis emarginatus, Myotis nattereri, Myotis myotis, Myotis blythii (see Chapter 7 species notes).

NOTES

We would like to acknowledge the support of Centro Pipistrelli Sardegna.

8.21 Noctule *Nyctalus noctula* (Schreber, 1774)

Jon Russ

© René Janssen

DISTRIBUTION

Widespread throughout much of Europe. In the north as far as southern Scotland and Scandinavia and as far south as northern Spain, with an isolated pocket in southern Spain, and to the Mediterranean. Not present in Ireland.

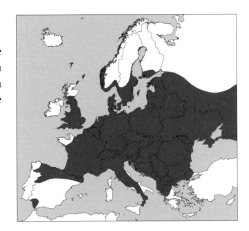

EMERGENCE

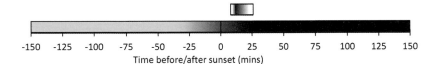

Time before/after sunset (mins)

FLIGHT AND FORAGING BEHAVIOUR

Very fast and straight but includes deep dives close to the ground or water to chase insects. Occasionally glides. Usually feeds up to 40 m, circling often, with scouting flights to 100 m and a maximum of 300 m (O'Mara *et al.* 2019). In flight, the wings nearly touch beneath the body. Not particularly agile. Early on in the evening noctules fly high, but they may descend to lower altitudes later in the night. Occasionally forage over streetlights. Predominantly forage by aerial hawking: prey is pursued and caught in flight.

HABITAT

A migratory bat throughout much of its range, although appears to be sedentary in the north. Found in a wide range of open habitats. Common over deciduous woodland, parkland, pasture, marshland, lakes and rivers. Not very common in larger cities unless there is a substantial amount of tree cover.

Predominantly a tree-roosting bat, often in woodpecker holes but also other cavities. Also roosts in buildings and caves, particularly in its southern distribution.

ECHOLOCATION

Measured parameters	Mean (range)	
	FM-qCF call	qCF call
Inter-pulse interval (ms)	216.9 (120.3–413.1)	372.2 (120.2–807.2)
Call duration (ms)	14.7 (8.8–23.4)	22.1 (13.2–29.9)
Frequency of maximum energy (kHz)	24.5 (22.4–33.6)	19.3 (17.5–23.6)
Start frequency (kHz)	37.9 (23.8–52.2)	23.2 (18.2–30.4)
End frequency (kHz)	23.7 (21.4–32.2)	18.3 (17.1–23.0)

Heterodyne

One of the loudest calls of European bats. In open habitats often produces two alternating call types. The first is a qCF call which is loudest at about 19 kHz and the second is a FM-qCF call with maximum energy at about 24 kHz. In sequence, this sounds like 'chip-chop-chip-chop'. The calls are similar to those of Leisler's bat, although Leisler's bat calls are loudest at 24 kHz and 27 kHz. Leisler's bats do not emit the chip-chop sequence quite as frequently. In closed habitats, such as when foraging over streetlights or next to woodland edge, only an FM-qCF call is produced and the loudest portion of the call may vary between 24 kHz and 28 kHz. In such situations, it may be difficult to separate this species from serotine and Leisler's bat. In extreme clutter, the loudest frequency may be as high as 33 kHz. At very high altitudes the FM portion of the call can be dropped so that only the 'chop' is heard (at around 19 kHz). Often, harmonics are produced, and these will be heard at double the frequency of the fundamental.

Eptesicus calls have a similar repetition rate, frequency range and tonal quality, but these do not usually alternate between two call types.

Other species' social calls (particularly pipistrelle) can sound superficially similar to a noctule's echolocation calls, but they are more sporadic with no obvious rhythmic quality.

Time expansion/full spectrum

Produces two main call types, an FM-qCF call and a qCF call. The FM-qCF call sweeps down from about 35 kHz to 24 kHz and is about 14 ms in duration (Figure 8.21.1). The qCF call starts at about 26 kHz, ends at about 18 kHz, and is on average 22 ms in duration. The frequencies containing maximum energy of the two calls are, on average, 24 kHz and 19 kHz respectively. Often these calls are produced alternately. However, when in a very open environment such as when flying at high altitude the FM-qCF call is omitted (Figure 8.21.2), but when flying in relative clutter the qCF call is omitted (Figure 8.21.3). When flying quite low, such as over and around streetlights, calls are variable (e.g. Figure 8.21.4). In very cluttered environments the FM-qCF call becomes much more broadband, FmaxE rises, and the call is of short duration (Figure 8.21.5), becoming more broadband as the degree of clutter increases (Figure 8.21.6).

In very extreme clutter, such as when flying in dense vegetation, calls become extremely FM and are sometimes mistaken for those of long-eared bats (e.g. Figure 8.21.7).

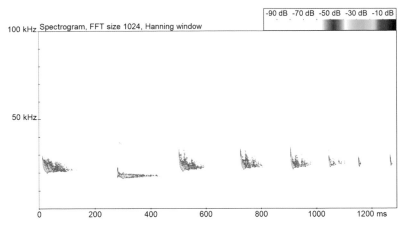

Figure 8.21.1   Echolocation calls of a noctule foraging above a riparian treeline (Jon Russ).

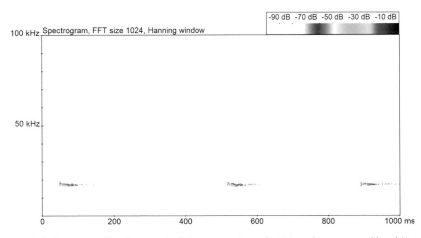

Figure 8.21.2   Echolocation calls of a noctule flying approximately 200 m above ground level (Jon Russ).

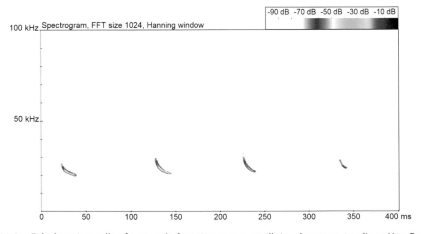

Figure 8.21.3   Echolocation calls of a noctule foraging over a small river between treelines (Jon Russ).

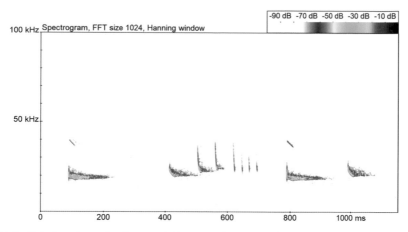

Figure 8.21.4   Echolocation calls of a noctule foraging above streetlights (Jon Russ).

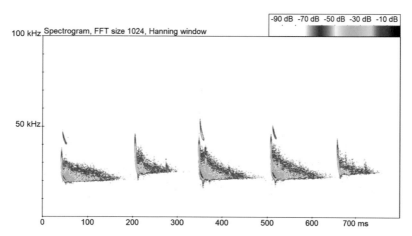

Figure 8.21.5   Echolocation calls of a noctule foraging low over buildings (Jon Russ).

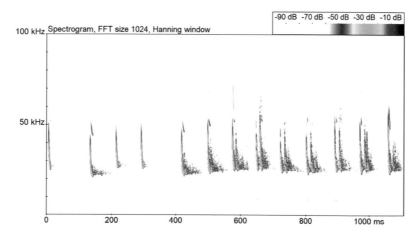

Figure 8.21.6   Echolocation calls of a noctule flying very low under the canopy of a tree (Jon Russ).

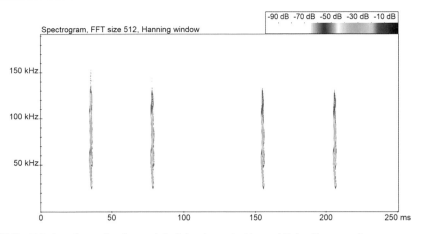

Figure 8.21.7   Echolocation calls of a noctule flying in a wind tunnel (Arjan Boonman).

SOCIAL CALLS

Heterodyne
One of the most vocal of the European bats. Males have a large call repertoire which they generally emit while stationary, although occasionally calls are emitted on the wing. A common call from a stationary male is a low, almost constant-frequency call which is loudest at about 11 kHz. This is repeated at a slow rate. When disturbed, a male will emit a loud, intense angry-sounding call.

Time expansion/full spectrum
A wide range of vocalisations are produced, usually by stationary males (Figures 8.21.8–8.21.16). These are low-frequency and may consist of rapid FM pulses, rapidly rising and falling warbles or long drawn-out CF calls. These are produced much more frequently during the autumnal mating season. When held in the hand, bats emit distress calls (Figures 8.21.17 and 8.21.18).

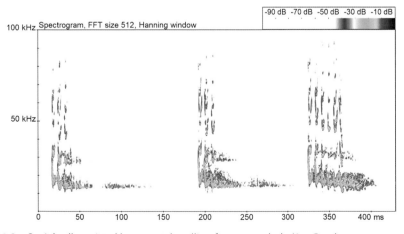

Figure 8.21.8   Social calls emitted by a noctule calling from a tree hole (Jon Russ).

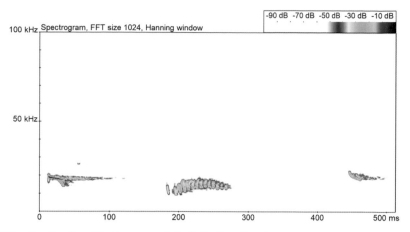

Figure 8.21.9 Social call emitted by a noctule in flight (David Lee).

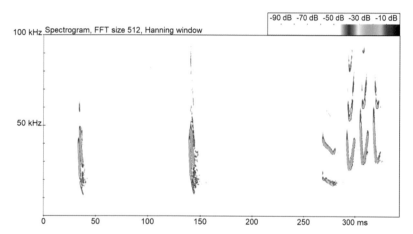

Figure 8.21.10 Social calls, though to be agonistic sounds, emitted by a noctule in flight (Jon Russ).

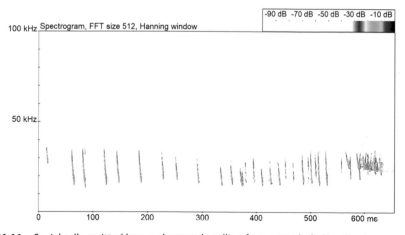

Figure 8.21.11 Social calls emitted by a male noctule calling from a tree hole (Jon Russ).

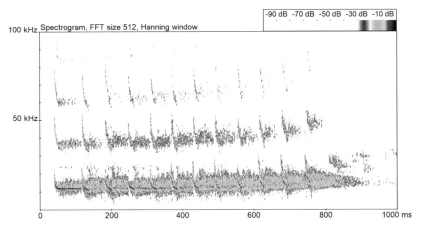

Figure 8.21.12   Social calls emitted by a male noctule calling from a tree hole during an encounter with a close-approaching bat (Jon Russ).

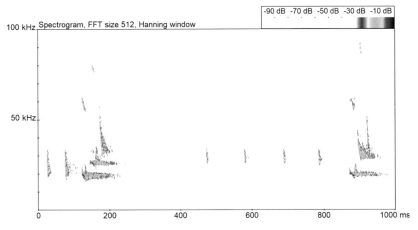

Figure 8.21.13   Social calls emitted by a male noctule calling from a tree hole (Jon Russ).

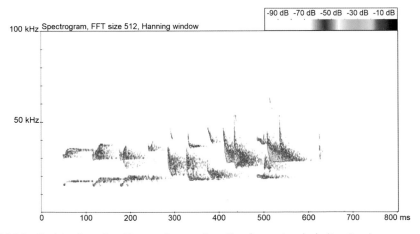

Figure 8.21.14   Social calls emitted by a male noctule calling from a tree hole (Jon Russ).

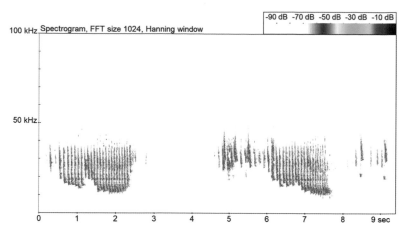

Figure 8.21.15   Social calls emitted by a male noctule calling from a tree hole. Example of variation in element frequency (Jon Russ).

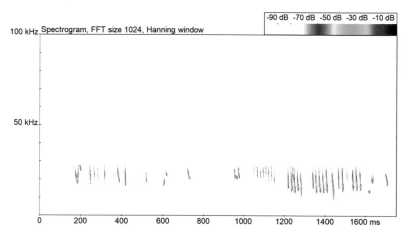

Figure 8.21.16   Social calls emitted by a male noctule calling from a tree hole (Jon Russ).

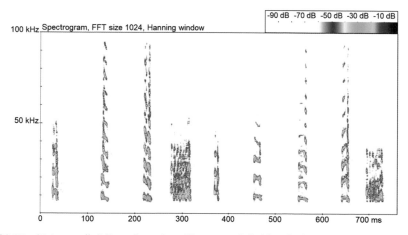

Figure 8.21.17   Distress calls (clipped) produced by a noctule held in the hand (Phil Redditt).

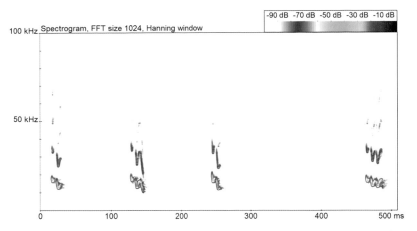

Figure 8.21.18   Distress calls emitted by a male noctule held in the hand following removal from a harp trap (Jon Russ).

SPECIES WITH SIMILAR OR OVERLAPPING ECHOLOCATION CALLS

Eptesicus spp., *Nyctalus leisleri*, *Nyctalus lasiopterus*, *Vespertilio murinus* (see Chapter 7 species notes).

8.22 Greater noctule *Nyctalus lasiopterus* (Schreber, 1780)

Péter Estók, Alex Lefevre, Marc Van De Sijpe and Jon Russ

© Péter Estók

DISTRIBUTION

In the western part of Europe, the greater noctule has a Mediterranean distribution, and in eastern Europe its range extends further to the north. In Europe, most records come from the Iberian peninsula, with other regular occurrences known from France, Italy and Hungary. In other areas of Europe, records are very scarce and the distribution appears patchy (Ibáñez *et al.* 2004).

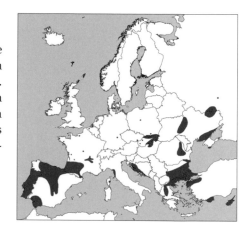

EMERGENCE

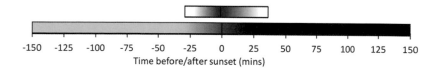

Time before/after sunset (mins)

-150 -125 -100 -75 -50 -25 0 25 50 75 100 125 150

FLIGHT AND FORAGING BEHAVIOUR

Very fast flight possibly at a considerable height (several hundred metres), but detailed direct observations are missing. Typical aerial hawking bat, catching prey in the open air. Very steep fast dives to capture prey or drink. Forages on large insects and preys on migrating songbirds during their migration season (Dondini and Vergari 2000, Ibáñez *et al.* 2001, Popa-Lisseanu *et al.* 2007).

HABITAT

The greater noctule is a forest-dwelling bat species, found in a variety of coniferous and deciduous forests. Very limited information is available on foraging habitat. In southern Spain, greater noctules select mainly marshland habitats during foraging (Popa-Lisseanu *et al.* 2009).

Most of the located roosts in natural habitats are in deciduous woodlands. In Spain, colonies have been found in city parks (Ibáñez *et al.* 2001). In Hungary, radio-tracked specimens roosted exclusively in beeches in old montane beech forest (Estók *et al.* 2007).

ECHOLOCATION

Measured parameters	Mean (range)	
	FM-qCF call	qCF call
Inter-pulse interval (ms)	290.0 (80.0–400.0)	532.0 (392.0–855.0)
Call duration (ms)	19.0 (9.0–24.0)	28.9 (20.0–35.0)
Frequency of maximum energy (kHz)	18.0 (16.0–30.0)	13.7 (11.4–14.9)
Start frequency (kHz)	27.8 (26.0–45.0)	13.8 (11.7–17.3)
End frequency (kHz)	17.1 (13.0–29.0)	12.3 (11.1–13.8)

Heterodyne

Very loud echolocation call. The sounds are similar to those of the noctule but are generally lower in frequency. In open habitats often produces two alternating call types. The first is a qCF call which is loudest at about 13 kHz and the second is an FM-qCF call with maximum energy at about 18 kHz. In sequence, this sounds like 'chip-chop-chip-chop'. In closed habitats, such as when foraging over streetlights or next to woodland edge, only an FM-qCF call is produced, and the loudest portion of the call may vary between 16 and 30 kHz. In extreme clutter, the loudest frequency may be as high as 29 kHz. At very high altitudes the FM portion of the call can be dropped so that only the 'chop' is heard (at around 13 kHz). Often, harmonics are produced, and these will be heard at double the frequency of the fundamental.

Time expansion/full spectrum

Greater noctules produce similar pulses to noctules but of lower frequency and longer duration (Estók and Siemers 2009). Produces two main call types, an FM-qCF call and a qCF call, but call types grade into each other. The qCF call starts at about 14 kHz, ends at about 12 kHz, and is on average 29 ms in duration (Figures 8.22.1 and 8.22.2). The FM-qCF call sweeps down from about 28 kHz to 17 kHz and is about 19 ms in duration (Figures 8.22.3 and 8.22.4). The frequencies containing maximum energy of the two calls are, on average, 18 kHz and 13 kHz respectively. Often these calls are produced alternately but there are also variations in frequency within call types (see for example Figures 8.22.2 and 8.22.4). In a very open environment such as when flying at high altitude the FM-qCF call is omitted, but when flying in relative clutter the qCF call is omitted with just FM-qCF calls being produced. When flying in very cluttered environments the FM-qCF call becomes much more broadband, the FmaxE rises, and the call is of short duration (Figure 8.22.5).

Feeding buzzes are typical of those of the *Nyctalus* bats (Figure 8.22.6).

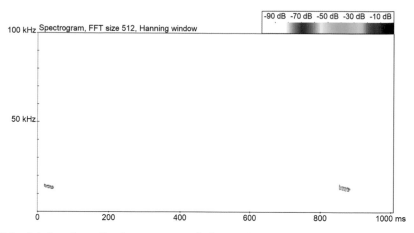

Figure 8.22.1 Echolocation calls of a greater noctule flying at high altitude (Marc Van De Sijpe).

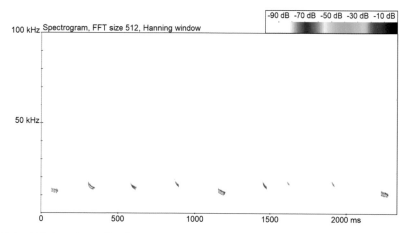

Figure 8.22.2 Echolocation calls of a greater noctule flying at high altitude, showing variation in call shape and FmaxE (Marc Van De Sijpe).

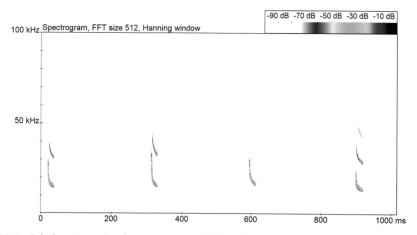

Figure 8.22.3 Echolocation calls of a greater noctule flying above a wooded area (Marc Van De Sijpe).

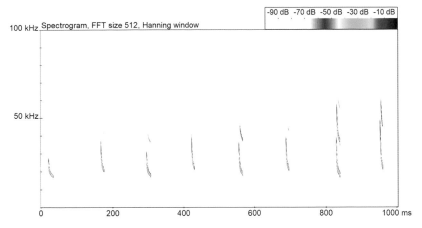

Figure 8.22.4 Echolocation calls of a greater noctule flying in semi-clutter above woodland, showing variation in call shape and frequency (Alex Lefevre).

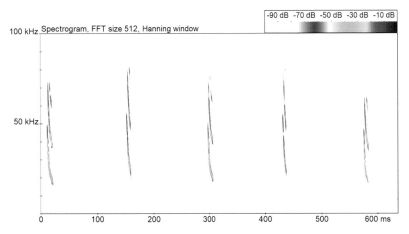

Figure 8.22.5 Echolocation calls of a greater noctule flying in clutter (Alex Lefevre).

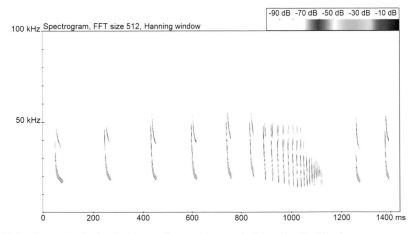

Figure 8.22.6 Example of a feeding buzz of a greater noctule (Marc Van De Sijpe).

SOCIAL CALLS

Heterodyne
Similar harsh-sounding calls to noctule at around 15–25 kHz. Very loud.

Time expansion/full spectrum
Social calls recorded from a maternity colony contain characteristic qCF pulses at low frequencies (FmaxE 9.3–18.16 kHz, mean 14.06 kHz, $n=6$) (Figure 8.22.7). Various other social calls have been recorded from bats in flight, the functions of which are unknown (Figures 8.22.8–8.22.12).

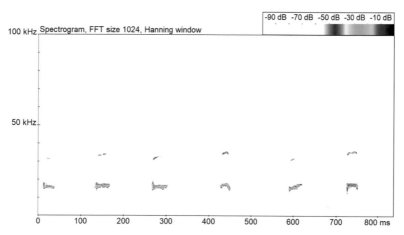

Figure 8.22.7   Example of a social call produced by a greater noctule in the roost (Péter Estók).

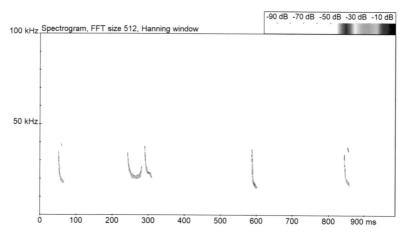

Figure 8.22.8   Social calls produced by a greater noctule in flight (Yannick Beucher).

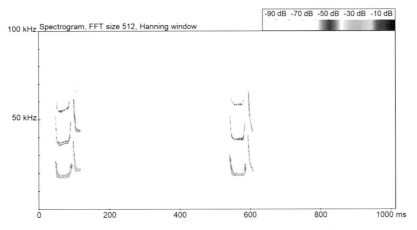

Figure 8.22.9   Social calls produced by a greater noctule in flight (Yannick Beucher).

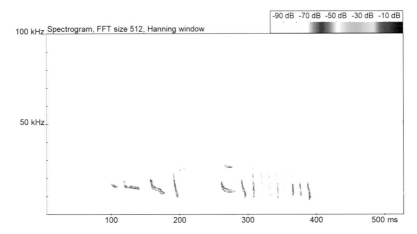

Figure 8.22.10   Social calls produced by a greater noctule in flight (Alex Lefevre).

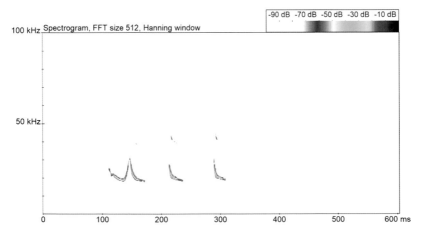

Figure 8.22.11   Social calls produced by a greater noctule in flight (Alex Lefevre).

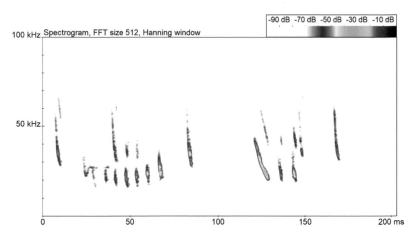

Figure 8.22.12   Social calls produced by a greater noctule in flight (Yannick Beucher).

SPECIES WITH SIMILAR OR OVERLAPPING ECHOLOCATION CALLS

Eptesicus spp., *Nyctalus leisleri*, *Vespertilio murinus*, *Tadarida teniotis* (see Chapter 7 species notes).

8.23 Leisler's bat *Nyctalus leisleri* (Kuhl, 1818)

Jon Russ and Sérgio Teixeira

© René Janssen

DISTRIBUTION

Widespread throughout Europe but varying greatly in abundance. In the UK and southern Sweden to the south of Spain, Portugal through to the Urals and the Caucasus. On the island of Madeira there is an endemic subspecies *Nyctalus leisleri verrucosus*, which is an abundant bat on the island (see notes).

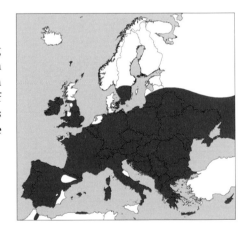

EMERGENCE

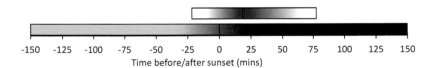

Time before/after sunset (mins)

-150 -125 -100 -75 -50 -25 0 25 50 75 100 125 150

FLIGHT AND FORAGING BEHAVIOUR

Often high (up to about 100 m) and relatively fast but also occasionally quite low. Usually forages in a straight line with shallow stoops and occasional steep dives, and sometimes seen circling. Predominantly forages by aerial hawking – prey is pursued and caught in flight.

HABITAT

Commonly observed above parkland, cattle pasture, meadows and tree canopies or flying in large circles over habitat boundaries, such as deciduous woodland and meadow. Also seen above water bodies, diving quite close to the water surface. In Madeira, this behaviour has been observed while chasing moths above the ocean surface. May be observed above white streetlights. Considered to be a typical woodland bat of hardwood forests in continental Europe.

In Ireland and some parts of Germany, roosts are commonly found in attic spaces. However, generally, the species utilises naturally occurring roosting features in trees. Occasionally found in crevices in walls and rocks.

ECHOLOCATION

Measured parameters	Mean (range)	
	FM-qCF call	qCF call
Inter-pulse interval (ms)	118.9 (107.3–313.1)	312.2 (100.2–801.2)
Call duration (ms)	8.3 (6.1–18.4)	17.1 (10.5–25.1)
Frequency of maximum energy (kHz)	27.1 (25.0–32.1)	23.5 (21.9–24.6)
Start frequency (kHz)	42.9 (29.8–61.7)	26.2 (23.5–29.9)
End frequency (kHz)	26.5 (24.2–30.7)	21.9 (20.9–24.1)

Heterodyne

Leisler's bat generally produces two very loud calls, often in sequence, resulting in a bubbly 'chip-chop' sound when tuned to about 25 kHz. As the detector is tuned up the range of frequencies, away from 25 kHz, the call develops into a 'click'. The pulse repetition rate is irregular and slow – much slower than many smaller species. The chip-chop is made up of two calls produced sequentially. The first is a low-frequency qCF call that is loudest at about 24 kHz and the second is an FM-qCF call that is loudest at about 27 kHz. The alternating chip-chop calls are produced less regularly than those of the noctule. In very open environments, when the bat is flying high, only the qCF call is produced. When the bat flies into a relatively cluttered habitat, for example when flying low over a river, only the FM-qCF call is produced. Generally, the greater the clutter, the higher the frequency containing maximum energy (FmaxE), sometimes reaching 31 kHz.

Eptesicus calls have a similar repetition rate, frequency range and tonal quality, but these species do not alternate between two call types.

Other species' social calls (particularly pipistrelle) can sound superficially similar to a Leisler's bat's echolocation calls, but they are more sporadic with no obvious rhythmic quality.

Time expansion/full spectrum

Leisler's bat often produced two different call types. An FM-qCF call of around 8 ms duration with FmaxE around 27 kHz, and a qCF call of around 17 kHz with FmaxE around 23–24 kHz (Figure 8.23.1). In open situations, such as when flying high above cattle pasture, uses primarily the qCF call with an inter-pulse interval of around 200 ms (Figures 8.23.2 and 8.23.3). In a closed environment such as when flying low over a river between two treelines, the call is a relatively steep FM and the FmaxE may rise to 29 kHz (Figure 8.23.4). Often harmonics are present. In very cluttered situations such as when released from the hand the calls become even more FM and the FmaxE can rise to as much as 32 kHz (Figure 8.23.5). Feeding buzzes show the typical transition from search phase to very steep FM (Figure 8.23.6).

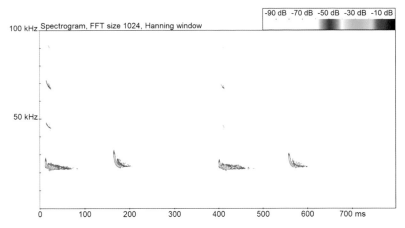

Figure 8.23.1   Echolocation calls of a Leisler's bat flying above woodland and pasture (Jon Russ).

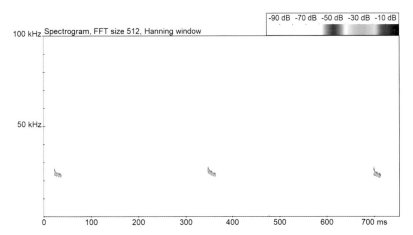

Figure 8.23.2   Echolocation calls of Leisler's bat flying high above cattle pasture (Jon Russ).

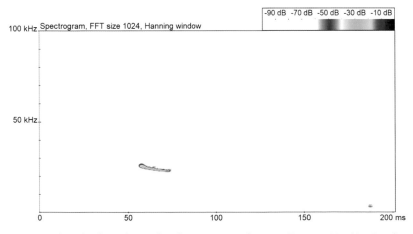

Figure 8.23.3   A single pulse from the Leisler's bat sequence shown in Figure 8.23.2 (Jon Russ).

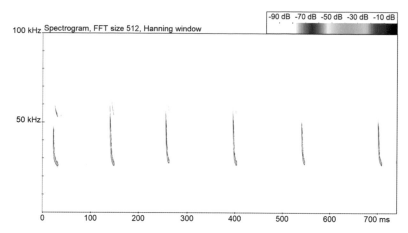

Figure 8.23.4   Echolocation calls of a Leisler's bat recorded from an individual foraging relatively low above a wooded river (Jon Russ).

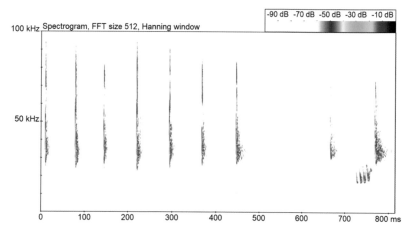

Figure 8.23.5   Echolocation calls of a Leisler's bat released from the hand. Note the presumably agonistic social call following release (Jon Russ).

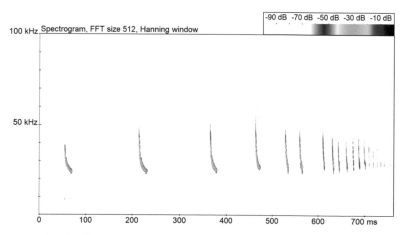

Figure 8.23.6   Leisler's bat feeding buzz (Jon Russ).

SOCIAL CALLS

Heterodyne

Social calls can be heard occasionally and sound similar to those of the pipistrelle species, but are much louder and sharper and can often be heard without the aid of a bat detector. The low-frequency call, produced by males, is easy to identify and is generally emitted by a stationary bat producing a series of audible (i.e. without a detector) sharp 'clicks' repeated every 0.5–1.0 seconds. The FmaxE of these calls is about 14 kHz. Other social calls are heard but it is very difficult to identify them as Leisler's bat calls using a heterodyne detector. Distress calls, produced from bats under physical duress, are harsh and are of long duration, with the loudest frequency at about 20 kHz

Time expansion/full spectrum

The main social call produced by males is an almost CF pulse of about 25 ms duration and with peak energy at about 14 kHz (Figures 8.23.7 and 8.23.8). These are usually produced by a stationary individual (perched in a tree, for example) and are repeated every 0.5–1.0 seconds. A variety of other social calls are produced, often in flight, and these usually consist of a combination of short trills with peak frequencies ranging from about 12 kHz to about 23 kHz (Figures 8.23.9–8.23.13). The functions of these calls are unknown, but it is likely that some of them play a role in territorial interactions.

Distress calls produced when an individual is under physical duress consist of a long series of FM pulses descending from around 55 kHz to around 15 kHz (Figures 8.23.14 and 8.23.15). These are generally quite intense, and multiple harmonics are usually present.

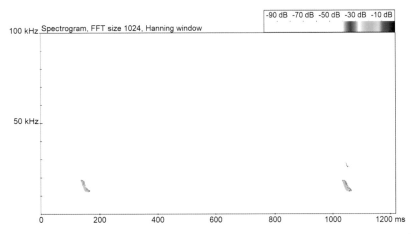

Figure 8.23.7   Social call emitted from a stationary male Leisler's bat located in a tree on the edge of woodland overlooking cattle pasture (Liat Wicks).

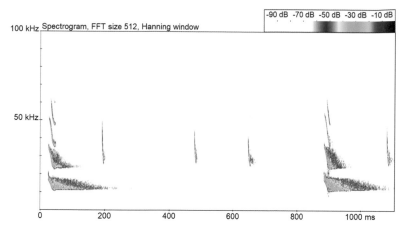

Figure 8.23.8   A sequence of Leisler's bat social calls similar to those shown in Figure 8.23.7 (Jochen Lueg).

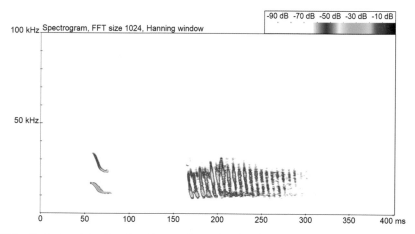

Figure 8.23.9   Typical social call emitted by a male Leisler's bat followed by a series of rapidly produced FM sweeps (Jochen Lueg).

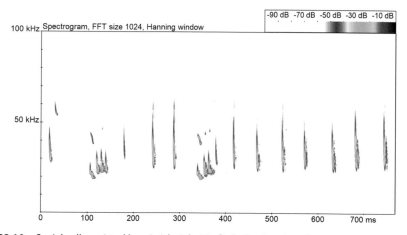

Figure 8.23.10   Social calls emitted by a Leisler's bat in flight (Jochen Lueg).

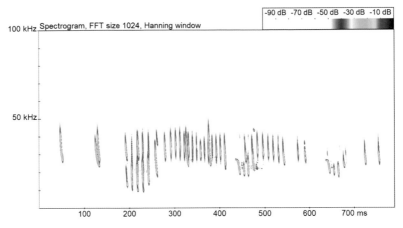

Figure 8.23.11 Social calls emitted by a Leisler's bat in flight (two bats present) (Ian Nixon).

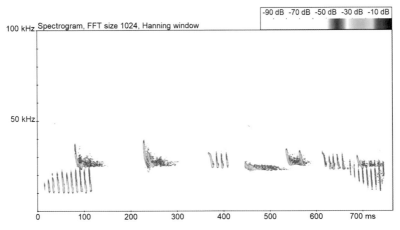

Figure 8.23.12 Social calls emitted by a Leisler's bat flying near to buildings on the edge of parkland (Jon Russ).

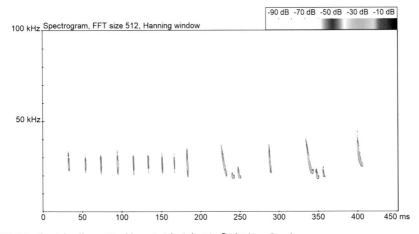

Figure 8.23.13 Social calls emitted by a Leisler's bat in flight (Jon Russ).

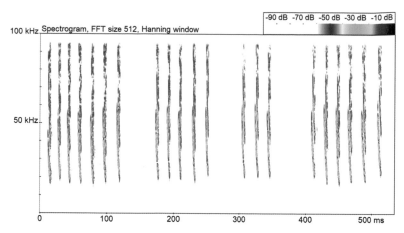

Figure 8.23.14   Distress calls of a Leisler's bat held in the hand (Jon Russ).

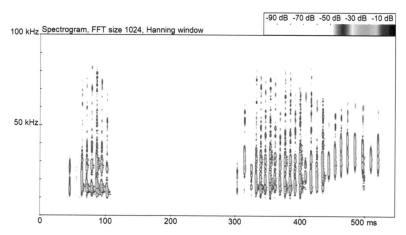

Figure 8.23.15   Distress calls of a Leisler's bat held in the hand (Anton Vlaschenko).

SPECIES WITH SIMILAR OR OVERLAPPING ECHOLOCATION CALLS

Eptesicus spp., *Nyctalus* spp., *Vespertilio murinus* (see Chapter 7 species notes).

NOTES

Madeira Leisler's bat *Nyctalus leisleri verrucosus* is an endemic subspecies of Madeira Island alone, being absent from other islands in the archipelago. This bat is smaller than mainland Leisler's bat. It is very common on the island and it is found predominantly flying above the canopy of humid laurel forests, sweet chestnut orchards and coastal and mountainous sheer cliffs. On Madeira, it has been seen foraging in mid-afternoon between eucalyptus forest and a farmland area. There is also a record of daylight flight under the canopy of large trees in a primary laurel forest area. On the island, barn owls predate Leisler's bats.

The echolocation calls of Madeira Leisler's bat appear to be similar in structure, frequency and duration to those of Leisler's bat (Figures 8.23.16–8.23.19). Social calls are also similar in structure, although there may be differences in the frequencies used (Figures 8.23.20–8.23.23).

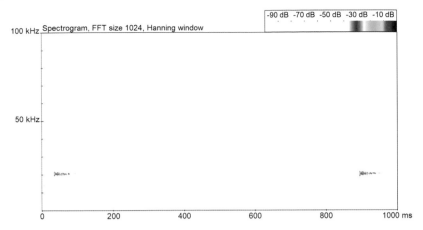

Figure 8.23.16 Echolocation calls of Madeira Leisler's bat flying high above a mountainous cliff (Sérgio Teixeira).

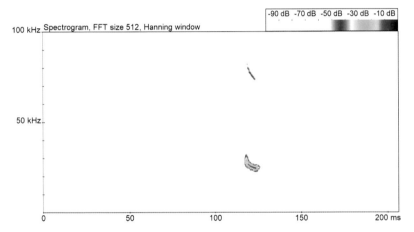

Figure 8.23.17 Echolocation calls of Madeira Leisler's bat flying in the open (Sérgio Teixeira).

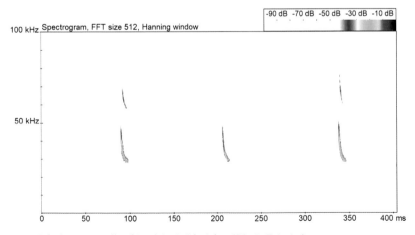

Figure 8.23.18 Echolocation calls of Madeira Leisler's bat (Sérgio Teixeira).

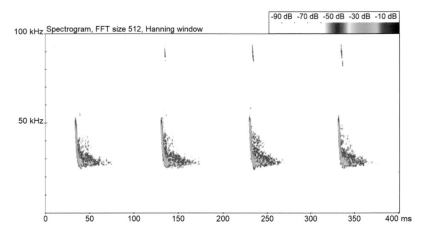

Figure 8.23.19   Echolocation calls of Madeira Leisler's bat flying in clutter (Sérgio Teixeira).

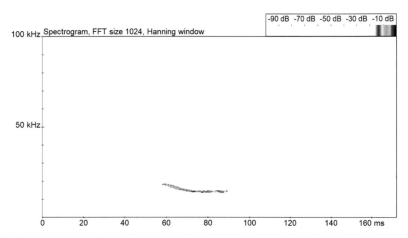

Figure 8.23.20   Social call of Madeira Leisler's bat, from a stationary bat (Sérgio Teixeira).

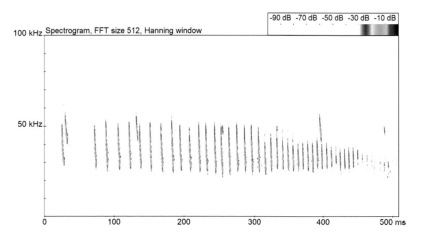

Figure 8.23.21   Social calls of Madeira Leisler's bat (Sérgio Teixeira).

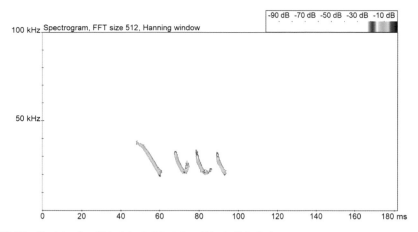

Figure 8.23.22   Social calls of Madeira Leisler's bat (Sérgio Teixeira).

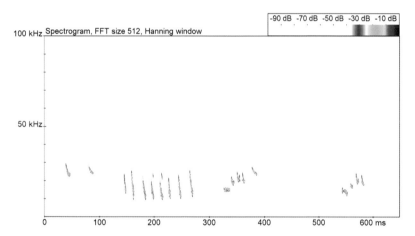

Figure 8.23.23   Social calls of Madeira Leisler's bat (Sérgio Teixeira).

8.24 Azorean noctule *Nyctalus azoreum* (Thomas, 1901)

Jorge M. Palmeirim and Ana Rainho

© Ana Rainho

DISTRIBUTION

The species is endemic to the Portuguese archipelago of the Azores. It is present on most islands (Santa Maria, São Miguel, Terceira, São Jorge, Graciosa, Pico and Faial) but is absent from the two westernmost islands, Flores and Corvo.

EMERGENCE

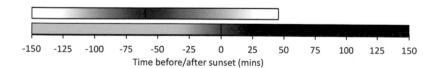

-150 -125 -100 -75 -50 -25 0 25 50 75 100 125 150

Time before/after sunset (mins)

FLIGHT AND FORAGING BEHAVIOUR

Although foraging activity of the Azorean noctule is most intense at night, it is often seen foraging during the daytime. Flies at a very variable height above the ground with a fast and manoeuvrable flight. Captures its prey by aerial hawking.

HABITAT

Uses most habitats available on the islands (Rainho *et al.* 2002). It is common in built-up areas, where it often forages around streetlamps, but it also frequently forages over farmland and wetlands. Pastures and woodlands are also used.

Its natural roosts are holes in trees and rock crevices, but the largest colonies, with up to a few hundred individuals, are currently found in buildings.

ECHOLOCATION

Measured parameters	Mean (range) FM-qCF call and qCF call
Inter-pulse interval (ms)	102.9 (23.5–211.0)
Call duration (ms)	7.2 (2.9–14.8)
Frequency of maximum energy (kHz)	32.1 (26.2–38.7)
Start frequency (kHz)	34.6 (28.8–57.1)
End frequency (kHz)	30.9 (24.5–36.3)

Heterodyne
Very loud calls with 'slaps' at around 32 kHz that turn into 'clicks' as the detector is gradually tuned up above this frequency. The frequency containing maximum energy (FmaxE) may reach 38 kHz in highly cluttered situations. In the open, the pulse repetition rate may be variable.

The Madeira pipistrelle is also present on some Azorean islands, but the two species can be separated by the pipistrelle's higher FmaxE (with 'slaps' at around 46 kHz) and more regular repetition rate in open environments.

Time expansion/full spectrum
Emits FM-qCF calls that typically start at about 35 kHz and end near 31 kHz, with the FmaxE around 32 kHz (Figure 8.24.1). Pulses have a typical duration of about 7 ms and are on average 100 ms apart. However, there is substantial call variation influenced by the level of clutter in the environment. As with many other bats in cluttered situations, pulses tend to become steeper and have higher frequencies; in the example shown in Figure 8.24.2 calls include a steep FM sweep ranging from over 50 kHz down to about 33 kHz and have an FmaxE of 37 kHz. In contrast, in open spaces calls can be narrowband qCF (29–26 kHz) and have an FmaxE of about 27 kHz (Figure 8.24.3). However, it is important to note that, at least in open spaces, bats often use irregular sequences of narrowband and FM sweeps (Figure 8.24.4). During the approach phase preceding feeding buzzes, pulses consist of very broadband FM sweeps (Figure 8.24.5).

The Madeira pipistrelle is also present on some Azorean islands, but local measurements show that it emits distinctively higher-frequency calls, with an FmaxE of about 45 kHz (43–50 kHz).

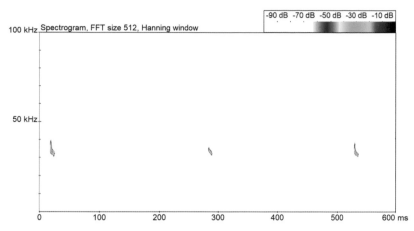

Figure 8.24.1   Echolocation calls of an Azorean noctule flying in an urban environment on the island of Terceira (Ana Rainho).

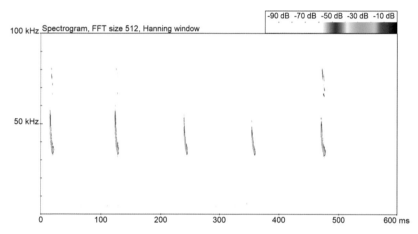

Figure 8.24.2   Echolocation calls of an Azorean noctule flying over farmland on the island of São Miguel (Ana Rainho).

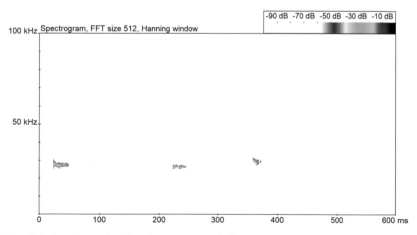

Figure 8.24.3   Echolocation calls of an Azorean noctule flying in an open space on the island of Terceira (Ana Rainho).

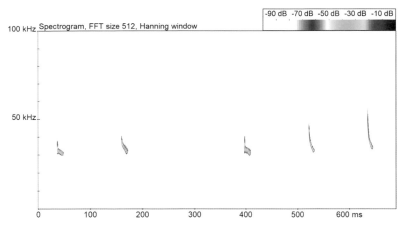

Figure 8.24.4   Sequence of echolocation calls of an Azorean noctule flying in an open space on the island of São Miguel, showing an irregular mixture of calls of different types (Alex Lefevre).

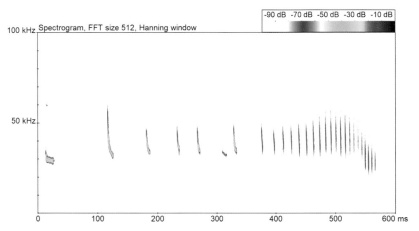

Figure 8.24.5   Feeding buzz of an Azorean noctule flying over a wooded ravine on the island of São Miguel (Alex Lefevre).

SOCIAL CALLS

Heterodyne
Social calls of the Azorean noctule have been heard during all the months sampled (March to August inclusive, several years).

Time expansion/full spectrum
The Azorean noctule produces a great variety of social calls, but there is little knowledge about their functions or the contexts in which they are emitted. Some of the calls are similar to those of its closest relative on the European mainland, the Leisler's bat (Pfalzer and Kusch 2003). That is the case of the trill in Figure 8.24.6, which consists of a repetition of downward pulses with FmaxE 18 kHz. It has been suggested that calls of island bats are subjected to stabilising selection and thus remain similar to those of their ancestral species (Russo *et al.* 2009).

In addition, the Azorean noctule produces a social call that consists of a rapid series of 3–4 FM sweeps, with an FmaxE of about 38 kHz (Figures 8.24.7 and 8.24.8). The structure of these calls is visually quite similar to that of the Madeiran pipistrelle, which is present on some Azorean islands, but the Azorean noctule's calls are emitted at much lower frequencies.

Another distinct social call, consisting of a series of broadband downward sweeps of variable steepness, has been recorded from bats flying near the entrance of a roost (Figure 8.24.9).

Further examples of calls emitted in the vicinity of maternity roosts are shown in Figures 8.24.10 and 8.24.11.

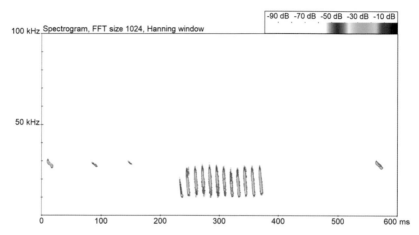

Figure 8.24.6   Social calls of Azorean noctule flying in a village on the island of São Jorge during March (Ana Rainho).

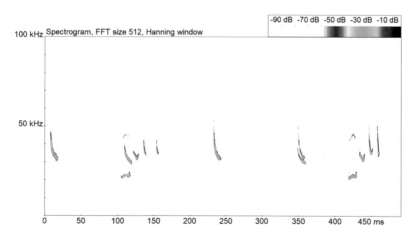

Figure 8.24.7   Social calls of Azorean noctule flying in a village on the island of São Jorge during March (Ana Rainho).

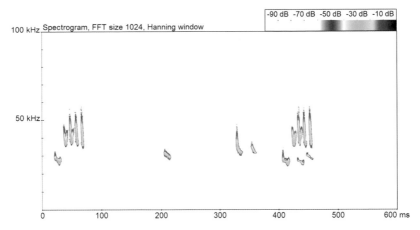

Figure 8.24.8   Social calls of Azorean noctule, emitted while flying in an urban environment, on the island of São Miguel, in July (Ana Rainho).

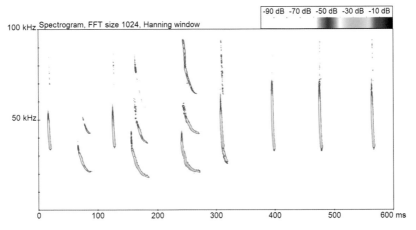

Figure 8.24.9   Social calls of Azorean noctule near the entrance of a roost, during emergence, on the island of São Miguel (Ana Rainho).

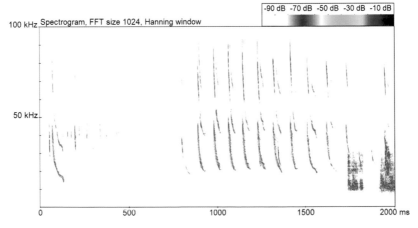

Figure 8.24.10   Social calls of Azorean noctule near the entrance of a roost, during emergence, on the island of São Miguel (Alex Lefevre).

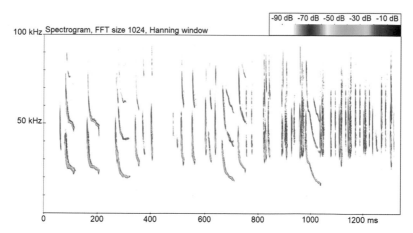

Figure 8.24.11   Social calls of Azorean noctule near the entrance of a roost, during swarming, on the island of São Miguel (Alex Lefevre).

SPECIES WITH SIMILAR OR OVERLAPPING ECHOLOCATION CALLS

Eptesicus spp., *Nyctalus* spp., *Vespertilio murinus* (see Chapter 7 species notes).

NOTES

The Azorean noctule is partly diurnal, so individuals can emerge from the roosts throughout the day, but most emerge in the late afternoon. The values shown under *Emergence*, above, are an indication of the times when emergence usually starts, based on data in Leonardo and Medeiros (2011) and Irwin and Speakman (2003).

8.25 Serotine *Eptesicus serotinus* (Scheber, 1774)

Pedro Horta, Helena Raposeira, Hugo Rebelo, Jon Russ, Erik Korsten
and Marc Van De Sijpe

© René Janssen

DISTRIBUTION

The serotine is mainly distributed along the Eurosiberian biogeographical region (Santos *et al.* 2014). It has a broad range covering most of Europe up to a latitude of 55°N, including the south of England, the southernmost areas of Sweden and the Baltic region. Despite some indications about the species spreading northwards, albeit with a fragmented distribution, it is apparently absent from Ireland, Norway, Finland and north of the Baltic region (Dietz *et al.* 2009). It is widely distributed in Mediterranean Europe (Santos *et al.* 2014), though with no records in the Atlantic islands (Dietz *et al.* 2009).

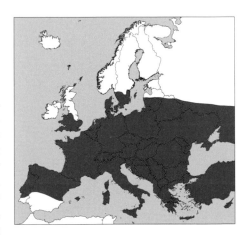

EMERGENCE

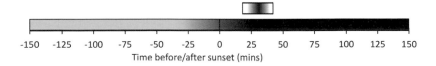

-150 -125 -100 -75 -50 -25 0 25 50 75 100 125 150
Time before/after sunset (mins)

FLIGHT AND FORAGING BEHAVIOUR

Agile and rapid flight along vegetation borders (Dietz *et al.* 2009), circling trees or capturing prey above the vegetation across open areas (flight speed 13.2 m·s⁻¹) (Bruderer and Popa-Lisseanu 2005). While hunting it flies between 6.8 and 10.7 m (occasionally 1–18 m) above the ground (Jensen and Miller 1999). It is also very common to detect serotines searching for prey at streetlamps for long periods of time (Dietz *et al.* 2009). This species can perform regular search flight circuits (frequently in groups), only interrupted when prey is found. It has also been reported that serotines can capture prey directly from the ground or from vegetation (Harbusch 2003).

HABITAT

Explores a large range of habitats in central Europe and the Mediterranean (Robinson and Stebbings 1997). It seems to be dependent on forests and open vegetation with deciduous trees as foraging habitats, where it hunts over hay meadows and grazed pasture, forest edges, parks and cleared agricultural areas with isolated trees and water edges (Catto *et al.* 1996). However, serotines are also commonly found in urban habitats (Baagøe 2001).

Each individual has an average hunting area of 4.6 km², and the maximum distance from the roost can be up to 48 km (Pérez and Ibáñez 1991, Robinson and Stebbings 1997, Harbusch 2003).

Typical roosts occur in buildings, especially in roof spaces (Dinger 1991), but some isolated bats or small groups can occupy bat boxes and tree holes (Dietz *et al.* 2009). In the southern area of their distribution, individuals particularly use bridge cornices, but also rock crevices and the entrances of caves (P. Horta, personal observation).

ECHOLOCATION

Measured parameters	Mean (range) FM-qCF call and qCF call
Inter-pulse interval (ms)	155.1 (96.8.3–291.0)
Call duration (ms)	9.75 (5.0–16.3)
Frequency of maximum energy (kHz)	24.8 (21.0–31.0)
Start frequency (kHz)	33.5 (28.5–60.0)
End frequency (kHz)	21.0 (19.0–24.5)

Heterodyne
Produces a relatively low-frequency 'click' detected between 20 and 35 kHz depending on the distance between the bat and the receiver, gradually becoming clearer near 23 kHz. The pulse repetition rate is relatively slow and constant, especially in open areas. In cluttered habitats such as dense forests, serotines usually echolocate using a larger range of frequencies (frequently with more frequency-modulated calls) and emit shorter pulses (Neuweiler 1989). In these situations, the frequency containing maximum energy (FmaxE) will rise and may reach 29 kHz.

There is an overlap in the FmaxE of serotines and meridional serotines between 23.4 and 28.8 kHz, making their separation very hard in regions where both species occur, namely in some regions of the central Iberian peninsula. In this sympatric region, the two species are known to interbreed and produce viable offspring (Centeno-Cuadros *et al.* 2019), with possible unknown introgression events that may modulate the traits of serotines, including their echolocation calls. Here, the existence of first-generation hybrids (at least), and serotine backcrosses (Centeno-Cuadros *et al.* 2019), may make it impossible to separate their calls.

The echolocation characteristics of serotines are very similar to those of Leisler's bat, especially in open areas, where serotines increase the temporal components of the calls. It is important to remember that the duration of pulses and the inter-pulse interval are longer in the *Nyctalus* genus, which is especially evident in cluttered environments where *Nyctalus* bats produce calls of longer duration, with longer inter-pulse intervals, than *Eptesicus* bats.

Barbastelles can sound similar to serotines in a cluttered environment – syncopated 'smacks' with peak frequencies in the range 31–34 kHz. However, serotine calls can usually still be heard when tuning above 50 kHz which is not normally the case with barbastelle.

Time expansion/full spectrum

Serotines produce FM-qCF calls that sweep down from about 33.5 kHz to around 21.0 kHz with FmaxE usually at 24.8 kHz (Figure 8.25.1). Mean call duration is about 10 ms and inter-pulse interval approximately 150 ms (Horta *et al.* 2015).

In open habitats, when for example serotines fly over the vegetation at great altitude, call duration becomes longer, inter-pulse interval increases and FmaxE drops (Figure 8.25.2), eventually reaching the lower extreme of the species' echolocation frequency range (Figure 8.25.3). Conversely, in cluttered habitats their call duration will shorten, the inter-pulse interval will decrease, and the FmaxE will increase (Figure 8.25.4), eventually reaching the highest extreme of their range (Figure 8.25.5) – and this, in some cases, may cause confusion with *Myotis* calls (which are characterised by FM pulses). Therefore, it is important to check carefully the end part of each pulse, which may provide further clues to allow distinction between these genera.

Feeding buzzes are typical of bats of the genus *Eptesicus* (Figures 8.25.6 and 8.25.7).

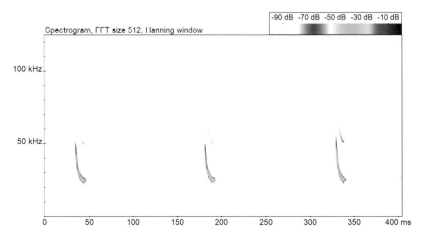

Figure 8.25.1   Echolocation calls of a serotine hunting along a woodland edge (Marc Van De Sijpe).

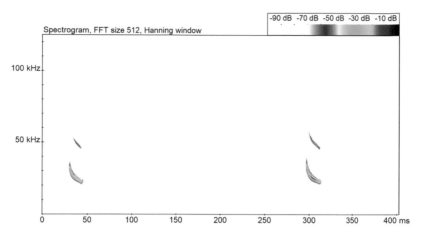

Figure 8.25.2   Echolocation calls of a serotine hunting over grassland (Marc Van De Sijpe).

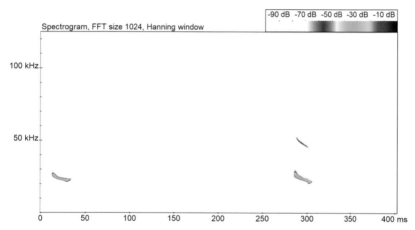

Figure 8.25.3   Echolocation calls of a serotine hunting high over an illuminated bridge by a canal (Marc Van De Sijpe).

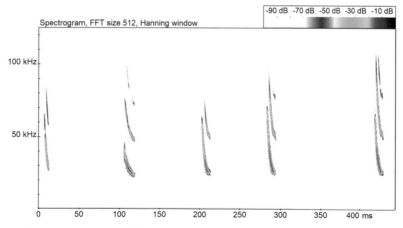

Figure 8.25.4   Further example of echolocation calls of a serotine hunting along a woodland edge (Marc Van De Sijpe).

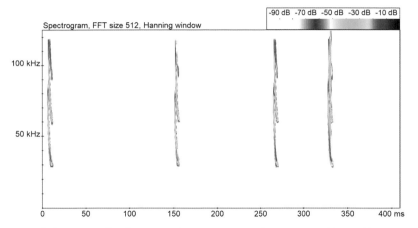

Figure 8.25.5   Echolocation calls of a serotine hunting in a pine forest. Note additional harmonics as a result of clipping (Marc Van De Sijpe).

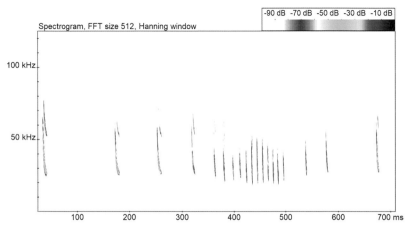

Figure 8.25.6   Feeding buzz of a serotine hunting along a woodland edge (Marc Van De Sijpe).

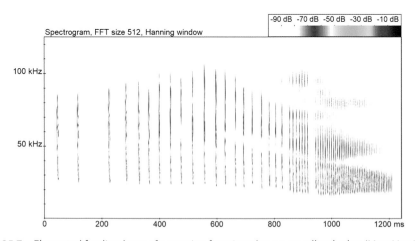

Figure 8.25.7   Elongated feeding buzz of a serotine foraging along a woodland edge (Marc Van De Sijpe).

SOCIAL CALLS

Heterodyne
Serotines are not particularly vocal bats, and social calls are rarely heard. Distress calls emitted by bats held in the hand are very loud with an FmaxE of about 15 kHz. Calls are produced prior to emergence and also occasionally on the wing.

Time expansion/full spectrum
Social calls are produced from the roost prior to emergence and these can be quite complex, consisting of a variety of low-frequency buzzes, n-shaped calls, sweeps and hooked calls (e.g. Figures 8.25.8–8.25.10). Elongated low-frequency hooked calls, possibly produced by juveniles, which are approximately half the frequency of the adult echolocation calls have been recorded (Figure 8.25.11). A call produced in flight has been described for this species which consists of three FM sweeps, similar in structure to those produced by soprano pipistrelles but with a longer inter-component interval and an FmaxE of about 17 kHz, often produced consecutively (Figure 8.25.12). Occasionally v-shaped calls are reported during emergence (Figure 8.25.13).

An interesting behaviour has been recorded on many occasions in the Netherlands which involves a single stationary bat calling from the facade of a building or a tree trunk (E. Korsten, personal observation). In each case, the calls resemble echolocation calls but are fixed in frequency, call length and call intervals (e.g. Figure 8.25.14). Bats remain in this position calling constantly for a few minutes to over 2 hours, occasionally flying away but often returning to the same location. Usually, when the bats produce these stationary calls, there are no other serotines around, although on the rare occasion that other serotines are nearby the caller starts to make buzz-like social calls (Figure 8.25.15). This behaviour has been observed in April, July and August.

Distress calls are long drawn-out calls which vary in duration and usually have an FmaxE of about 15 kHz (Figure 8.25.16).

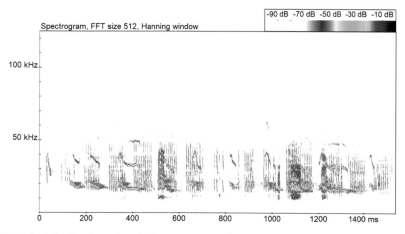

Figure 8.25.8   Social calls of serotines before emergence from a roost exit (Marc Van De Sijpe).

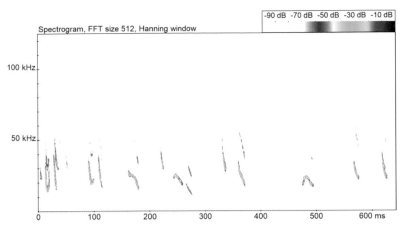

Figure 8.25.9   Social calls of serotines at a roost entrance (Marc Van De Sijpe).

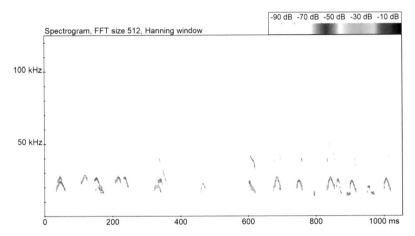

Figure 8.25.10   Social calls of serotine in a roost (Marc Van De Sijpe).

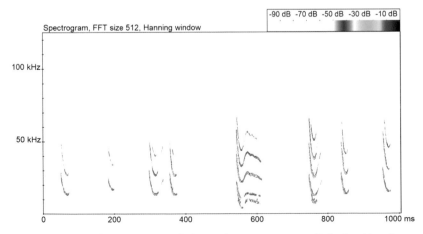

Figure 8.25.11   Social calls of a serotine recorded outside a roost entrance (Sally-Ann Hurry).

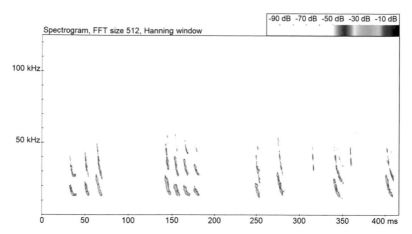

Figure 8.25.12   Social calls of a serotine recorded outside a roost entrance (Sally-Ann Hurry).

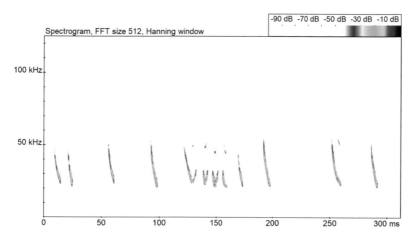

Figure 8.25.13   Social calls of a serotine recorded during emergence from a church (Erika Dahlberg).

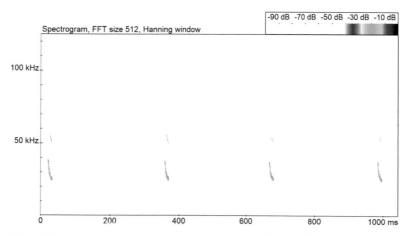

Figure 8.25.14   Calls produced by a serotine while stationary on the west facade of an apartment block (Erik Korsten).

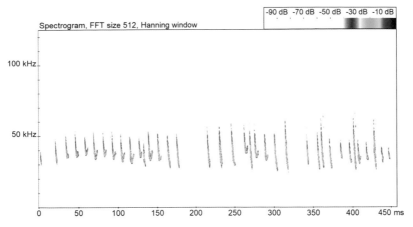

Figure 8.25.15   Social calls produced by a serotine immediately upon flying from a stationary perch (see Figure 8.25.14) (Erik Korsten).

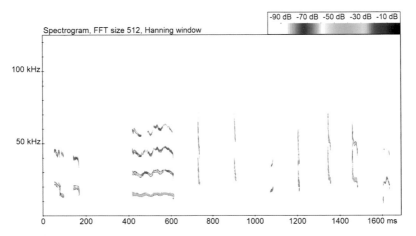

Figure 8.25.16   Distress calls produced by a serotine held in the hand (Jon Russ).

SPECIES WITH SIMILAR OR OVERLAPPING ECHOLOCATION CALLS

Eptesicus spp., *Nyctalus* spp., *Vespertilio murinus* (see Chapter 7 species notes).

NOTES

Only recently was the meridional serotine promoted to species level in the Iberian peninsula. The separation from the serotine was supported by genetic analyses showing a level of genetic divergence up to 16% (Ibáñez *et al.* 2006). More studies are needed regarding the ecology of both species in sympatric zones, especially focusing on their acoustic behaviour.

8.26 Meridional serotine
Eptesicus isabellinus (Temminck, 1840)

Helena Raposeira, Hugo Rebelo and Pedro Horta

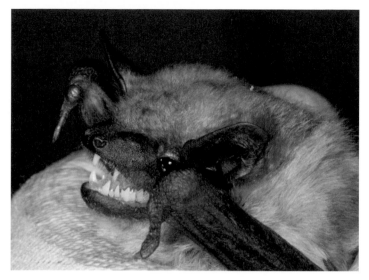

© Martyn Cooke

DISTRIBUTION

The meridional serotine occurs over the Mediterranean biogeographical region of the Iberian peninsula and north Africa (Ibáñez *et al.* 2006). The species is limited to the south by the Sahara Desert, with the Atlas Mountains being the southern limit of its distribution (Dietz *et al.* 2009). It is present to the west in the Canary Islands (Trujillo 1991), occurring from Morocco, Algeria, Tunisia to Libya, the eastern extreme of their distribution (Ibáñez *et al.* 2006). In the north, the meridional serotine is found only in the centre and south of the Iberian peninsula, the only European region where it is present (Santos *et al.* 2014).

EMERGENCE

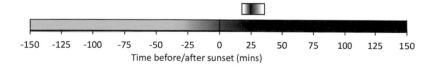

-150   -125   -100   -75   -50   -25   0   25   50   75   100   125   150

Time before/after sunset (mins)

FLIGHT AND FORAGING BEHAVIOUR

The meridional serotine is an agile and fast-flying bat, similar to the serotine, foraging around trees or capturing prey above vegetation, especially across open habitats (Brosset 1955). Being smaller than the serotine (Dietz *et al.* 2009), the meridional serotine is probably better able to explore cluttered habitats. This species also uses temporary roosts during the night, where it rests for long periods of time between hunting sessions (Helena Raposeira, personal observation).

HABITAT

This species is adapted to hot and dry macrohabitats, exploring open agricultural and forest landscapes of north Africa (especially along the coast) and the Iberian Mediterranean (Dietz *et al.* 2009). In semi-desert areas, it is associated with water bodies, usually with dense surrounding vegetation (Dalhoumi *et al.* 2016). Some individuals have been frequently detected hunting under streetlamps (Dalhoumi *et al.* 2017).

Their natural and typical roosts are probably rock crevices, but currently the majority of the colonies found occur in human structures, particularly in bridges over permanent rivers, but also in box-culverts, under-passages and cracks in ruins (Helena Raposeira, personal observation). Some isolated bats or small groups can occupy human buildings (Aulagnier and Thevenot 1986), especially in the attics or gaps between roof tiles (Helena Raposeira, personal observation).

ECHOLOCATION

Measured parameters	Mean (range) FM-qCF call and qCF call
Inter-pulse interval (ms)	126.9 (59.5–263.0)
Call duration (ms)	11.3 (3.3–22.5)
Frequency of maximum energy (kHz)	29.5 (22.0–35.8)
Start frequency (kHz)	52.0 (20.4–65.5)
End frequency (kHz)	22.0 (18.5–28.0)

Heterodyne
Produces a relatively low-frequency 'click' detected between 20 and 35 kHz depending on the distance between the bat and the receiver, gradually becoming louder around 29 kHz. As for the serotine, the pulse repetition rate is relatively slow and constant, especially in open areas.

In cluttered habitats such as dense forests, the meridional serotine usually echolocates using a larger range of frequencies (frequently with more FM calls) and shorter pulses (Neuweiler 1989).

There is a large overlap in FmaxE between the meridional serotine and serotine between 23.4 kHz and 28.8 kHz, making it difficult to separate them in regions where both species occur, namely some regions of central Iberia. Particular attention should be paid to the fact that in this sympatric region the two species viably interbreed (Centeno-Cuadros *et al.* 2019) with possible unknown introgressive hybridisation, which can modulate the traits of meridional serotines, including their echolocation calls, despite being asymmetrically biased in the serotine's direction. Even so, in sympatric regions, the existence of first-generation hybrids (at least) and backcrosses (Centeno-Cuadros *et al.* 2019) means that it may be impossible to separate their calls.

Similarly to the serotine acoustic identification, there is also a possible confusion with Leisler's bat when present, especially in open areas, where both species tend to produce similar calls.

Time expansion/full spectrum

Produces FM-qCF calls that sweep down from about 52 kHz to about 24 kHz with FmaxE usually around 29 kHz (Figure 8.26.1). Mean call duration is about 7 ms and inter-pulse interval around 130 ms (Horta *et al.* 2015).

In habitats with cluttered vegetation the call duration shortens, the inter-pulse interval decreases and the FmaxE also increases to the highest extreme of the species' range (Figure 8.26.2). As the bat moves to a more open habitat the call duration increases and the bandwidth decreases, and there is often a sharp kink in the call at the lowest frequency of the FM portion of the call (Figure 8.26.3). As the bat moves into a completely open habitat the bandwidth decreases further and the call becomes qCF (Figure 8.26.4). An example of a group of foraging bats utilising a variety of call shapes is shown in Figure 8.26.5.

Occasionally single echolocation calls are of lower frequency in a sequence with a 'warbling' structure (Figure 8.26.6).

Feeding buzzes are typical of those of *Eptesicus* species (Figure 8.26.7).

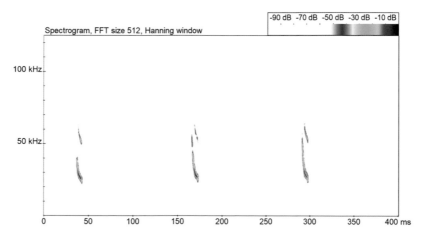

Figure 8.26.1   Typical echolocation calls of meridional serotine (Helena Raposeira).

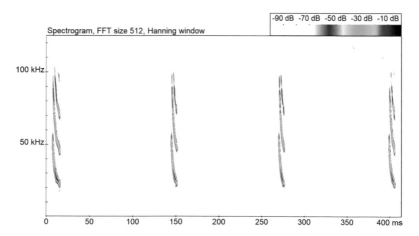

Figure 8.26.2   Echolocation calls of meridional serotine flying in clutter under a roost in a bridge (Marc Van De Sijpe and Alex Lefevre).

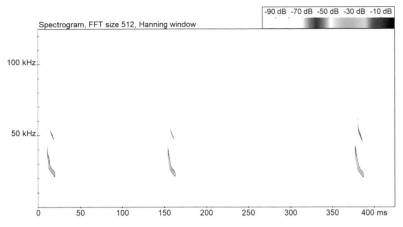

Figure 8.26.3   Echolocation calls of meridional serotine flying in a semi-cluttered environment near a roost in a bridge (Marc Van De Sijpe and Alex Lefevre).

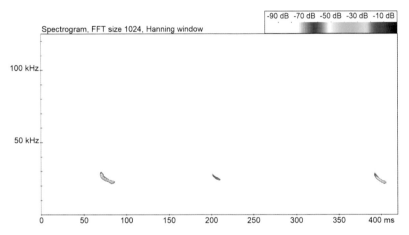

Figure 8.26.4   Echolocation calls of meridional serotine flying in the open (Marc Van De Sijpe and Alex Lefevre).

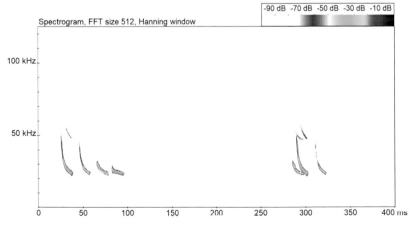

Figure 8.26.5   Echolocation calls of hunting meridional serotines, showing calls from different bats in varying degrees of clutter (Marc Van De Sijpe and Alex Lefevre).

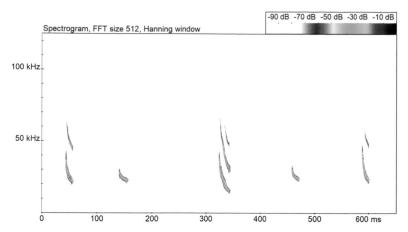

Figure 8.26.6   Unusual echolocation calls of meridional serotine, recorded near a roost in a bridge (Marc Van De Sijpe and Alex Lefevre).

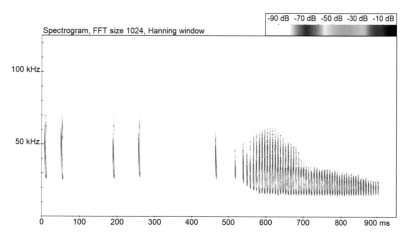

Figure 8.26.7   Feeding buzz of a meridional serotine (Marc Van De Sijpe and Alex Lefevre).

SOCIAL CALLS

Heterodyne
Calls heard from roosts are harsh and loud.

Time expansion/full spectrum
Social calls have been rarely recorded. Calls from maternity roosts are of long duration with multiple harmonics (Figure 8.26.8)

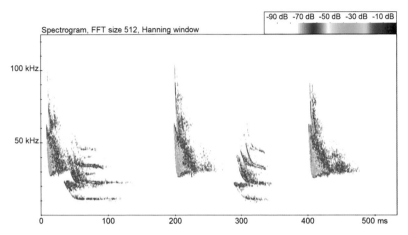

Figure 8.26.8 Social calls of a meridional serotine from a maternity roost (Marc Van De Sijpe and Alex Lefevre).

SPECIES WITH SIMILAR OR OVERLAPPING ECHOLOCATION CALLS

Eptesicus spp., *Nyctalus* spp., *Vespertilio murinus* (see Chapter 7 species notes).

NOTES

Only recently was the meridional serotine promoted to species level. The separation from the serotine was supported by genetic analyses showing a level of genetic divergence up to 16% from serotine (Ibáñez *et al.* 2006). Therefore, the knowledge gap for this species is quite large and more studies are needed regarding the ecology of the meridional serotine, especially focusing on their acoustic behaviour and description of their acoustic parameters.

8.27 Anatolian serotine *Eptesicus anatolicus* (Felten, 1971)

Alex Lefevre and Panagiotis Georgiakakis

© Panagiotis Georgiakakis

DISTRIBUTION

This recently described European serotine species is distributed along the coast of the Aegean Sea from Rhodes (Greece) in the west, along the Levantine Sea coast of Turkey to Lebanon in the southeast. So far, in Europe, this species is found only on the Greek islands off the Anatolian coast such as Rhodes, Cyprus and possibly Samos.

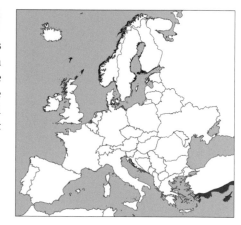

EMERGENCE

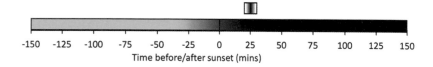

Time before/after sunset (mins)

FLIGHT AND FORAGING BEHAVIOUR

The Anatolian serotine has a fast and agile flight, alternating with some rapid dives to capture prey. Prefers foraging at heights between 2 and 5 m, depending on wind strength. Forages back and forth above Mediterranean vegetation such as dry calcareous grassland, shrubland, garrigue and maquis along coastal zones and cliffs. Foraging animals have also been observed around streetlamps and in parks with ancient ruins.

HABITAT

The Anatolian serotine seems to be associated with Mediterranean landscapes, and it seems to avoid forested areas. On Rhodes (Greece) and Cyprus, it hunts over sparsely vegetated coastal zones with bare rocky slopes and dry streams. It appears that this species is a typical rock-crevice dweller, some individuals having been observed leaving crevices in coastal cliffs and flying above beaches. Other roosts have been found in crevices on ancient ruins, in other buildings, and under bridges. The species is absent from mountain areas of Rhodes and mainland Turkey.

ECHOLOCATION

Measured parameters	Mean (range) FM-qCF call and qCF call
Inter-pulse interval (ms)	164.5 (114.0–458.0)
Call duration (ms)	10.0 (4.37–14.0)
Frequency of maximum energy (kHz)	29.6 (27.5–31.9)
Start frequency (kHz)	40.9 (36.1–65.0)
End frequency (kHz)	27.3 (26.1–29.4)

Heterodyne

The species can be recorded with a heterodyne bat detector at a distance of up to 25–35 m. Produces loud 'smacks', loudest at around 28 kHz, which is similar to the northern bat. The pulse repetition rate is irregular and very similar to other serotines species. Sounds like a marble dropped on a stone floor.

Time expansion/full spectrum

Echolocation calls from the Anatolian serotine have lower FmaxE (mean 29.5 kHz) than Botta's serotine *Eptesicus bottae* (32.5 kHz; Holderied *et al.* 2005). Our measured parameters are based on calls from Anatolian serotines recorded on the island of Rhodes. So far, this is the only registered *Eptesicus* species on the island and cannot be confused with Botta's serotine. In 1998 von Helversen reported the presence of Botta's serotine in Rhodes and described its echolocation calls, but it was later proven that they belonged to the Anatolian serotine (Juste *et al.* 2013). The end frequency (26–29.4 kHz) is higher than in the serotine, more similar to the northern bat (around 28 kHz) (Figures 8.27.1 and 8.27.2). In clutter, the start frequency is higher, occasionally up to 65 kHz (Figure 8.27.3). In open areas almost CF calls with a reduced FM initial part and an end frequency of 27–28 kHz and call duration of about 10–14 ms (Figure 8.27.4). The call structure is very similar to that of the northern bat.

Feeding buzzes are typical for those of the genus (Figure 8.27.5).

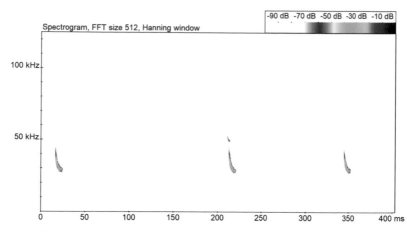

Figure 8.27.1   Echolocation calls of an Anatolian serotine flying above a sparsely vegetated coastal zone (Alex Lefevre).

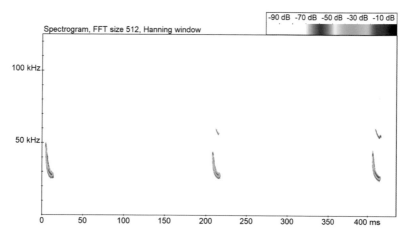

Figure 8.27.2   Further example of echolocation calls of an Anatolian serotine flying above a sparsely vegetated coastal zone (Alex Lefevre).

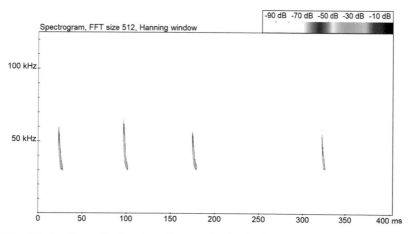

Figure 8.27.3   Echolocation calls of an Anatolian serotine briefly flying in clutter above a sparsely vegetated coastal zone (Alex Lefevre).

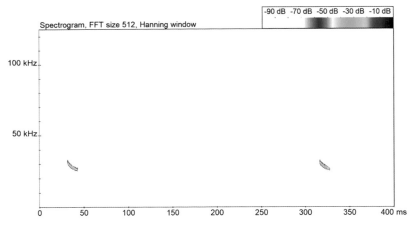

Figure 8.27.4   Echolocation calls of an Anatolian serotine flying in the open above a sparsely vegetated coastal zone (Alex Lefevre).

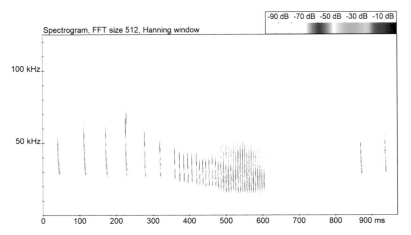

Figure 8.27.5   Feeding buzz of an Anatolian serotine (Alex Lefevre).

SOCIAL CALLS

Heterodyne
No data.

Time expansion/full spectrum
No data.

SPECIES WITH SIMILAR OR OVERLAPPING ECHOLOCATION CALLS

Eptesicus spp., *Nyctalus* spp., *Vespertilio murinus* (see Chapter 7 species notes).

NOTES

Until 2013, *Eptesicus anatolicus* was seen as a subspecies of *E. bottae*. DNA-based approaches have recently shown that *E. anatolicus* is a fully distinct species (Juste *et al.* 2013).

8.28 Northern bat

Eptesicus nilssonii (Keyserling and Blasius, 1839)

Jens Rydell and Johan Eklöf

© Jens Rydell

DISTRIBUTION

From northernmost Fennoscandia south through Scandinavia, the Baltic States and Russia to central Europe and Ukraine. Further southeast in Europe the distribution is patchy and restricted to mountain areas. The species is rare or absent from the lowland areas of western Europe such as Denmark, Netherlands and the UK.

EMERGENCE

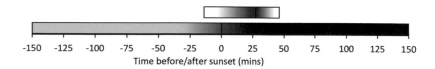

Time before/after sunset (mins)

FLIGHT AND FORAGING BEHAVIOUR

Flight is rather rapid and straight. Foraging takes place in open or semi-open habitats 3–50 m above the ground, sometimes higher (e.g. 100 m at wind turbines). Catches prey by aerial hawking, typically through an agile vertical dive (Jensen *et al.* 2001).

HABITAT

Foraging habitats include rivers and lakes, gardens, forest edges, along rows of trees or lit roads, over pastures and fields, and sometimes also in the open air high above forest. In the far north, it usually forages in gaps in spruce forests and over water. Frequently roosts in attics of heated houses, at least in the north, but probably also in tree holes. In winter in houses, mines and cellars and in Norway also in scree and rock crevices.

ECHOLOCATION

Measured parameters	Mean (range) FM-qCF call and qCF call
Inter-pulse interval (ms)	222.7 (85.0–440.0)
Call duration (ms)	12.6 (6.0–16.7)
Frequency of maximum energy (kHz)	28.7 (26.0–36.6)
Start frequency (kHz)	38.0 (33.3–71.0)
End frequency (kHz)	27.0 (25.7–32.1)

Heterodyne

Typically emits search-phase calls ending with a narrowband component which sweeps from about 50 kHz to 30 kHz or slightly lower. Powerful 'tocks' or 'slaps' are heard when the detector is tuned to the frequency containing maximum energy (FmaxE) at 28–32 kHz. The frequency band used by this species is almost unique in Europe, and it is easy to recognise in most situations on the frequency alone. The pulse repetition rate is usually one per wingbeat (10 per second) in low flight or near clutter and about one per two or three wingbeats (3–5 per second) in more open space. Most often, the pulse repetition rate is somewhere in between and typically includes longer inter-pulse intervals here and there, resulting in an easily recognised irregular pulse rhythm. A strong harmonic sometimes occurs at ca 60 kHz, and to avoid confusion with soprano pipistrelles the detector should be tuned down to 30 kHz as well.

Time expansion/full spectrum

Typically emits search-phase FM-qCF calls which sweep from 50–60 kHz to 28–32 kHz and with FmaxE at 28–32 kHz (Figure 8.28.1). The variation relates to the distance to background or clutter-producing objects. The most frequently observed and most typical foraging situation is in semi-open situations such as gaps in forest and along treelines, roads or lake shores. The bat typically flies straight and rather fast, while patrolling back and forth along the same path 5–10 m above the ground. The pulses are usually 10–12 ms long, quite powerful, and typically begin with a short broadband component. The FmaxE is usually 29–31 kHz but sometimes higher. In such cases, the pulse rhythm is usually irregular but on average has inter-pulse intervals of about 200 ms. This pulse rhythm is the most common, and is characteristic of the species.

In search flight in open air space high above the ground, powerful, ca 15 ms long, narrowband pulses (Figure 8.28.2) are the most common, but occasionally the pulses are longer, up to 20 ms. The FmaxE is 26–30 kHz, most frequently 27–28 kHz. The pulse rhythm under such conditions corresponds to one pulse every two or three wingbeats, hence with the inter-pulse intervals around 200 or 300 ms, usually changing irregularly.

When patrolling low (3–4 m) over an open place or under the tree canopy, the search pulses are typically 6–8 ms long, occasionally as short as 4 ms, including a broadband sweep and with FmaxE well above 30 kHz (Jensen *et al.* 2001). The pulse rhythm is usually very regular in such cases, with inter-pulse intervals of about 100 ms, corresponding to one pulse per wingbeat (Figure 8.28.3).

The frequency band used by this species is almost unique in Europe and it is easy to recognise in most situations from the call frequency alone (see also Ahlén 1981, Barataud 2015).

Feeding buzzes are typical of most European vespertilionids (Figure 8.28.4).

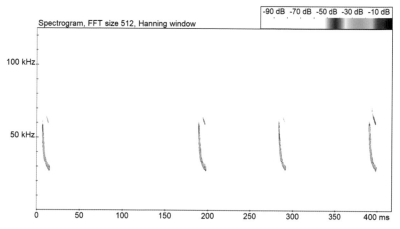

Figure 8.28.1 Typical echolocation calls of the northern bat recorded in semi-open space below treetop level (Jens Rydell).

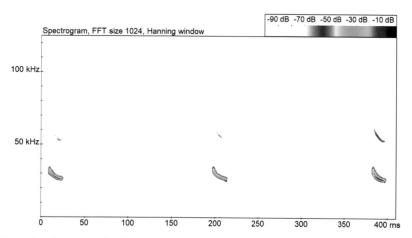

Figure 8.28.2 Echolocation calls of a northern bat flying in the open around 10 m above a lake (Jens Rydell).

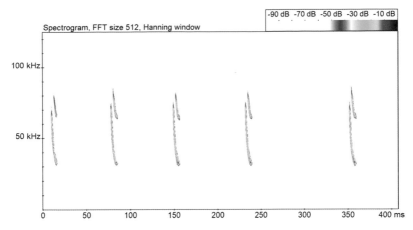

Figure 8.28.3   Echolocation calls of a northern bat patrolling low over a grass field searching for dung beetles (Jens Rydell).

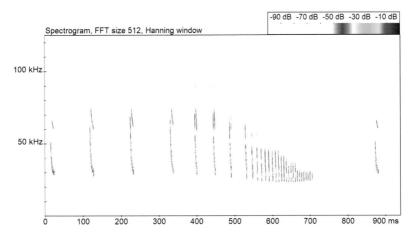

Figure 8.28.4   Feeding buzz of a northern bat foraging along a woodland edge (Jens Rydell).

SOCIAL CALLS

Heterodyne
The communication calls or 'social' calls of this species have not been well studied, and few such calls have been described. Some social calls are quite complex and variable and can be heard as rather powerful 'scratches' with FmaxE somewhere in the 20–40 kHz range and sometimes much lower.

Time expansion/full spectrum
Social calls are heard, for example, when two foraging individuals encounter each other in a feeding place in what seems to be a territorial conflict. The calls are often emitted within echolocation call sequences (Rydell 1986, Middleton *et al.* 2014).

There are also other quite complex and more broadband calls, sometimes heard during similar situations and/or during the swarming period in late July and August.

Social calls consisting of shallow sweeps at about 20 kHz have been recorded near maternity roosts, and these may perhaps represent mother–young communication (Figure 8.28.5).

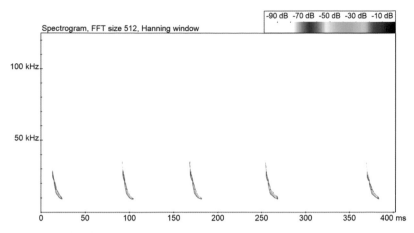

Spectrogram, FFT size 512, Hanning window

-90 dB -70 dB -50 dB -30 dB -10 dB

Figure 8.28.5   Social calls of a northern bat recorded outside a roost (Marc Van De Sijpe).

SPECIES WITH SIMILAR OR OVERLAPPING ECHOLOCATION CALLS

Eptesicus spp., *Nyctalus* spp., *Vespertilio murinus* (see Chapter 7 species notes).

NOTES

Time of emergence relative to sunset depends strongly on latitude and time of year. In southern Sweden in summer, 20–45 minutes after sunset. In the far north may feed in daylight in summer (while the sun remains above the horizon all night). The echolocation calls of *E. nilssonii* are quite variable but still easy to recognise in most situations. The frequency band does not overlap with that of any other species in continental Europe, as long as the same type of calls are compared. Search phase calls are about 2–4 kHz higher in frequency than in *E. serotinus* and *V. murinus* in similar situations (Dietz and Kiefer 2014). The FmaxE of *E. nilssonii* calls is never below 26 kHz and seldom below 27 kHz, even in the long narrowband calls. Also, *E. nilssonii* is normally easy to recognise based on its irregular pulse rhythm, particularly when compared to the more regular rhythms of patrolling *E. serotinus*, *V. murinus* and *Nyctalus* spp. In addition to the calls, *E. nilssonii* and *E. serotinus* can be distinguished easily on size, whenever they are seen. *E. nilssonii* is much smaller.

8.29 Parti-coloured bat *Vespertilio murinus* (Linnaeus, 1758)

Jens Rydell and Arjan Boonman

© René Janssen

DISTRIBUTION

From central Scandinavia and southern Finland and eastwards through Russia, southwards to the Alps, the Balkans and Greece, westwards to Belgium and eastern France. Some populations are migratory, and the distribution is complex and poorly understood. Vagrants may turn up far away from the normal breeding range, such as in northern Norway, in England and on North Sea oil rigs.

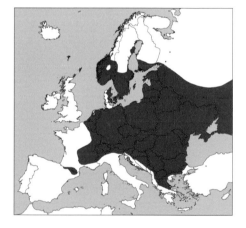

EMERGENCE

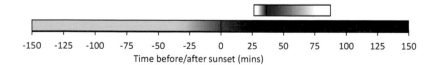

Time before/after sunset (mins)

FLIGHT AND FORAGING BEHAVIOUR

Flight is rather rapid and straight. Foraging takes place in open or semi-open habitats 10–50 m or more above the ground and sometimes much more (e.g. at least 100 m at wind turbines). Catches prey by aerial hawking, often in short vertical dives after small flies, but also in deep dives or spiralling flights after tympanate moths.

HABITAT

Foraging habitats often in openings in forest or woodland and along treelines but also much higher along cliffs and high buildings. Also frequently feeds over rivers and lakes in the north, but probably most often in the open air and without any obvious preference for particular habitats on the ground. Probably feeds at high altitude above urban areas, including major city centres (Rydell and Baagøe 1994, Zhigalin and Moskvitina 2017).

In the north, colonies typically live in attics of houses in summer but individuals may also roost in scree and cliff crevices, particularly in winter. In autumn and winter males often roost in the ventilation system of high town and city buildings, and they also feed in urban habitats.

ECHOLOCATION

Measured parameters	Mean (range) FM-qCF call and qCF call
Inter-pulse interval (ms)	276.0 (112.0–492.0)
Call duration (ms)	14.2 (5.9–21.6)
Frequency of maximum energy (kHz)	25.0 (22.4–27.5)
Start frequency (kHz)	34.8 (22.9–50.7)
End frequency (kHz)	22.8 (20.7–25.8)

Heterodyne

Near the canopy of trees, the search-phase calls usually include a steep FM component starting at ca 50 kHz and ending in a narrowband component with the frequency containing maximum energy (FmaxE) at 24–25 kHz. In the open air, the steep FM component is usually omitted and the entire pulse is narrowband or qCF. FmaxE is usually at 23–25 kHz, a little higher near clutter.

Powerful 'slaps' are heard when the detector is tuned to the FmaxE. The pulse repetition rate near vegetation is usually one per wingbeat (8 per second) or less and one every second, third or fourth wing beat (2–4 per second) in more open space. The pulse repetition rate is usually but not always regular and without the 'chip-chop' typical of noctules, for instance, flying in open places.

Time expansion/full spectrum

Typical search-phase calls start at around 40 kHz and sweep down to around 24 kHz, ending with a qCF tail (Figures 8.29.1 and 8.29.2). In open habitats the pulses are more narrowband, usually starting below 30 kHz, ending at 21–23 kHz, and with FmaxE at 22–25 kHz. In more open habitats, such as when flying above a high cliff, the pulses are sometimes very narrowband, sweeping from c.23 kHz to 21 kHz with FmaxE at 22 kHz (Figure 8.28.3) (Ahlén 1981, Barataud 2015). When flying near clutter-producing objects, the search-phase calls are usually 6–12 ms long, with a steep sweep from about 50 kHz down to about 23 kHz and ending with a narrowband component. The FmaxE is usually at 24–25 kHz in such cases, sometimes a little higher (Figure 8.29.4). In extreme clutter, such as during hand release or flying inside a room, the pulses are broadband and of very short duration (Figure 8.29.5). The start frequency is never higher than 58 kHz, whereas the serotine can start at 68 kHz.

Pulses are usually 6–12 ms long when the bat searches for prey near clutter-producing structures, using FM-CF calls. The qCF calls used in open situations are typically 15–22 ms long, perhaps longer. The pulse repetition rate is usually one every other wing beat (pulse interval 110–130 ms) near clutter-producing objects and about one every two, three or even four wingbeats (pulse interval 250 ms or more) in open space. Strong second harmonics are sometimes heard as well.

The frequency band used by this species has a broad overlap with other species, most notably with the serotine and Leisler's bat, which means that confusions can occur easily (see *Notes*).

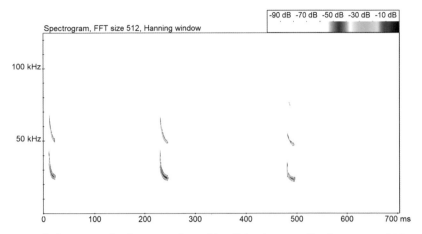

Figure 8.29.1 Echolocation calls of a parti-coloured bat flying in a woodland gap around 15 m above the ground (Jens Rydell).

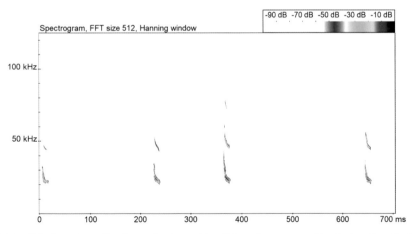

Figure 8.29.2 Echolocation calls of a parti-coloured bat flying above a woodland edge (Jens Rydell).

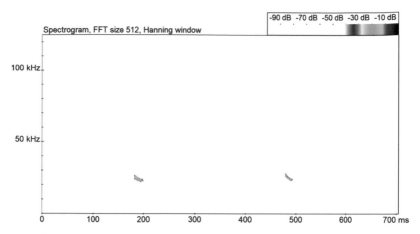

Figure 8.29.3   Echolocation calls of a parti-coloured bat flying in the open in front of a high cliff (Bengt Edqvist).

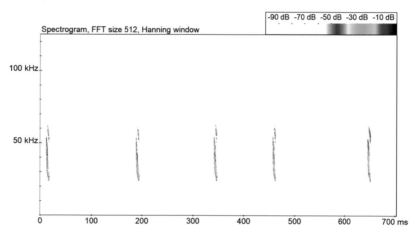

Figure 8.29.4   Echolocation calls of a parti-coloured bat flying in clutter (Bengt Edqvist).

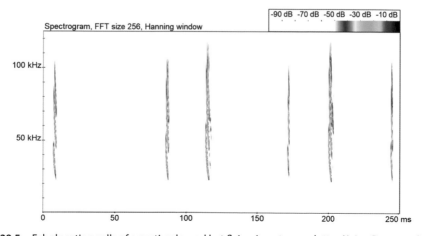

Figure 8.29.5   Echolocation calls of a parti-coloured bat flying in extreme clutter (Arjan Boonman).

SOCIAL CALLS

Heterodyne

This species is known for its highly diagnostic and audible (to young people) mating calls emitted by the males on mild autumn and winter evenings. These calls are emitted very regularly about 4–6 times per second and sound like a 'tock' when the detector is tuned to 12–14 kHz. More complex calls or call sequences, which may be agonistic, are sometimes heard when calling males encounter other individuals (Middleton *et al.* 2014).

Other communication calls or 'social' calls of this species have not been well studied, and few such calls have been described.

Time expansion/full spectrum

Distress calls emitted by parti-coloured bats under physical duress, such as when being held in the hand, are broadband and harsh (Figure 8.29.6).

The songs emitted by males in autumn are rather complex and consist of a steep FM sweep with a qCF part in the middle, occasionally the FM components being omitted (Figures 8.29.7 and 8.29.8). The entire sweep extends from 30 to *c.*10 kHz and the narrowband part, and hence the FmaxE, is usually at 12–14 kHz. The call often includes a second harmonic. In addition, a series of weaker sweeps that sound like a 'buzz' is heard when the bat is close. The song is very characteristic of the species and highly diagnostic (Zagmajster 2003).

Other social calls are also occasionally produced, possibly during interactions during song flight (e.g. Figure 8.29.9).

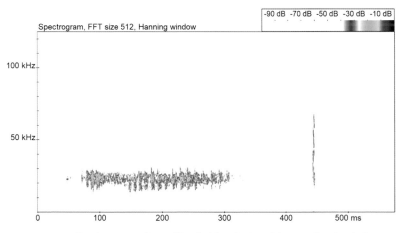

Figure 8.29.6   Distress calls of a parti-coloured bat held in the hand (Anton Vlaschenko).

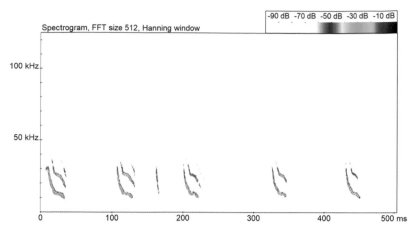

Figure 8.29.7   Social calls of a male parti-coloured bat (Bengt Edqvist).

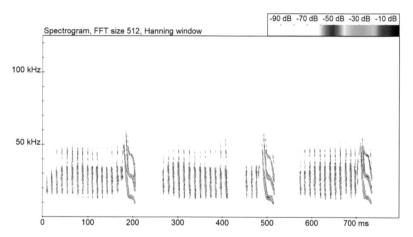

Figure 8.29.8   Typical social calls of a male parti-coloured bat (Bengt Edqvist).

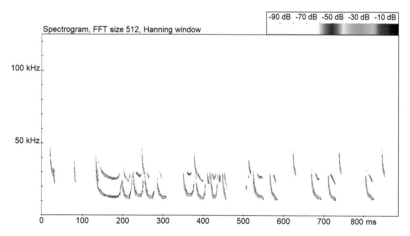

Figure 8.29.9   Uncommon social calls of a parti-coloured bat (Bengt Edqvist).

SPECIES WITH SIMILAR OR OVERLAPPING ECHOLOCATION CALLS

Eptesicus spp., *Nyctalus* spp. (see Chapter 7 species notes).

NOTES

Time of emergence relative to sunset depends on latitude and time of year. In southern Sweden in summer, it is about 35 minutes after sunset. In late autumn and winter, singing males may feed in daylight or early dusk for a while before they start the song-flight display.

The frequency band used by this species overlaps broadly with other species, notably *Eptesicus serotinus* and *Nyctalus leisleri* (Barataud 2015), which means that confusion can occur easily. However, *Vespertilio murinus* can be recognised if the echolocation calls are complemented with some other information. For example, its light belly immediately distinguishes it from the *Nyctalus* spp., which all are brown beneath, whenever this trait can be seen, and it is also easily distinguished from *E. serotinus* by its much smaller size. However, *V. murinus* is similar to *E. nilssonii* in size and general appearance, but the two use different frequency bands when flying in open situations.

Safe recognition of *V. murinus* in flight is often tricky and may require several criteria in combination. Therefore, to minimise the risk of misidentification in this account, the recordings were made when the bats could be seen clearly and hence identified independently by using a combination of size, wing shape and colour. In some cases, a male was recorded as it stopped singing intermittently to feed for a while before resuming the song flight. Hence in both cases, the recorded bats were unambiguously assigned to species by independent criteria (Rydell *et al.* 2017).

8.30 Common pipistrelle
Pipistrellus pipistrellus (Schreber, 1774)

Jon Russ

© René Janssen

DISTRIBUTION

Widespread throughout much of Europe. Up to the north of Ireland and Scotland, the southern tip of Sweden and Finland, and as far south as the Mediterranean.

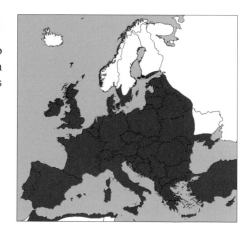

EMERGENCE

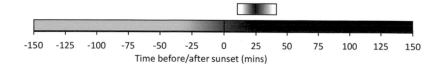

-150   -125   -100   -75   -50   -25   0   25   50   75   100   125   150
Time before/after sunset (mins)

FLIGHT AND FORAGING BEHAVIOUR

Flight is fast and agile with frequent rapid changes of direction. Hunts 3–10 m above the ground. Catches prey by aerial hawking. Forages along linear structures on fixed flight paths.

HABITAT

The common pipistrelle is found in a wide variety of habitats and can be seen foraging within gardens, farmland, parkland, urban areas, deciduous woodland rides and edges, rivers, streams and lakes. Frequently forages along edges such as treelines, hedgerows and water edge. Occasionally it can be found in coniferous woodland, and over beaches. In relatively mountainous areas it may occur up to the point at which the hedgerows and other linear features end. Tends to avoid foraging in very open areas.

Nursery roosts in a wide variety of structures containing crevices such as under tiles, in cavity walls, behind fascias, in tree splits. Male roosts and hibernation roosts also found in a wide variety of crevices.

ECHOLOCATION

Measured parameters	Mean (range) FM-qCF call and qCF call
Inter-pulse interval (ms)	102.5 (59.9–211.0)
Call duration (ms)	5.9 (3.2–8.6)
Frequency of maximum energy (kHz)	46.6 (41.6–50.6)
Start frequency (kHz)	68.8 (50.8–95.2)
End frequency (kHz)	45.9 (41.2–49.9)

Heterodyne
Produces a frequency-modulated sweep which terminates in a constant-frequency tail. When the bat detector is tuned to around 46 kHz a relatively loud irregular slapping noise is heard. As the detector is gradually tuned up the range of frequencies from this frequency, the 'slaps' develop into 'clicks'. The pulse repetition rate is quite fast and erratic, like a badly tuned car. In open or uncluttered situations, the frequency containing maximum energy (FmaxE) may drop to around 43 kHz, whereas in closed or cluttered situations such as an enclosed woodland path the FmaxE may rise to about 48 or 49 kHz.

50 kHz pipistrelles – occasionally the FmaxE is about 50 kHz. This may be due to geographical variation, age, environment or sex. Although soprano pipistrelles have been recorded echolocating at 50 kHz (identified by their social calls), common pipistrelles have also been recorded at 50 kHz. Therefore it is not possible to assign these calls to either species.

Nathusius's pipistrelle near clutter can be confused with common pipistrelle flying in the open, as repetition rate and FmaxE can be similar (around 42 kHz).

In a cluttered environment, common pipistrelle calls tend to sound like dry 'clicks' which can be confused with *Myotis* calls, so it is important to tune down and check for end frequency, which may provide further clues.

Time expansion/full spectrum
Produces FM-qCF calls that sweep down from about 70 kHz to about 43 kHz with FmaxE usually at around 46 kHz (Figure 8.30.1). Mean call duration is about 6 ms and inter-pulse interval approximately 100 ms. In open habitats, such as when foraging more than 6 m away from the nearest structure, calls become longer, FmaxE may drop to about 43 kHz or occasionally lower, and inter-pulse interval increases (Figure 8.30.2). In closed habitats FmaxE may rise to 48 or 49 kHz, the call duration will decrease, and the inter-pulse interval will decrease to 60 ms (Figure 8.30.3). When the bat is foraging in extreme clutter, such as within a building or when drinking (Figure 8.30.4), the echolocation calls are sometimes

confused by inexperienced surveyors with those of the *Myotis* species. Feeding buzzes are similar to those produced by other vespertilionid bats (Figure 8.30.5).

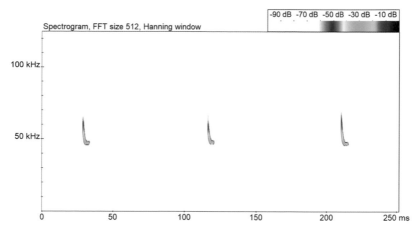

Figure 8.30.1   Echolocation calls of common pipistrelle foraging along a treeline (Jon Russ).

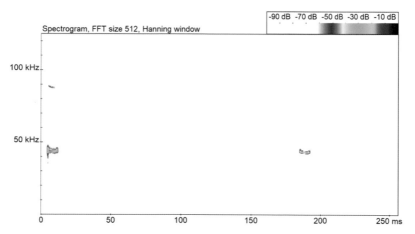

Figure 8.30.2   Echolocation calls of a common pipistrelle flying high above a pond (Jon Russ).

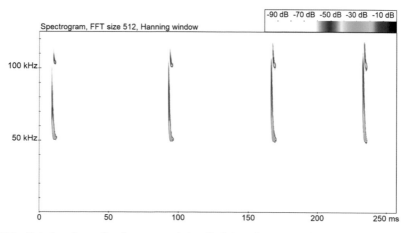

Figure 8.30.3   Echolocation calls of common pipistrelle flying along an enclosed woodland path (Jon Russ).

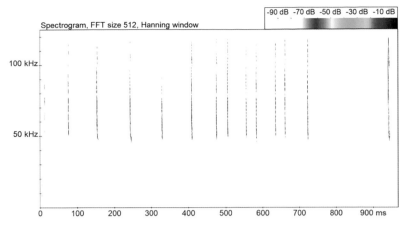

Figure 8.30.4   Echolocation calls produced by a common pipistrelle while drinking from a small pond (Jon Russ).

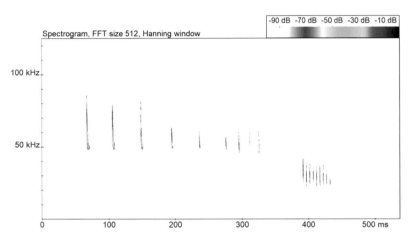

Figure 8.30.5   Feeding buzz produced by a common pipistrelle (Jon Russ).

SOCIAL CALLS

Heterodyne
Social calls are heard throughout the active period, but more often during the autumnal mating period. The majority of common pipistrelle social calls are loudest at around 22 kHz, although they can be heard from about 45 kHz down to 15 kHz. With a heterodyne detector, the call is like an extremely rapid grating sound or 'chonking'. This can be mistaken for noctules by inexperienced surveyors. These calls are almost always produced on the wing, rarely from a perch. In the mating season, males produce social calls at a very rapid rate for long periods of time while flying within their territories.

Time expansion/full spectrum
The most commonly produced social call consists of a series of 3–5 downward FM sweeps starting at about 42 kHz and ending at around 20 kHz (Figure 8.30.6). These calls are very similar to the social calls produced by soprano pipistrelles. However, generally, the calls of soprano pipistrelles consist of three components (FM sweeps) on average, whereas those of the common pipistrelle consist on average of four components (Barlow and Jones 1997a).

In addition, the mean FmaxE is around 18 kHz, whereas for the soprano pipistrelle it is around 21 kHz. Note that the individual components of the common pipistrelle's social call are quite similar in structure, whereas for the soprano pipistrelle the last component is structurally very variable. These social calls are almost always produced between echolocation pulses. However, echolocation calls are much quieter than social calls and will only appear on the sonogram if the bat is close to the microphone. The calls can be heard throughout the active period, and during periods of low insect density they may be used to defend patches of prey (Barlow and Jones 1997b). However, the calls (which appear to be similar in structure) are also produced by males defending territories (Lundberg and Gerell 1986), with frequency of calling being very high during the autumnal mating period (Russ *et al.* 2003). Males always produce these calls on the wing as they patrol their territories, but regularly returning to their mating roosts where they 'false-land' before flying off again (Figure 8.30.7).

Pre-emergence calls are regularly heard at maternity roosts and can be seen as a series of low-frequency calls (e.g. Figure 8.30.8). They are often very distorted except when the bat is calling at the roost entrance.

Calls, thought to be produced by juveniles, can often be heard outside maternity roost entrances while the bats are in flight (e.g. Figures 8.30.9–8.30.11). Similar calls have been recorded from juveniles calling to their mothers (e.g. Figure 8.30.12)

High-frequency calls, integrated into the calls sequence, up to around 62 kHz (Pfalzer and Kusch 2003), can occasionally be heard which are emitted by a resident at a foraging site while chasing an intruder or during frontal encounters (Götze *et al.* 2020) (Figure 8.30.13).

Distress calls emitted by common pipistrelles held in the hand consist of a series of rapid downward FM pulses each of about 2.4 ms duration beginning at an average frequency of 25.7 kHz and falling to an average frequency of 17.8 kHz (Russ *et al.* 1998) (Figure 8.30.14). The median number of components is five but it has been recorded to be as high as seven. The FmaxE is around 22 kHz. Distress calls probably function in attracting conspecifics to perform mobbing behaviour as an anti-predator response. Distress calls are structurally convergent between congeners (and other species) and common pipistrelles, soprano pipistrelles and Nathusius's pipistrelles respond to each other's distress calls, indicating that attracting heterospecifics increases the chance of repelling predators by mobbing (Russ *et al.* 2004).

In captivity, distress or agonistic calls have also been recorded which differ in structure and may have an alternative function (Figure 8.30.15).

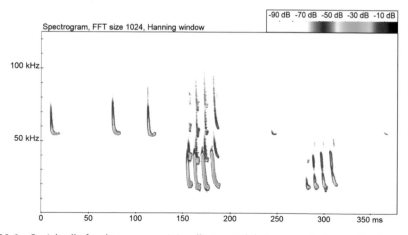

Figure 8.30.6   Social call of male common pipistrelle emitted during song flight (Jon Russ).

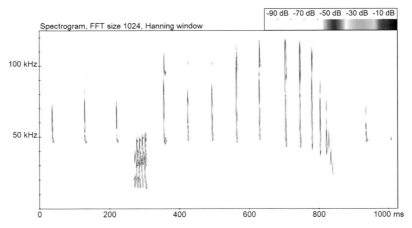

Figure 8.30.7   Social calls and echolocation calls of a male common pipistrelle during a false-landing event (Jon Russ).

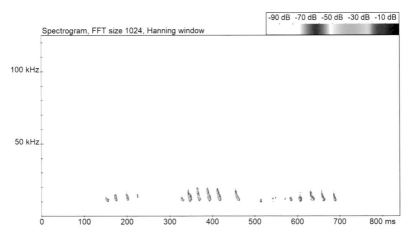

Figure 8.30.8   Common pipistrelle pre-emergence calls at a maternity roost (Jon Russ).

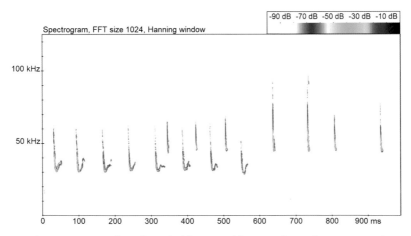

Figure 8.30.9   Common pipistrelle calls probably emitted by juveniles at the entrance of a maternity roost (Jon Russ).

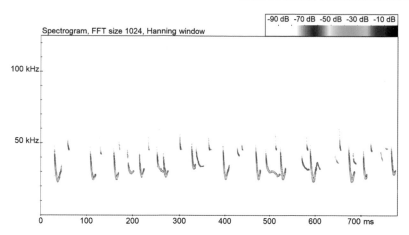

Figure 8.30.10   Further example of common pipistrelle calls probably emitted by juveniles at the entrance to a maternity roost (Jon Russ).

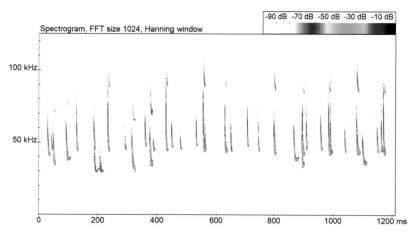

Figure 8.30.11   Calls emitted by a common pipistrelle in flight during high activity outside a maternity roost containing juveniles (Jon Russ).

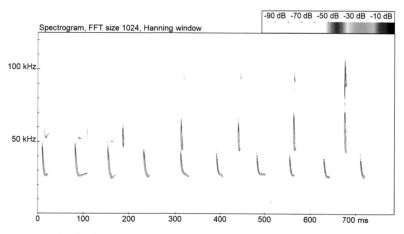

Figure 8.30.12   Social calls of a juvenile common pipistrelle being circled by the mother immediately prior to being picked up by her (Tricia Scott).

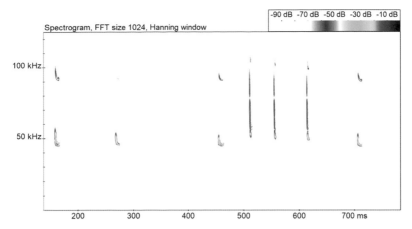

Figure 8.30.13   High-frequency calls produced by a common pipistrelle in flight which possibly have a social function (Jon Russ).

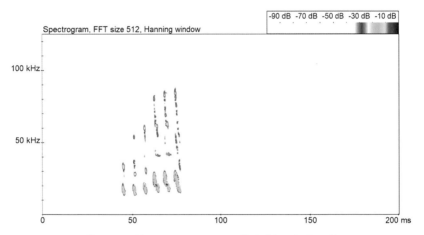

Figure 8.30.14   Distress calls emitted by a common pipistrelle held in the hand (Jon Russ).

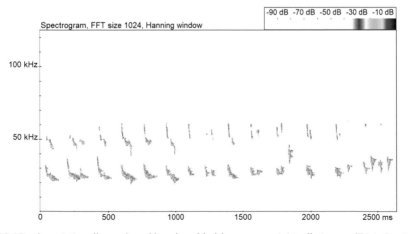

Figure 8.30.15   Agonistic calls produced by a hand-held common pipistrelle in care (Tricia Scott).

SPECIES WITH SIMILAR OR OVERLAPPING ECHOLOCATION CALLS

Pipistrellus pygmaeus, Pipistrellus nathusii, Pipistrellus hanaki, Miniopterus schreibersii (see Chapter 7 species notes).

Note that in the absence of conspecifics, bats can expand the range of their frequencies. For example, in parts of Greece such as Crete where soprano pipistrelles do not occur, common pipistrelles have a greater frequency range.

8.31 Soprano pipistrelle *Pipistrellus pygmaeus* (Leach, 1825)

Jon Russ

© Jon Russ

DISTRIBUTION

Occurs from European Mediterranean and western Asia Minor throughout southern and central Europe to Scotland and southern Norway and Sweden.

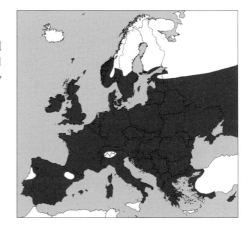

EMERGENCE

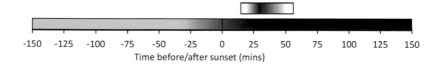

-150   -125   -100   -75   -50   -25   0   25   50   75   100   125   150

Time before/after sunset (mins)

FLIGHT AND FORAGING BEHAVIOUR

Rapid flight and quite agile. Flies 3–10 m from the ground. Very similar flight behaviour to the common pipistrelle. Catches prey by aerial hawking.

HABITAT

Compared to the common pipistrelle, which is found in a wide variety of habitats, the soprano pipistrelle is more of a specialist in its choice of habitats. It is common in areas containing bodies of water such as rivers, canals, lakes, ponds and reservoirs, with associated riparian vegetation. Habitats such as broadleaf and mixed woodland edge and parkland are also used frequently for hunting. Generally, farmland (improved grassland/arable land) is used to a much lesser extent.

Typically, a crevice dweller roosting under roof tiles, in cavity walls, under flat roofs, behind fascias on buildings, in crevices in stone walls and rock faces, and in holes in trees. Maternity roosts under tiles, in cavity walls, behind soffits, in deep crevices in walls, and in cracks, splits and holes in trees. Males roost in similar locations, as do hibernating bats.

ECHOLOCATION

Measured parameters	Mean (range) FM-qCF call and qCF call
Inter-pulse interval (ms)	89.1 (51.0–217.1)
Call duration (ms)	5.5 (2.1–8.2)
Frequency of maximum energy (kHz)	55.1 (49.8–64.1)
Start frequency (kHz)	79.6 (63.8–108.6)
End frequency (kHz)	54.7 (49.6–63.9)

Heterodyne

Produces very loud 'slaps' which are similar in sound to the common pipistrelle's calls but are loudest at around 55 kHz. As the detector is gradually tuned up the range of frequencies above 55 kHz the slaps develop into 'clicks'. The pulse repetition rate is very fast and erratic, like a badly tuned car. In open areas, such as when flying high above a pond, the frequency containing maximum energy (FmaxE) may drop to about 52 kHz. In cluttered situations, such as when flying along a woodland path, the FmaxE will rise, and may even reach 61 kHz.

50 kHz pipistrelles – occasionally the FmaxE is about 50 kHz. This may be due to geographical variation, age, environment or sex. Although soprano pipistrelles have been recorded echolocating at 50 kHz (identified by their social calls) common pipistrelles have also been recorded at 50 kHz. Therefore it is not possible to assign these calls to either species.

In a cluttered environment, soprano pipistrelle calls tend to sound like dry 'clicks' which can be confused with *Myotis* calls, so it is important to tune down and check for end frequency, which may provide further clues.

Time expansion/full spectrum

Produces FM-qCF calls that sweep down from about 80 kHz to about 53 kHz with FmaxE usually at around 55 kHz (Figure 8.31.1). Mean call duration is about 6 ms and inter-pulse interval approximately 90 ms. In uncluttered situations, for example when flying high above the ground, call duration becomes longer, calls become more narrowband, inter-pulse interval increases, and FmaxE may drop to about 52 kHz (Figure 8.31.2). Conversely, in cluttered situations such as in woodland or when chasing, call duration will shorten, calls become very broadband, the inter-pulse interval will decrease, and the FmaxE will increase, sometimes going as high as 61 kHz (Figure 8.31.3). Feeding buzzes are an extreme example of bats in clutter as they home in on an insect (Figure 8.31.4). Very occasionally, when flying in the open, soprano pipistrelles may produce upward-sloping qCF signals (Figure 8.31.5).

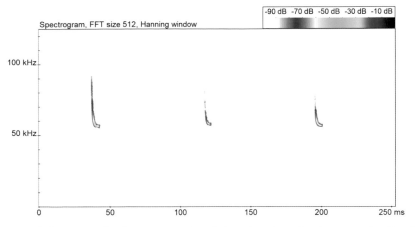

Figure 8.31.1   Echolocation calls of a soprano pipistrelle foraging along riparian vegetation (Jon Russ).

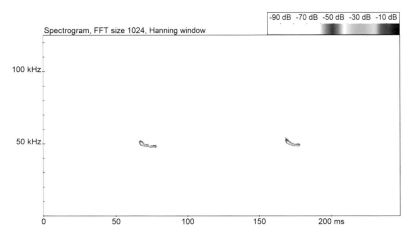

Figure 8.31.2   Echolocation calls of a soprano pipistrelle flying high in parkland (Jon Russ).

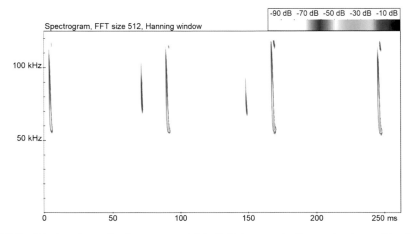

Figure 8.31.3   Echolocation calls of a soprano pipistrelle flying near to the roost entrance (Jon Russ).

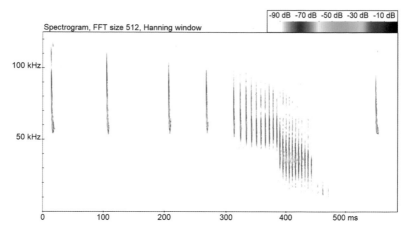

Figure 8.31.4   Feeding buzz of a soprano pipistrelle as it catches an insect (Jon Russ).

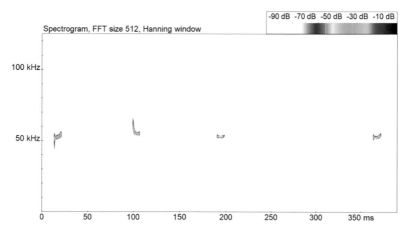

Figure 8.31.5   Unusual upward-sloping echolocation calls produced by a soprano pipistrelle flying in the open (Marc Van De Sijpe).

SOCIAL CALLS

Heterodyne

Social calls from soprano pipistrelles can be heard all year round except during hibernation. The most common type of social call heard with a heterodyne detector is a rapid 'chonking' sound which is loudest at about 22 kHz, although it can be heard from 15 to 45 kHz. This can be mistaken for noctules by inexperienced surveyors. Occasionally these calls are repeated in very quick succession, especially during the autumnal mating period when social activity reaches a peak. As many as 80 calls in a one-minute period have been counted. These calls are often associated with aerial chasing. Social calls such as these may be emitted to attract a mate or defend a territory. These social calls are always emitted during flight and never from a perch.

As with most bat species, distress calls are emitted when the bat is under physical duress, such as being held in the hand, and these are loud scolding calls that are loudest at about 22 kHz.

Time expansion/full spectrum

The most commonly heard social call consists of a series of 2–4 downward FM sweeps (Figure 8.31.6). In comparison with the common pipistrelle, which has a similarly structured call, the soprano pipistrelle has on average three components to its social call whereas the common pipistrelle has about four, and the mean FmaxE is around 21 kHz whereas for the common pipistrelle it is around 18 kHz (Barlow and Jones 1997a). Kuhl's pipistrelles also have a similarly structured call which also consists of three components on average, but the FmaxE is around 17 kHz (Russo and Jones 1999). Note that the individual components of the common pipistrelle's social call are quite similar in structure, whereas for the soprano pipistrelle the last component is structurally very variable.

The occurrence of these social calls, produced by the males, increases dramatically during the autumnal mating period (August to early October), indicating that they are involved in mate attraction, female choice and/or territorial defence. During May males begin to set up individual territories around their day roosts, and the male monopolises its roost throughout the summer and the subsequent mating period, when the females join the males (Gerell and Lundberg 1985*). Males intruding into a territory are chased away by the territory owner while emitting these social calls, thought to be agonistic (Lundberg and Gerell 1986*). 'Song-flight display' is performed by males during the mating period and consists of a fixed circular or elliptical path with a length of 100–200 m which touches the roost site, during which time the males constantly emit social calls (Lundberg and Gerell 1986*). At the beginning of the mating period, song flights of both common and soprano pipistrelle were recorded in the second third of the night, whereas from the end of September, the peak of display activity moved to the first third of the night (Bartoničková et al. 2016). Males that spend the greatest proportion of their time in song-flight display are visited by the greatest number of females (Lundberg and Gerell 1986*). The structure of song-flight calls varies between individuals. A series of rapid frequency-modulated sweeps can occasionally be heard when two bats are interacting during chasing behaviour (Figure 8.31.7).

Similarly structured social calls irregularly produced by soprano pipistrelles may function in agonistic interactions to repel other bats from a food patch when insects are scarce (Barlow and Jones 1997b).

Some social calls of unknown function are emitted in flight, and these variable calls consist of a long-duration broadband FM sweep often with a long qCF tail (Figures 8.31.8 and 8.31.9). These are characteristic for interactions of females and pups and may also function as contact calls (Pfalzer and Kusch 2003).

Calls emitted by juveniles calling from the roost are generally low-frequency with FmaxE being about half that of the echolocation calls of the adult, i.e. about 27 kHz (Figure 8.31.10).

Distress calls emitted by soprano pipistrelles held in the hand consist of a series of rapid downward FM pulses each of about 3 ms duration beginning at an average frequency of 39.8 kHz and falling to an average frequency of 18.2 kHz (Russ et al. 1998**) (Figure 8.31.11). The median number of components is six but has been recorded to be as high as 22. The FmaxE is around 26.8 kHz. Playbacks of soprano pipistrelle distress calls have been shown to result in an 80-fold increase in bat activity over the loudspeaker (Russ et al. 1998**). Distress calls probably function in attracting conspecifics to perform mobbing behaviour as an anti-predator response. Distress calls are structurally convergent between congeners (and other species), and common pipistrelles, soprano pipistrelles and Nathusius's pipistrelles respond to each other's distress calls, indicating that attracting heterospecifics increases the chance of repelling predators by mobbing (Russ et al. 2004).

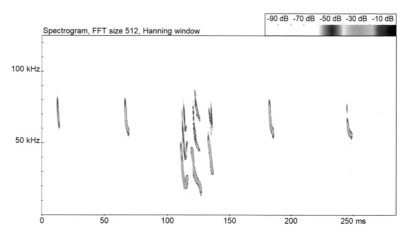

Figure 8.31.6 Social calls emitted by a male soprano pipistrelle during song flight. Note the echolocation calls before and after the social call (Jon Russ).

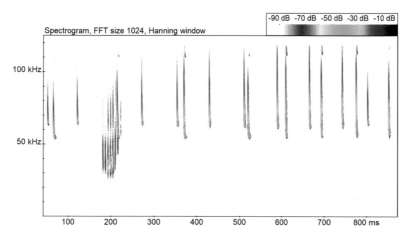

Figure 8.31.7 Soprano pipistrelle chasing and emitting a rapid frequency-modulated buzz (Jon Russ).

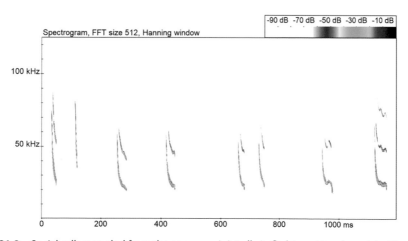

Figure 8.31.8 Social calls recorded from the soprano pipistrelle in flight next to a large lake (Jon Russ).

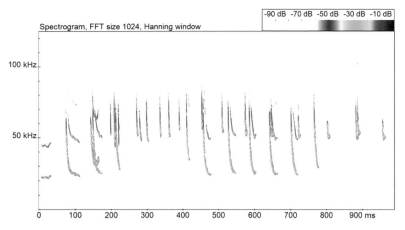

Figure 8.31.9   Social calls recorded from the soprano pipistrelle in flight on the edge of a woodland (Jon Russ).

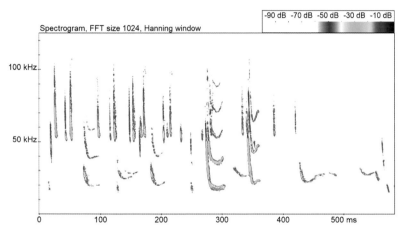

Figure 8.31.10   Social calls probably produced by a juvenile soprano pipistrelle calling from the roost entrance (Jon Russ).

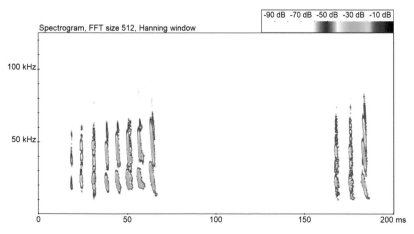

Figure 8.31.11   Distress calls recorded from a soprano pipistrelle held in the hand (Jon Russ).

SPECIES WITH SIMILAR OR OVERLAPPING ECHOLOCATION CALLS

Pipistrellus pipistrellus, Pipistrellus hanaki, Miniopterus schreibersii (see Chapter 7 species notes).

NOTES

* Gerell and Lundberg's work refers to the common pipistrelle *Pipistrellus pipistrellus*. However, in recent years it has been shown that their observations were in fact of soprano pipistrelle *P. pygmaeus*.
** Although Russ *et al.*'s work refers to the common pipistrelle *Pipistrellus pipistrellus*, it is very likely that the species was the soprano pipistrelle *P. pygmaeus*.

8.32 Hanak's pipistrelle

Pipistrellus hanaki (Hulva and Benda, 2004)

Panagiotis Georgiakakis

© Panagiotis Georgiakakis

DISTRIBUTION

The European distribution of the species is limited to the island of Crete (Greece; Benda *et al.* 2009). Outside Europe, it occurs only in the Cyrenaica peninsula of northeastern Libya (Hulva *et al.* 2007).

EMERGENCE

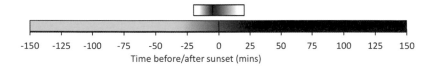

Time before/after sunset (mins)

FLIGHT AND FORAGING BEHAVIOUR

Agile and erratic flight. Flies both below and above the tree canopy, also in small forest openings and around trees close to fresh water. Prey is caught by aerial hawking. Tracked bats have been observed foraging in small areas for up to 1.5 hours.

HABITAT

Hanak's pipistrelle has foraging habitats similar to those of the soprano pipistrelle. It prefers forested areas and old tree cultivations (e.g. olive groves), as well as above water bodies (streams, rivers, ponds, lakes, dams) with riparian vegetation.

It uses a variety of roost types, including rock fissures, tree cavities, roof tiles, even electricity posts (Georgiakakis *et al.* 2018).

ECHOLOCATION

Measured parameters	Mean (range) FM-qCF call and qCF call
Inter-pulse interval (ms)	136.3 (26.2–378.0)
Call duration (ms)	5.7 (2.5–11.5)
Frequency of maximum energy (kHz)	48.4 (44.8–52.1)
Start frequency (kHz)	70.7 (46.8–109.5)
End frequency (kHz)	48.2 (44.3–51.0)

Heterodyne

Echolocation sounds are similar to those of the other pipistrelle species when using a heterodyne detector. For Hanak's pipistrelle, loud slaps are heard when the detector is tuned to around 49 kHz which turn into dry clicks as the frequency dial is tuned upwards. In clutter, the slaps may rise to as much as 52 kHz and drop to around 45 kHz in the open.

Time expansion/full spectrum

Although Hanak's pipistrelle is genetically closer to the soprano pipistrelle (Hulva *et al.* 2007), its echolocation calls resemble those of common pipistrelle. It produces FM-qCF calls that sweep down from about 70 kHz to 48 kHz, with the frequency containing maximum energy (FmaxE) usually at around 49 kHz (Figure 8.32.1). Mean call duration is about 6 ms and the inter-pulse interval is quite variable with a mean value of around 136 ms. In open areas calls become more 'flat' and longer, with longer inter-pulse intervals, and FmaxE may be less than 45 kHz (Figure 8.32.2). On rare occasions, when the bat is flying very close to trees, calls become more 'steep' and shorter, with a higher repetition rate. FmaxE may go as high as 51 kHz (Figure 8.32.3). When approaching prey, calls initially become 'steeper' and shorter with higher frequencies, as in cluttered conditions, but at the final stage of the feeding buzz end frequency falls below 40 kHz (Figure 8.32.4). Since common pipistrelles are not present on Crete (or in Libya), all FM-qCF calls with FmaxE between 44.8 and 50.4 kHz belong to this species. A small percentage of calls have an FmaxE above 50 kHz and can be confused with those of Schreiber's bent-winged bat, and some statistical methods may be needed for accurate identification.

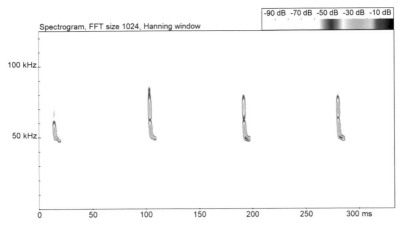

Figure 8.32.1   Echolocation calls of a Hanak's pipistrelle foraging in a forest opening (Panagiotis Georgiakakis).

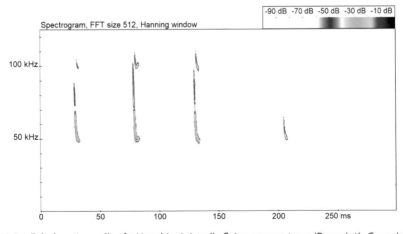

Figure 8.32.2   Echolocation calls of a Hanak's pipistrelle flying over a provincial road (Panagiotis Georgiakakis).

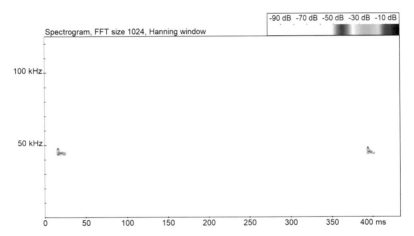

Figure 8.32.3   Echolocation calls of a Hanak's pipistrelle flying among trees (Panagiotis Georgiakakis).

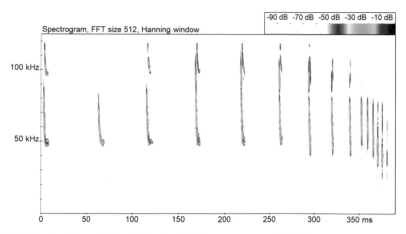

Figure 8.32.4   Feeding buzz of Hanak's pipistrelle (Panagiotis Georgiakakis).

SOCIAL CALLS

Heterodyne
Agonistic and mating calls sound like a dry 'chonk' when the detector is tuned between 70 kHz and 22 kHz, and they are loudest at around 22 kHz.

Time expansion/full spectrum
Unlike most other pipistrelle species, the agonistic and mating calls of Hanak's pipistrelle have clear structural differences (Georgiakakis and Russo 2012) and great variability has been observed within each category. During spring and summer, when flying bats spend most of their time foraging, single-component, multi-harmonic FM-qCF agonistic calls are emitted. These calls occasionally resemble the echolocation calls of serotines and Leisler's bats (Fmax up to 77 kHz, Fmin 22.0 ± 3.34 kHz, mean FmaxE 28 kHz), having similar spectral characteristics, but they are considerably longer, since they may last up to 40 ms (20 ms on average). Agonistic calls with different spectral and temporal characteristics can be found even in a single sequence (Figure 8.32.5).

Beginning from August, mating calls are increasingly emitted, until November, when the activity of the species falls dramatically. Mating calls have on average lower start and end frequencies, but their most striking feature is their fluctuating frequency (Figure 8.32.6). Usually, the mating calls of Hanak's pipistrelle have a single component, but in some instances they have two or three 'parts', giving the impression of a 'fractured' call (Figure 8.32.7).

In the hand, Hanak's pipistrelle emits variable distress calls with 1–8 components. Often, the first 3–4 components are short and steep, and they are followed by one or more components with fluctuating frequency (Figure 8.32.8). They begin at a frequency of around 39 kHz, fall to a frequency of 14 kHz or even less, and they last up to 60 ms.

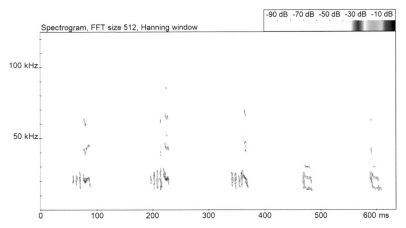

Figure 8.32.5   Agonistic calls of Hanak's pipistrelle. Note the echolocation calls emitted by several specimens (Panagiotis Georgiakakis).

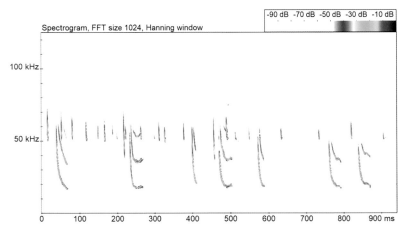

Figure 8.32.6   Uni-component mating calls of Hanak's pipistrelle. Echolocation calls of Hanak's pipistrelle (two individuals) are also displayed (Panagiotis Georgiakakis).

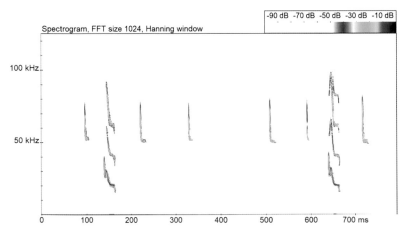

Figure 8.32.7   Multi-component mating call of Hanak's pipistrelle. Note also the echolocation calls of the species (Panagiotis Georgiakakis).

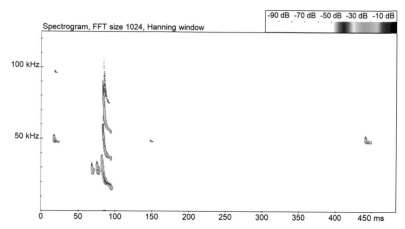

Figure 8.32.8 Distress calls of Hanak's pipistrelle held in the hand (just after being taken out of a mist net) (Panagiotis Georgiakakis).

SPECIES WITH SIMILAR OR OVERLAPPING ECHOLOCATION CALLS

Pipistrellus pygmaeus, Pipistrellus pipistrellus, Pipistrellus maderensis, Miniopterus schreibersii (see Chapter 7 species notes).

8.33 Nathusius's pipistrelle

Pipistrellus nathusii (Keyserling and Blasius, 1839)

Jon Russ

© René Janssen

DISTRIBUTION

Widespread throughout Europe from northeast Scotland and southern Scandinavia to the Mediterranean, and from northwest Spain to the Ural Mountains. In Britain and Ireland, records are widespread but relatively scarce. In autumn and winter, it migrates in a southwesterly direction (Strelkov 1969) and returns during the late spring (Aellen 1983). Males establish mating territories along the migration routes, and some remain in areas along the route, where they mate throughout the year (Kapteyn and Lina 1994). Many records from along the Baltic coast and offshore oil platforms in the North Sea.

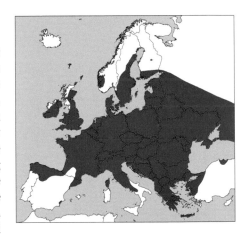

EMERGENCE

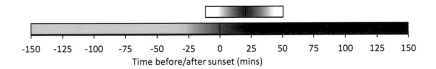

-150 -125 -100 -75 -50 -25 0 25 50 75 100 125 150
Time before/after sunset (mins)

FLIGHT AND FORAGING BEHAVIOUR

Rapid flight – slightly faster than the smaller common and soprano pipistrelles and not quite as manoeuvrable. Relatively deep wing beats when flying in a straight line. Usually observed 4–15 m above the ground. Captures prey by aerial hawking.

HABITAT

Lowland woodland rides and edge (deciduous and occasionally coniferous), damp lowland forests, meadows, and also frequently observed over or near water such as canals, rivers, lakes and waterlogged areas. In Britain and Ireland, the majority of records are within a few kilometres of large water bodies, and this is typical throughout the range, particularly during the migration period.

Maternity roosts in buildings in Northern Ireland and England but elsewhere in Europe often in crevices in bark, tree holes and bat boxes. Single males roost in most available crevices: tree holes, cracks in walls, under soffits and tiles, in bridges etc.

ECHOLOCATION

Measured parameters	Mean (range) FM-qCF call and qCF call
Inter-pulse interval (ms)	129.0 (88.6–237.0)
Call duration (ms)	5.88 (3.0–7.9)
Frequency of maximum energy (kHz)	39.4 (35.5–46.1)
Start frequency (kHz)	62.8 (40.6–95.1)
End frequency (kHz)	38.6 (35.0–45.1)

Heterodyne

Similar in sound to common pipistrelle and soprano pipistrelle except that the loudest frequency (slaps) typically occurs at 36–41 kHz. The repetition rate is noticeably slower and more regular than for the common and soprano pipistrelles. As the detector is tuned to higher frequencies the 'slaps' turn into clicks. In cluttered situations, such as when foraging along a woodland path, the frequency containing maximum energy (FmaxE) may rise to 42 kHz. In open situations, such as when flying high above a reservoir, the FmaxE may drop to about 36 kHz and the repetition rate will increase.

Nathusius's pipistrelle can be confused with a common pipistrelle flying in the open, as repetition rate and loudest frequency can be similar (around 42 kHz).

In a cluttered environment, Nathusius's pipistrelle calls tend to sound like dry 'clicks' which can be confused with *Myotis* calls, so it is important to tune down and check for end frequency, which may provide further clues.

Time expansion/full spectrum

Produces a typical FM-qCF call which starts at around 51 kHz and ends at around 36 kHz (Figure 8.33.1). The average duration is about 6 ms and the inter-pulse interval is about 130 ms. The FmaxE is on average about 39 kHz. As with most bats, in cluttered situations the FmaxE will rise slightly (up to 42 kHz) and the call duration and inter-pulse interval will decrease (Figure 8.33.2). Conversely, in uncluttered situations, the FmaxE will decrease (down to 36 kHz) and the inter-pulse interval and duration will increase (Figure 8.33.3). Occasionally produces a low-bandwidth terminal FM sweep (Figure 8.33.4). Nathusius's pipistrelles have also been observed to emit a 'whispering echolocation' when swarming around a tree, and this is likely to be similar to the calls produced in a wind tunnel (Figure 8.33.5).

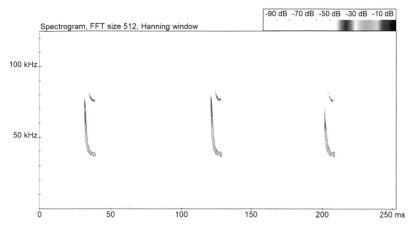

Figure 8.33.1 Echolocation calls of Nathusius's pipistrelle flying along riparian vegetation (Jon Russ).

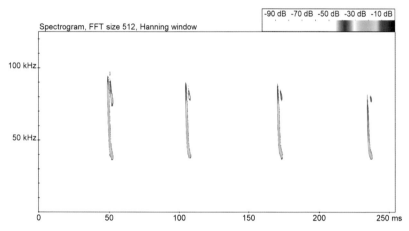

Figure 8.33.2 Echolocation calls of Nathusius's pipistrelle flying in circles outside a roost (Jon Russ).

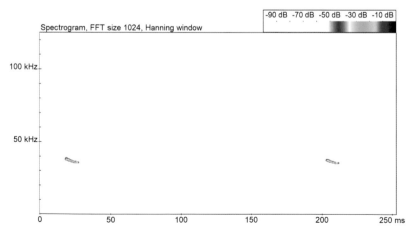

Figure 8.33.3 A Nathusius's pipistrelle flying high over a lake towards a woodland edge (Jon Russ).

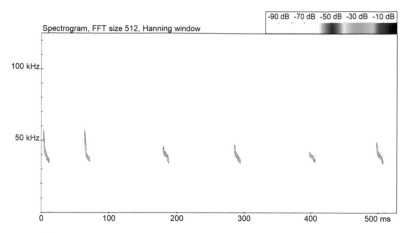

Figure 8.33.4 Nathusius's pipistrelle flying in the open above the edge of a lake (Jon Russ).

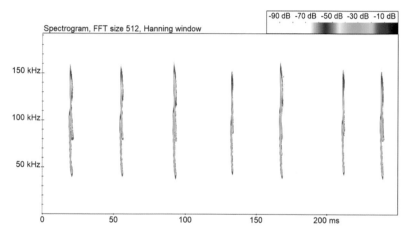

Figure 8.33.5 Nathusius's pipistrelle flying in a wind tunnel (Arjan Boonman).

SOCIAL CALLS

Heterodyne

Social calls sound similar to those of the other pipistrelle species when using a heterodyne detector. The most common type of social call is emitted by males and may function in attracting a mate. Nathusius's pipistrelles are the most vocal of the European pipistrelles. Males call from April to October but more frequently during the autumnal mating periods, when males often call continually throughout the night. In contrast to other pipistrelle species, male Nathusius's pipistrelles tend to emit these calls from a perch, such as a tree or building, particularly during August and September. Social calls are loudest at about 20 kHz. Distress calls, heard while the bat is under physical duress such as being held in the hand or in a net, are often of long duration compared to the rapid and harsh 'chonk' of advertisement and patch-defence calls.

Time expansion/full spectrum

Nathusius's pipistrelles produce a wide variety of social calls. The calls emitted by males can be very complex, usually consisting of four different call types emitted in sequence (Figure 8.33.6). The last call type is much quieter than the first three and often is not detected unless the microphone is quite close. However, it may occasionally be omitted from the sequence, particularly outside the mating season when the calls are more often emitted in flight. Occasionally other calls in the sequence are dropped or call types are repeated several times (e.g. Figure 8.33.7). The frequency of occurrence of mating calls gradually increases through the summer, but there is almost a nine-fold increase in the autumn mating period (Furmankiewicz 2003). During May, mating calls are occasionally emitted in flight (usually without the last call type, which does not appear to be emitted in flight), in June bats gradually begin to emit vocalisations from a perch, and in August nearly 80% of calls are emitted from a perch, sustained for long periods (Jahelková and Horáček 2011). For example, males in Northern Ireland have been observed calling continually for 4.5 hours without a break. Interactions between territorial males can be very aggressive, and often the number of components within the calls can become very high (e.g. Figure 8.33.8).

Low-frequency calls are often produced from the roost during pre-emergence, and these consist of FM sweeps often with variable terminal frequencies (e.g. Figures 8.33.9 and 8.33.10). Occasionally variable calls of juveniles can be heard from a roost, and these are usually of long duration with FmaxE being on average about half that of the echolocation calls of the adults, at about 20 kHz (Figure 8.33.11).

Distress calls emitted by Nathusius's pipistrelles held in the hand consist of a series of rapid downward FM pulses each of about 4 ms duration, beginning at an average frequency of 28 kHz and falling to an average frequency of 20 kHz (Russ *et al.* 1998) (Figure 8.33.12). The median number of components is five but has been recorded to be as high as seven. The FmaxE is around 22 kHz. Distress calls probably function in attracting conspecifics to perform mobbing behaviour as an anti-predator response. Distress calls are structurally convergent between congeners (and other species), and common pipistrelles, soprano pipistrelles and Nathusius's pipistrelles respond to each other's distress calls, indicating that attracting heterospecifics increases the chance of repelling predators by mobbing (Russ *et al.* 2004).

Unusual social calls can also sometimes be recorded during flight and interactions between bats (e.g. Figure 8.33.13, and see Jahelková 2011).

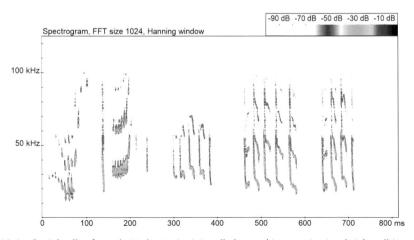

Figure 8.33.6   Social calls of a male Nathusius's pipistrelle located in a crevice in a brick wall (Jon Russ).

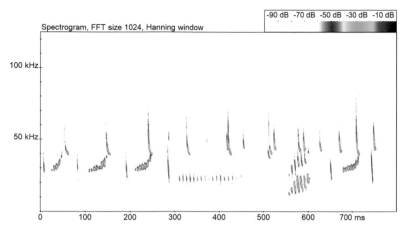

Figure 8.33.7   Social calls of a male Nathusius's pipistrelle located under a tile, showing repeated call types (Jon Russ).

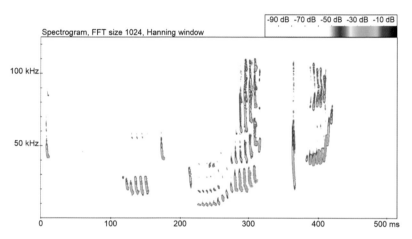

Figure 8.33.8   A social call produced by a male Nathusius's pipistrelle during an interaction with a territorial male from outside the calling area (Jon Russ).

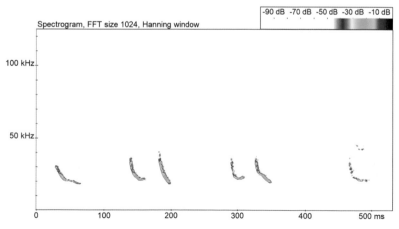

Figure 8.33.9   Nathusius's pipistrelle social calls emitted during pre-emergence from a maternity roost (Jon Russ).

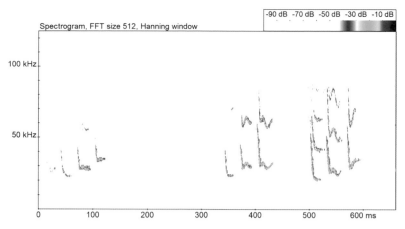

Figure 8.33.10   A further example of Nathusius's pipistrelle social calls emitted during pre-emergence from a maternity roost (Graeme Smart).

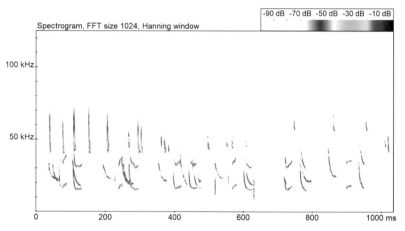

Figure 8.33.11   Social calls of Nathusius's pipistrelle recorded outside a maternity roost entrance probably produced by juvenile bats (Jon Russ).

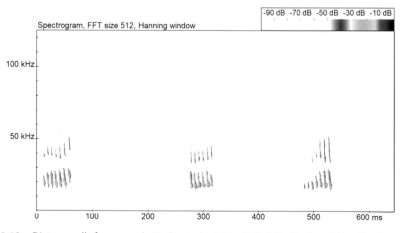

Figure 8.33.12   Distress calls from a male Nathusius's pipistrelle held in the hand (Jon Russ).

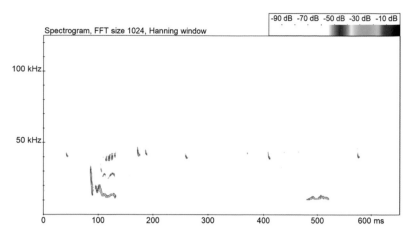

Figure 8.33.13   Unusual calls emitted by Nathusius's pipistrelle during flight (Tina Wiffen).

SPECIES WITH SIMILAR OR OVERLAPPING ECHOLOCATION CALLS

Pipistrellus kuhlii, Hypsugo savii (see Chapter 7 species notes).

8.34 Kuhl's pipistrelle *Pipistrellus kuhlii* (Kuhl, 1817)

Sérgio Teixeira and Danilo Russo

© Leonardo Ancillotto

DISTRIBUTION

Kuhl's pipistrelle occurs across the whole of the Mediterranean, Black Sea and Caspian Sea basins. To the west, it is present from the Canary Islands northwards to France. On the eastern limit of its range, it extends from Russia in the north to Oman in the south, and eastwards to Afghanistan, Pakistan, Kazakhstan and India. Occasionally recorded in the Czech Republic and China. The European range of Kuhl's pipistrelles has been expanding to the north and east since the mid-1990s, and records have recently been reported for the UK. Modelling work shows that range expansion is likely driven by climate change (Ancillotto *et al.* 2016).

EMERGENCE

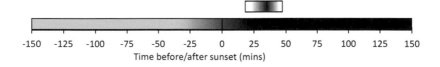

-150 -125 -100 -75 -50 -25 0 25 50 75 100 125 150

Time before/after sunset (mins)

FLIGHT AND FORAGING BEHAVIOUR

With a rapid and agile flight, Kuhl's pipistrelle flies between 2 and 7 m above the ground, usually at 3 or 4 m. In rare cases it is seen diving, almost touching the ground, probably to feed on emerging insects. Generally, Kuhl's pipistrelles follow straight trajectories with sudden turns to track prey. It is commonly seen flying between streetlamps in urban and suburban areas. Forages by aerial hawking, using its wing or tail membrane as a pouch to scoop insects, which are then brought to the mouth.

HABITAT

Considering its wide distribution, this species is very adaptable to a diversity of habitat types. Kuhl's pipistrelles have been observed in Saharan oases, Mediterranean shrublands, olive groves, conifer and broadleaf forests, and over rivers and lakes. It is one of the species best adapted to anthropogenic habitats ranging from urban areas to farmland, orchards and tree plantations, and prefers artificially illuminated sites to feed on insects attracted by lights.

Maximum altitude recorded for this species is 2,000 m above sea level. Kuhl's pipistrelles roost in a variety of roost types. Favoured natural roosts include tree and rock crevices, while in urban areas, where the species is frequent, they may roost under tiles, beneath rain gutters, in the narrow spaces between bricks, or in cracks in walls.

ECHOLOCATION

Measured parameters	Mean (range) FM-qCF call and qCF call
Inter-pulse interval (ms)	109.8 (45.6–284.7)
Call duration (ms)	5.7 (2.9–9.5)
Frequency of maximum energy (kHz)	39.4 (34–46.2)
Start frequency (kHz)	68.0 (41.8–97.2)
End frequency (kHz)	36.9 (35.3–45.2)

Heterodyne
When tuned at 41 kHz the bat detector emits very loud 'slaps' similar to an electronic clapping sound. As in other pipistrelle species, when the ultrasound detector is tuned to higher frequencies the sound heard becomes a 'tick'. As for other edge-space foragers, the repetition rate is fast and irregular, changing with clutter distance and variation in flight direction. When close to clutter the repetition rate escalates and becomes more consistent. In open areas, Kuhl's pipistrelles drop the frequency containing maximum energy (FmaxE) down to about 36 kHz, while in closed areas or near obstacles, the FmaxE can increase up to about 46 kHz.

Time expansion/full spectrum
Kuhl's pipistrelles produce FM-qCF calls that sweep down from about 70 kHz to about 35–40 kHz, with FmaxE usually at around 40–41 kHz (Figure 8.34.1). Mean call duration is about 6 ms and the inter-pulse interval is around 110 ms. In uncluttered situations, for example when flying high or away from obstacles, call duration becomes longer, reaching up to about 9 ms, and inter-pulse interval increases to around 300 ms, while frequency becomes narrower and FmaxE may drop to approximately 35 kHz (Figures 8.34.2 and 8.34.3). Conversely in cluttered situations such as in forest areas, or when pursuing prey,

call duration will decrease while frequency increases, resulting in wide broadband calls. In such situations, the inter-pulse interval typically decreases to 45–50 ms and call duration decreases to about 3 ms, while FmaxE typically increases to around 45 kHz (Figures 8.34.4 and 8.34.5). Duration is reduced to a minimum and frequency bandwidth is broadest in feeding buzzes, when the bat homes in on an insect, or in drinking buzzes, when the bat approaches the water surface to drink (Figure 8.34.7).

Calls occasionally terminate with an FM component (Figures 8.34.2 and 8.34.6).

Feeding buzzes consists of terminal buzzes I and II. Terminal buzz II is characterised by a higher call repetition rate and a reduction in frequency, which is thought to be required to broaden the acoustic beam and make it possible to track last-moment movements of prey. Buzz II is absent in drinking buzzes because water is a static target, which renders alterations of the acoustic beam unnecessary.

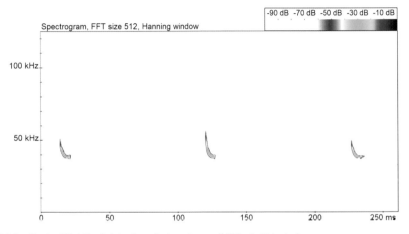

Figure 8.34.1 Typical Kuhl's pipistrelle echolocation call (Sérgio Teixeira).

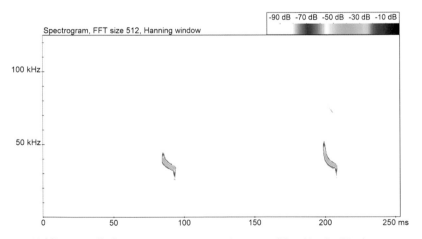

Figure 8.34.2 Kuhl's pipistrelle flying in a semi-open environment (Marc Van De Sijpe).

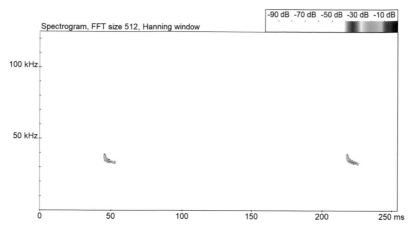

Figure 8.34.3 Echolocation calls of Kuhl's pipistrelle flying in the open (Sérgio Teixeira).

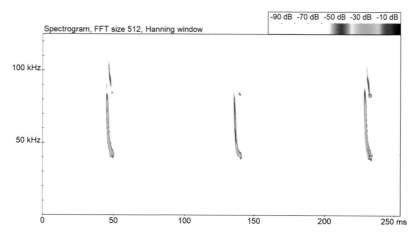

Figure 8.34.4 Echolocation calls of Kuhl's pipistrelle flying in clutter during emergence from a maternity roost (Sérgio Teixeira).

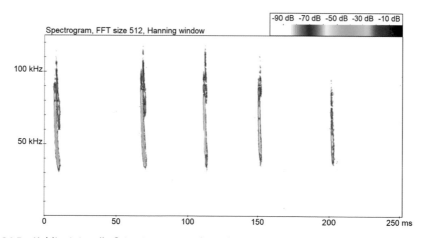

Figure 8.34.5 Kuhl's pipistrelle flying in extreme clutter between tall trees and a house (Marc Van De Sijpe).

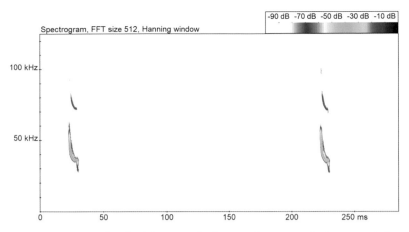

Figure 8.34.6   Echolocation calls of Kuhl's pipistrelle foraging by trees and white streetlights along a road, showing terminal FM sweep (Marc Van De Sijpe).

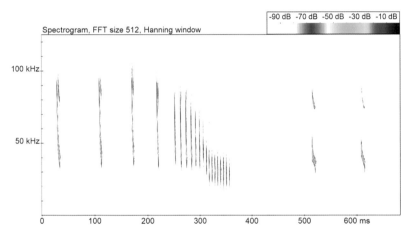

Figure 8.34.7   Feeding buzz of Kuhl's pipistrelle (Marc Van De Sijpe).

SOCIAL CALLS

Heterodyne
Rapid, strong trills, which sound like a very short, powerful buzz, best heard at around 15–16 kHz. Often broadcast in sequences at foraging sites with echolocation calls.

Time expansion/full spectrum
Kuhl's pipistrelle produces a range of social calls, only some of which have been described. The most typical calls are agonistic calls, emitted at foraging sites (Figures 8.34.8–8.34.10). When played back experimentally at foraging sites, the activity of Kuhl's pipistrelles declines, suggesting that they probably serve to warn off conspecifics. Such calls are more commonly heard later at night when temperature drops and bats compete over the same prey or feeding area, and they are often broadcast during chases. The structure of these calls is identical to that of song-flight calls emitted by males in flight in the mating season, so their function appears to be context-specific. These social calls comprise trills of 2–5 components, rarely more, with three being more common. Single-component social calls are occasionally recorded at foraging sites; their function is unclear, but they are probably also agonistic.

The call's total duration is around 35 ms, ranging from 20 ms to over 55 ms. Call components are broadband (on average, from 40 kHz to 11 kHz, but maximum and minimum frequencies may reach extreme values of 9 kHz and >60 kHz, respectively) and spaced out by around 15 ms. The low minimum frequencies typical of these calls make them audible to the unaided ear. The mean FmaxE is around 18 kHz. Within the social call, call components are around 15 ms apart, which means their repetition rate is slower than in social calls of common and soprano pipistrelles, as can be noticed by listening to time-expanded recordings. Social calls in this species are highly diagnostic and provide an effective way to tell Kuhl's pipistrelle apart from species that emit similar echolocation calls, such as Nathusius's pipistrelle.

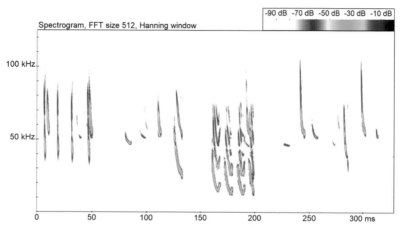

Figure 8.34.8   Kuhl's pipistrelle four-component social calls (along with echolocation calls from common and soprano pipistrelles) (Marc Van De Sijpe).

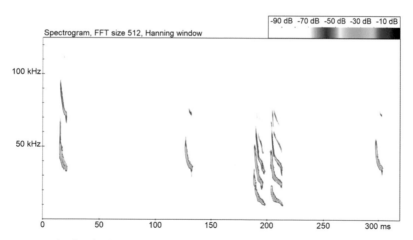

Figure 8.34.9   Social calls of Kuhl's pipistrelle with two components (Marc Van De Sijpe).

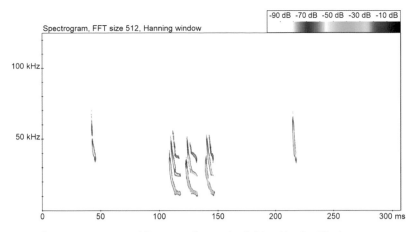

Figure 8.34.10   Three-component Kuhl's pipistrelle social call (Marc Van De Sijpe).

SPECIES WITH SIMILAR OR OVERLAPPING ECHOLOCATION CALLS

Pipistrellus nathusii, Hypsugo savii (see Chapter 7 species notes).

NOTES

Heterodyne is not suitable for identifying the species in flight in regions where other similarly calling species such as the common pipistrelle and Nathusius's pipistrelle are present.

Occasional daylight flights are observed, especially at the end of winter.

8.35 Madeira pipistrelle *Pipistrellus maderensis* (Dobson, 1872)

Sérgio Teixeira

© Sérgio Teixeira and José Jesus

DISTRIBUTION

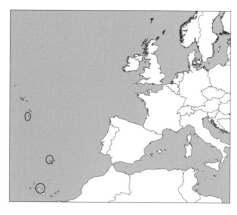

The Madeira pipistrelle inhabits six of the 31 Macaronesian Islands and is restricted to two of the five Macaronesian archipelagos: Madeira and the Canary Islands. In the Madeira archipelago it is present in the islands of Madeira and Porto Santo, and is occasionally spotted in Deserta Grande, although no roosts have been found so far on this island. In the archipelago of the Canaries, it is present on the islands of Tenerife, El Hierro, La Palma and La Gomera. In the Azores, pipistrelles have been detected on the islands of Santa Maria, São Jorge, Graciosa, Flores and Corvo, although further studies are required to determine specific status. Absent from the Selvagens and Cape Verde.

EMERGENCE

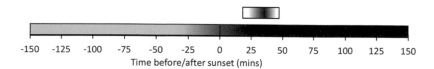

Time before/after sunset (mins)

FLIGHT AND FORAGING BEHAVIOUR

With a rapid and extremely agile flight, the Madeira pipistrelle flies between 1.5 and 15 m above the ground, most commonly at 6 m, which is mainly due to the presence of artificial lights at that height. Extremely manoeuvrable, the Madeira pipistrelle is commonly seen flying over the island's water canals locally known as levadas, foraging at 1.5 m above ground level and turning in circles of 2–3 m diameter. Forages by aerial hawking, sometimes with short dives, as typically performed by Leisler's bat.

HABITAT

This species is fairly adaptable to a wide range of habitat types. Madeira pipistrelle is commonly observed in anthropogenically altered areas such as mixed exotic forests, as well as broadleaf woodland and rural areas where contact exists between forests or thickets and humanised areas. As for many other species, it takes advantage of the presence of artificial light to forage. However, in its natural and unaltered habitat, the Madeira pipistrelle prefers water bodies such as rivers, ponds, lakes and streams located on the native dry and humid evergreen laurel forests. In areas of banana plantation or other managed crops, Madeira pipistrelle is mostly absent. Maximum altitude recorded for this species in the Canary Islands is 2,150 m, while on Madeira it is 1,550 m.

As for other pipistrelles, the Madeira pipistrelle roosts in crevices. Preferences are grey hollow bricks, wall cavities, basalt fractures, drystone walls and roof tiles.

ECHOLOCATION

Measured parameters	Mean (range) FM-qCF call and qCF call
Inter-pulse interval (ms)	123.1 (59.3–358.0)
Call duration (ms)	5.8 (2.1–9.5)
Frequency of maximum energy (kHz)	46.2 (41.3–51.7)
Start frequency (kHz)	59.4 (44.6–106.6)
End frequency (kHz)	44.9 (40.9–49.8)

Heterodyne
Similar to other pipistrelles and FM-qCF-emitting species – when the Madeira pipistrelle's frequency of maximum energy (FmaxE) (47 kHz) is tuned on the detector, the sound produced is a loud 'slap', comparable to an electronic hand-clap. As the detector is tuned to frequencies above 47 kHz the slap/clapping sound becomes a 'tick'-like sound. Comparable to other species that forage against edge-space backgrounds, the repetition rate is fast and irregular, but always adjusted as the distance to echo-producing background changes. When echolocating closer to objects, echoes are bounced back with a shorter time lag, and so the bat increases the repetition rate, with a regular inter-pulse interval, to be able to gather more information about the background and hence avoid collisions. In open areas, such as when flying across a valley, Madeira pipistrelles drop FmaxE down to 41 kHz with a lower repetition rate. In closed spaces, FmaxE can reach about 51 kHz.

Time expansion/full spectrum
Madeira pipistrelles are edge foragers, hence typically producing FM-qCF calls that sweep down from about 80–85 kHz to about 45 kHz, with FmaxE usually between 46 and 48 kHz, more commonly at about 47 kHz, and of an average duration of 5 ms. For these typical calls, the FM sweep portion ranges between 85 kHz and 50 kHz with a duration of 1.2–2.2 ms, while the qCF portion of the signal is between 50 and 45 kHz with a duration of 3.5–4.5 ms. This is the typical flight call for this species (Figure 8.35.1). Females have lower FmaxE, normally closer to 46 kHz, while males are closer to 47 kHz.

In the case of Madeira pipistrelle pups, the frequency range of FM-qCF calls is higher in all acoustic parameters, due to their smaller size. In juveniles, the FmaxE typically ranges from 49 kHz to 53 kHz, but the call duration and pulse interval are similar to those of adults (Figure 8.35.2). Madeira pipistrelles always have more energy in the qCF portion, except when flying close to objects, when qCF is mostly absent or reduced.

When flying further away from echo-producing objects, Madeira pipistrelles emit longer broadband qCF pulses lacking the FM portion with a shape resembling a boomerang, and this is typically used when commuting or foraging at greater distances from vegetation. These signals are much narrower in bandwidth than typical FM-qCF, normally with a start frequency of 60–50 kHz, end frequency between 43 and 46 kHz, FmaxE between 44 and 45 kHz, and a duration of 6 ms (Figure 8.35.3), but still much broader than the qCF calls used when flying in the open. In these situations, when flying and echolocating in the open, far from any echo-producing objects, Madeira pipistrelles emit narrowband flattish qCF signals, commonly with a start frequency of 47 kHz, FmaxE of 45 kHz, end frequency of 43 kHz, inter-pulse interval of 220 ms and duration of 7 ms, but in extremes reaching 9 ms duration and 350 ms inter-pulse interval. In these echolocation pulses, the FM sweep is absent and the qCF portion composes the entire signal. On the island of Madeira, these signals are commonly observed when Madeira pipistrelles forage in open areas such as beside vertical cliffs, or when they are commuting in high-altitude flights over river valleys (Figure 8.35.4).

In slightly cluttered situations, such as when close to vegetation, Madeira pipistrelles emit calls of higher bandwidth and shorter duration (Figure 8.35.5). In much more cluttered situations, such as within a forest or other forest-like habitat, echolocation pulses resemble *Myotis* FM sweeps but lack the typical heel curve produced by *Myotis* bats. In these conditions, the call sweeps from around 85 kHz to 60 kHz, the inter-pulse interval typically decreases to 60 ms, and call duration is 3 ms or less (Figure 8.35.6).

Feeding buzzes are an extreme example of FM calls such as when bats fly in cluttered-background spaces, but in this case bats narrow the bandwidth and put all the energy into the frequency most suited to gathering information about the prey and increasing the chance of success (Figure 8.35.7).

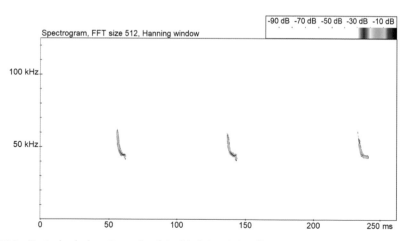

Figure 8.35.1   Typical echolocation calls of the Madeira pipistrelle (Sérgio Teixeira).

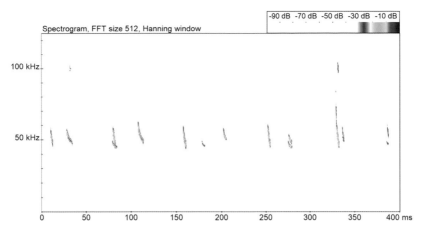

Figure 8.35.2   Two juvenile Madeira pipistrelles flying close to vegetation (Sérgio Teixeira).

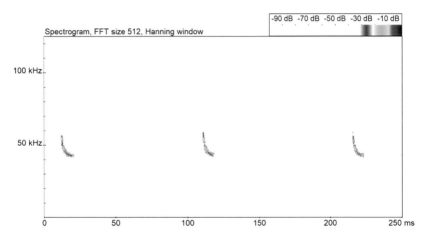

Figure 8.35.3   Madeira pipistrelle flying above a tree canopy on a steep mountain (Sérgio Teixeira).

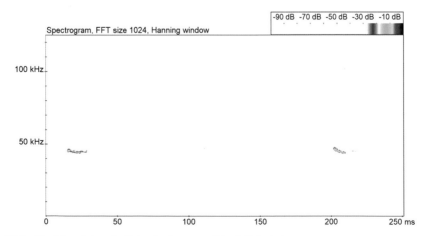

Figure 8.35.4   Madeira pipistrelle flying in the open (Sérgio Teixeira).

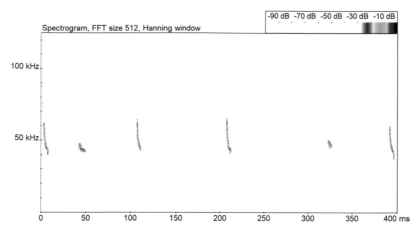

Figure 8.35.5   Madeira pipistrelle foraging in a laurel forest (Sérgio Teixeira).

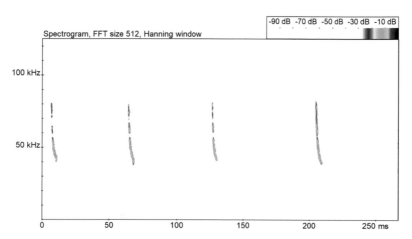

Figure 8.35.6   Madeira pipistrelle flying in a cluttered environment (Sérgio Teixeira).

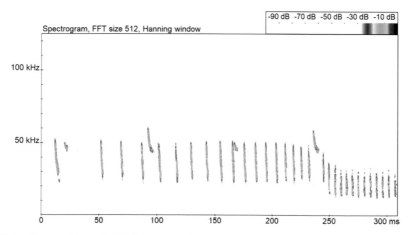

Figure 8.35.7   Feeding buzz of a Madeira pipistrelle (Sérgio Teixeira).

SOCIAL CALLS

Heterodyne
The most common social calls emitted by Madeira pipistrelles are agonistic calls related to food-patch or territorial defence, which frequently involves pursuit flights and even physical aggression such as biting the tail membrane of the fleeing bat (S. Teixeira, personal observation). In the Macaronesian Islands these calls are produced all year round because, in the mild climate of these islands, bats do not hibernate (Teixeira 2014). With a heterodyne detector, these calls sound like a rapid 'chonking' sound which is loudest at about 39 kHz, although it can be heard from 12 to 40 kHz (Russo *et al.* 2009, Teixeira and Jesus 2009). During the mating season, these calls are produced much more frequently when territorial defence and mate attraction are paramount for reproductive success. The rate of production of these calls during winter increases when temperatures are lower, but more importantly when food items are scarce and bats fight for the few insects they can find. These social calls are always emitted during flight and never from a roost or post, and are more frequently emitted closer to feeding areas than during commuting flight between roosts and feeding areas.

Curved social calls are the second most common type of social call. These calls are long curved FM-qCF calls, and sound like a rapid click on the detector.

Madeira pipistrelle distress calls are emitted only when under physical duress and are similar to those of other bat species. These calls are extravagant reproach sounds with continuous or high repetition rate and are like a rapid 'rasping' sound. Typically, the FmaxE is about 22 kHz.

Time expansion/full spectrum
Similar to other pipistrelles in Europe, Madeira pipistrelles emit single-element curved and complex songs. Also, squawk-like calls are produced during distress situations.

Single-element curved calls are more commonly produced during mother–infant tandem flights. These calls have not been recorded in any other situation. During March and April when juveniles start to follow their mothers in tandem flights, mother–infant interaction social calls are produced frequently. These calls are long broadband FM-qCF calls with an FmaxE of about 36 kHz. Calls are very variable in the FmaxE, which varies between 31 kHz, 33 kHz, 36 kHz and 38 kHz. Typically Madeira pipistrelle juveniles produce sequences of three individual components in rapid sequence, where the first component has a higher frequency (38 or 36 kHz) and a duration of about 6.5 ms, the second component is a bit lower in frequency (averaging 36 or 33 kHz) and about 9 ms in duration, and the last component is the lowest in frequency (about 31 kHz) and 15 ms in duration (Figure 8.35.8).

The most frequently emitted social call consists of a series of similar FM sweeps with a short qCF signal termination. The mean component number is three, equal to Kuhl's pipistrelle and soprano pipistrelle. The number of FM sweeps within this type of social call range between two (Figure 8.35.9), three (Figure 8.35.10) and five (Figure 8.35.11). In Madeira pipistrelles, these calls have an FmaxE of around 15 kHz, a minimum frequency of 13.6 kHz and a maximum frequency of 24 kHz, while the most closely comparable species, Kuhl's pipistrelle, has a mean FmaxE of 17 kHz and ranges between 11.5 kHz and 38 kHz (Russo and Jones 1999, Russo *et al.* 2009, Teixeira and Jesus 2009). In the common pipistrelle, the FmaxE is around 18 kHz (Barlow and Jones 1997a, Russo and Jones 1999). In Kuhl's pipistrelle the most common calls have three components, followed by the two-component song, four-component and lastly five-component (Russo and Jones 1999), whereas in the Madeira pipistrelle the three-component is also the most common, but the second most used variation is the four-component, with the two-component and five-component being the least common (S. Teixeira, personal observation). The variation within individual components of the song varies between species. In the songs of the Madeira pipistrelle, the first component is very variable and normally has a higher frequency and narrower bandwidth than the following components, while in soprano pipistrelle the variable component is the last, and in common pipistrelle the individual components are rather parallel in structure. In Kuhl's pipistrelle, as in the Madeira pipistrelle, the first component is the most variable and dissimilar, but frequency-wise it is similar to the other individual components in the song, unlike the Madeira pipistrelle.

The occurrence of these social calls, produced by the males, increases dramatically during the autumnal mating period (August to early October), indicating that they are involved in mate attraction, female choice and/or territorial defence. During May males begin to set up individual territories around their day roosts, and the male monopolises its roost throughout the summer and the subsequent mating period, when the females join the males. Males intruding into a territory are chased away by the territory owner while emitting these social calls, thought to be agonistic. 'Song-flight display' is performed by males during the mating period and consists of a fixed, circular or elliptical path with a length of 100–200 m which touches the roost site, during which time the males constantly emit social calls.

Madeira pipistrelle distress calls are extravagant reproach sounds, continuous or with a high repetition rate (Figure 8.35.12). Typically, the FmaxE is about 22 kHz. Another type of call emitted during distress is a trill-like signal, with an FmaxE of around 21 kHz. These are short FM sweeps with a maximum/start frequency of 23 kHz and a minimum/end frequency of 18 kHz and an average duration of 4 ms (Figure 8.35.13). Inter-pulse interval is on average 120 ms.

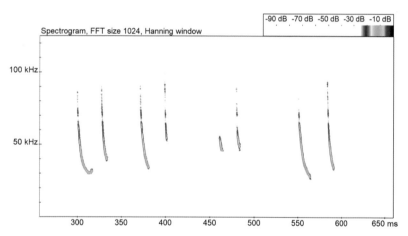

Figure 8.35.8   Social calls of a Madeira pipistrelle produced during mother–infant tandem flights (Sérgio Teixeira).

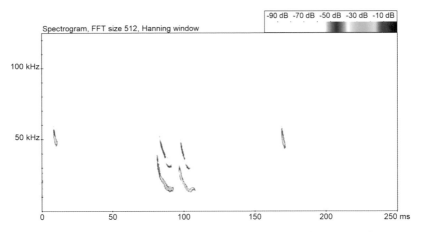

Figure 8.35.9   Social calls of a Madeira pipistrelle with two components (Sérgio Teixeira).

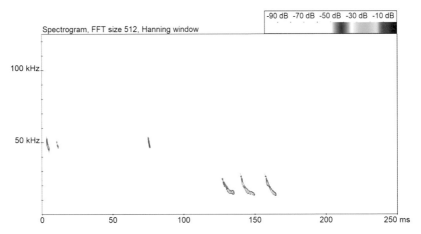

Figure 8.35.10   Social call of a Madeira pipistrelle with three components (Sérgio Teixeira).

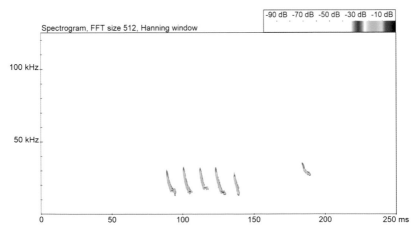

Figure 8.35.11   Social calls of a Madeira pipistrelle with five components (Sérgio Teixeira).

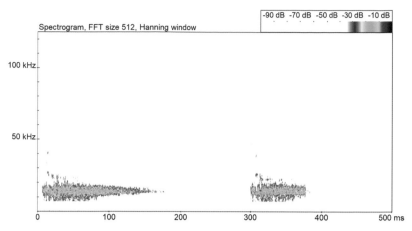

Figure 8.35.12   Distress calls of a Madeira pipistrelle held in the hand (Sérgio Teixeira).

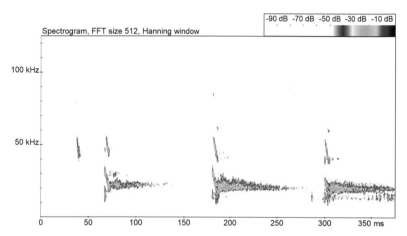

Figure 8.35.13   Further example of distress calls of a Madeira pipistrelle held in the hand (Sérgio Teixeira).

SPECIES WITH SIMILAR OR OVERLAPPING ECHOLOCATION CALLS

Pipistrellus pipistrellus (see Chapter 7 species notes).

NOTES

Diurnal activity – On Madeira Island, *Pipistrellus maderensis* has been observed foraging during mid-afternoon, taking advantage of warm foggy days. Diurnal activity in *P. maderensis* has always been recorded after 15.00 h, hence in this situation emergence may happen much earlier than its normal evening emergence.

Mother–infant calls have been observed in August, as well as tandem flight behaviour, indicating that on Madeira there might be two breeding seasons in the warmest years.

Some Madeira pipistrelles have been observed foraging over a 3 × 3 m fenced water reservoir, exhibiting FM signals and with an extremely fast repetition rate and an inter-pulse interval of about 40 ms. FM sweeps sounded like the *Myotis* click sound but lacked the heel typical of *Myotis* bats calls. This shows that this species is well adapted to forage in close-edge habitats, as this behaviour was observed during foraging and not commuting.

This behaviour was observed many times, over a long period, indicating that this is typical behaviour for the species.

Interestingly, sometimes when flying in tight spaces such as levada gaps, Madeira pipistrelles go very close to the vegetation or fly into foliage gaps and produce extreme high-frequency FM sweeps, but with low amplitude. In this case, FM sweeps are faint, resembling *Plecotus* whispers, and the frequencies are much higher than the typical FM used by the Madeira pipistrelle in these background situations as described above. This behaviour is still not well understood.

Although a paper was published reporting the presence of *Pipistrellus maderensis* in the island of Santa Maria (Trujillo and Gonzalez 2011), this population should be studied in more detail since the fur and membrane colours are much darker than in *P. maderensis* bats found both on Madeira and in the Canary Islands. The presence of this species in the Azores is still doubtful. Several researchers have found pipistrelles on several islands (Skiba 1996, Rainho *et al.* 2002), but none has given any certainty about which species they belong.

8.36 Savi's pipistrelle *Hypsugo savii* (Bonaparte, 1837)

Panagiotis Georgiakakis, Danilo Russo and Leonardo Ancillotto

© Panagiotis Georgiakakis

DISTRIBUTION

Occurs from the Middle East and Asia Minor throughout southern Europe to Iberia. The northern limit of its distribution is central France, southern Switzerland and Germany east to Crimea, but individual animals have been found in northern France and Germany. It is present in almost all Mediterranean islands. The north African populations may belong to a separate species (Mayer *et al.* 2007).

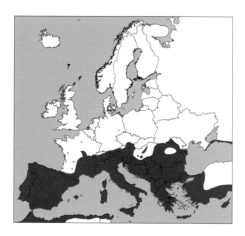

EMERGENCE

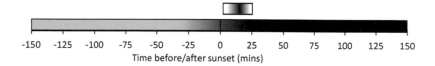

FLIGHT AND FORAGING BEHAVIOUR

Savi's pipistrelle is a fast flier, even when foraging, and performs fewer manoeuvres than species of the genus *Pipistrellus*. It prefers flying well above the ground, up to 100 m. Prey is caught on the wing.

HABITAT

Savi's pipistrelle can be found from sea level up to the alpine zone. Generally, it prefers open and semi-open areas such as meadows, scrublands, structured landscapes, wetlands and in some regions, including Italy and the Balkans, large cites. In other regions, such as Crete, it is quite common in forests and tree cultivations, but not in settlements.

Usually roosts in rock and wall crevices, also under roof tiles. Occasionally it is found in bridges and bird nests.

ECHOLOCATION

Measured parameters	Mean (range) FM-qCF call and qCF call
Inter-pulse interval (ms)	225.8 (33.77–549)
Call duration (ms)	10.2 (4.7–14.7)
Frequency of maximum energy (kHz)	34.6 (31.6–40.0)
Start frequency (kHz)	41.6 (34.0–71.4)
End frequency (kHz)	33.6 (31.0–36.0)

Heterodyne

Calls sound similar to those of bats of the genus *Pipistrellus* but tend to be louder and of a lower frequency, with the frequency containing maximum energy (FmaxE) around 35 kHz.

Time expansion/full spectrum

The echolocation calls of Savi's pipistrelles resemble those of Kuhl's pipistrelle, but they are usually longer and have lower frequencies. They produce FM-qCF calls that sweep from c.70 kHz to 33–34 kHz with FmaxE of 33–35 kHz (Figure 8.36.1). Mean call duration is about 10 ms, and the inter-pulse interval is very variable with a mean value of around 226 ms. In open areas calls become flatter and longer, with longer inter-pulse intervals, and FmaxE may fall below 32 kHz (Figure 8.36.2). Conversely, when flying very close to trees, calls become steeper and shorter, with a higher repetition rate and FmaxE as high as 37 kHz (Figure 8.36.3). When approaching prey, calls initially become 'steeper' and shorter with higher frequencies, but then, in feeding buzz II, they drop to much lower values (Figure 8.36.4). Occasionally, especially when more than one individual is foraging together, it may be difficult to discriminate Savi's pipistrelle calls from those of Kuhl's pipistrelle or Nathusius's pipistrelle where such species occur in sympatry.

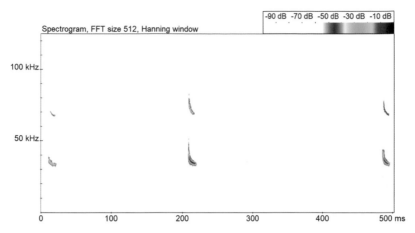

Figure 8.36.1   Echolocation calls of Savi's pipistrelle foraging above a forest road (Panagiotis Georgiakakis).

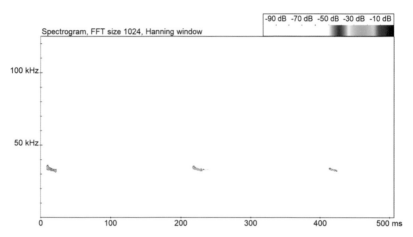

Figure 8.36.2   Echolocation calls of a Savi's pipistrelle flying over a dam (Panagiotis Georgiakakis).

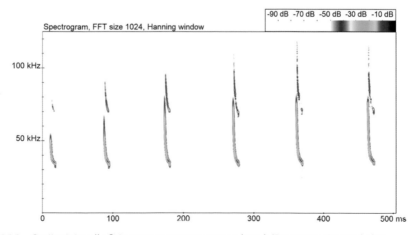

Figure 8.36.3   Savi's pipistrelle flying among trees next to a brook (Panagiotis Georgiakakis).

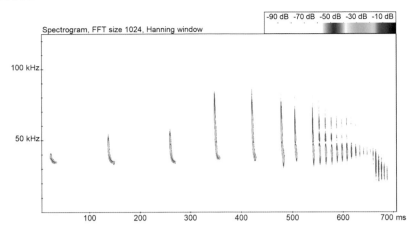

Figure 8.36.4   Feeding buzz of Savi's pipistrelle (Panagiotis Georgiakakis).

SOCIAL CALLS

Heterodyne
Social calls produced by this species range in FmaxE between about 10 kHz and 40 kHz, most peaking at 25–30 kHz, thus resembling noctules when heard in heterodyne, but louder and with a higher repetition rate.

Time expansion/full spectrum
Savi's pipistrelle emits a wide range of social calls both in flight and from the roost. Social calls emitted in flight have been studied in detail by Nardone *et al.* (2017), who described at least five different call types, each with extremely high variability, which can be emitted both as single calls and in sequence (between two and nine repetitions) or in combination with others.

Social calls emitted by Savi's pipistrelle while foraging closely resemble those of the genus *Pipistrellus* (Figure 8.36.5); i.e. they consist of 3–4 FM-qCF components, with a frequency range between 15 and 28 kHz. Therefore, they are similar in composition to those of Kuhl's pipistrelle and closer to the common pipistrelle in terms of frequencies.

Other calls emitted in flight show a high degree of variation. Most of these single- and multi-component calls consist of steep downward FM signals that end with an ascending shorter portion or a longer, lower part with evident frequency 'oscillations'. Both single- and multi-component calls can be found in longer sequences of calls, i.e. songs. During songs, repeated motifs can also be often found within the same sequence, each consisting of initial FM signals followed by one or more trills (Figure 8.36.6). This complex call is often recorded in the proximity of roosts and/or water sites, mainly in mid/late summer.

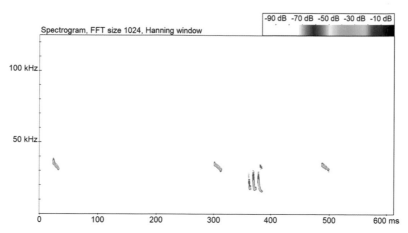

Figure 8.36.5 Social call of Savi's pipistrelle during foraging close to streetlights in open space (with echolocation calls) (Leonardo Ancillotto).

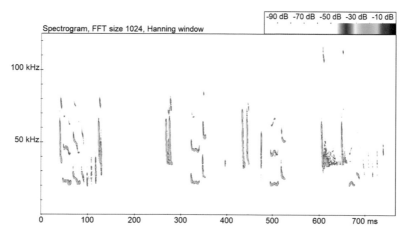

Figure 8.36.6 Song of Savi's pipistrelle recorded close to a water site. At least four motifs can be seen in this sequence (see text) (Leonardo Ancillotto).

SPECIES WITH SIMILAR OR OVERLAPPING ECHOLOCATION CALLS

Pipistrellus kuhlii, *Pipistrellus nathusii* (see Chapter 7 species notes).

8.37 Western barbastelle
Barbastella barbastellus (Schreber, 1774)

Jon Russ

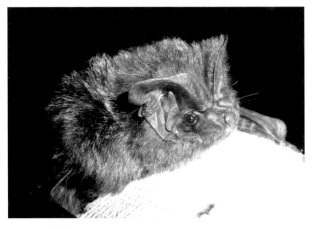

© James Shipman

DISTRIBUTION

Widespread throughout Europe but missing from parts of the eastern Mediterranean. Found in the Balearics, Corsica, Sardinia and the Canary Islands. North as far as Sweden and a single record in southern Norway. In the UK generally limited to southern and central England and Wales.

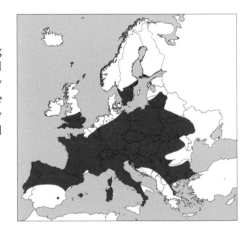

EMERGENCE

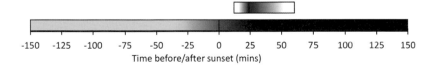

-150   -125   -100   -75   -50   -25   0   25   50   75   100   125   150

Time before/after sunset (mins)

FLIGHT AND FORAGING BEHAVIOUR

In between periods of hunting the flight is fast, and the species is quite agile. Forages under the canopy in the early evening, but later at 2–4 m above tree crowns. Also along vegetation in regular paths 4–5 m above the ground, around mercury vapour streetlamps and low over water. There is strong evidence that barbastelles are aerial hawkers, but they also glean prey from surfaces.

HABITAT

Prefers wooded countryside and hunts in wooded river valleys, often in the tree canopy. Also forages over wet meadows, ponds and rivers. Generally, forages where bodies of water are nearby. They also forage in more open countryside with scrubby trees and gorse and have even been recorded foraging alongside coastal cliffs and over beaches. Also known to forage around mercury vapour streetlights.

Roosts in summer are located under peeling bark, in tree crevices and in certain types of bat box. They are also found in buildings, usually timber-framed barns. Also behind window shutters and timber cladding. Winter roosts are located in trees, but also caves, mines, bridges, tunnels, bunkers and rock crevices.

ECHOLOCATION

Measured parameters	Mean (range)	
	qCF-FM call	FM call
Inter-pulse interval (ms)	72.4 (43.2–144.9)	108.4 (41.8–229.0)
Call duration (ms)	4.3 (2.0–6.6)	3.4 (2.5–5.1)
Frequency of maximum energy (kHz)	41.6 (33.5–43.8)	32.9 (29.2–44.7)
Start frequency (kHz)	44.1 (36.8–47.3)	39.4 (35.2–49.0)
End frequency (kHz)	28.9 (25.4–31.9)	28.0 (23.8–36.8)

Heterodyne
Typically, the barbastelle alternates between two call types: one with frequency containing maximum energy (FmaxE) at about 32 kHz sweeping down from about 40 kHz to 28 kHz (FM call), and a quieter one with FmaxE at about 42 kHz sweeping down to about 29 kHz (qCF-FM call). The overall effect of these alternating calls is a rattling sound similar to castanets. This is best heard when the detector is tuned to about 32 kHz, when the sound is unmistakable. In some situations, either call may be omitted.

Time expansion/full spectrum
When hunting and commuting the barbastelle often alternates between two call types (Figure 8.37.1). The first call type is loud, has a mean duration of about 2.5 ms and sweeps downwards (FM) from about 36 kHz to about 28 kHz. The second call type consists of a qCF-FM pulse starting of about 44 kHz and ending at about 29 kHz, with a duration of about 4.5 ms. The qCF-FM call is generally quieter than the FM call. Occasionally either call type is omitted. In cluttered situations, such as leaving the roost or flying in a barn, both call types can be more FM (Figure 8.37.2), and then in very cluttered habitats bats produce steep FM echolocation calls that start at about 50 kHz and end at about 27 kHz (Figures 8.37.3 and 8.37.4). The duration is about 2.8 ms and the FmaxE is about 40 kHz. As the bats move away from the highly cluttered situation these calls change into the second call type (Figure 8.37.5). In very open situations the CF portion of the qCF-FM call becomes longer.

When capturing an insect, feeding buzzes are produced (Figure 8.37.6).

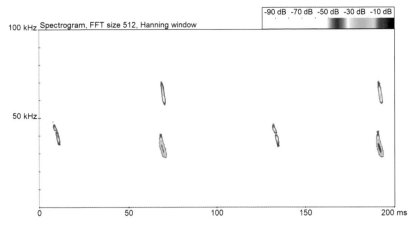

Figure 8.37.1 Echolocation calls of barbastelle recorded flying across a woodland clearing. Although harmonics are visible in this sonogram they are not always present (Jon Russ).

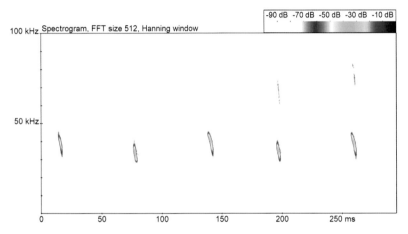

Figure 8.37.2   Barbastelle flying within a large barn (Jon Russ).

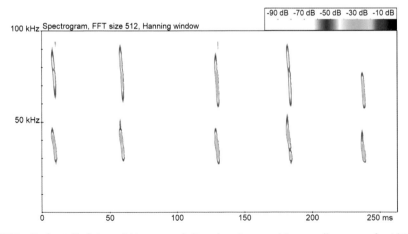

Figure 8.37.3   Barbastelle flying within a very cluttered environment in a small open roof void (Jon Russ).

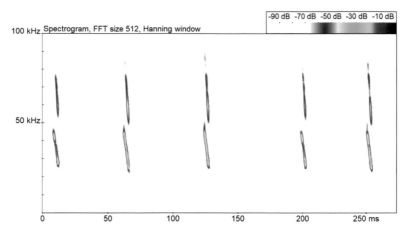

Figure 8.37.4   Barbastelle released from the hand (Jon Russ).

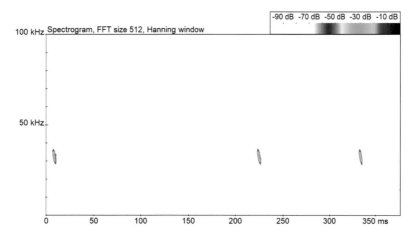

Figure 8.37.5   Barbastelle recorded flying along hedgerow leading from a woodland edge (Jon Russ).

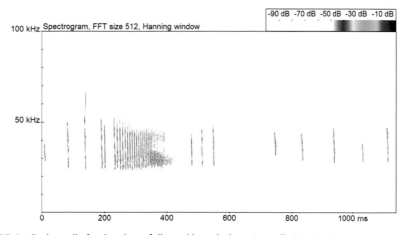

Figure 8.37.6   Barbastelle feeding buzz followed by echolocation calls (Jon Russ).

SOCIAL CALLS

Heterodyne
Few social calls have been reported for this species. Distress calls, produced by individuals under physical duress, are long and scolding and are loudest at around 40 kHz. Social calls are also produced during the autumnal mating period; these are quite variable and loudest at 30–35 kHz.

Time expansion/full spectrum
Distress calls consist of a series of rapidly downward-sweeping FM pulses with FmaxE at about 40 kHz (Figure 8.37.7). Social chatter can be recorded at roost sites or occasionally when held in the hand (Figure 8.37.8)

During late summer and the autumnal mating period, social calls are occasionally recorded, more often when two or more bats are flying together (Figures 8.37.9–8.37.11). The functions of these social calls are unknown.

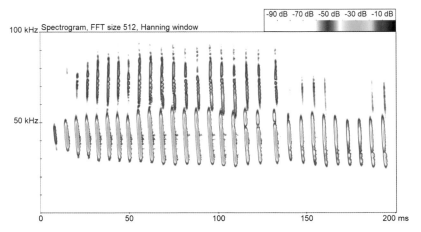

Figure 8.37.7   Distress calls of barbastelle held in hand (Jon Russ).

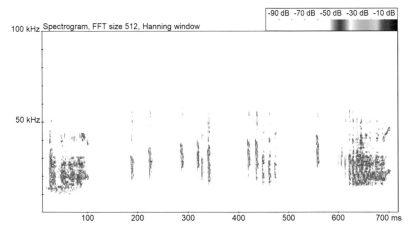

Figure 8.37.8   Social chatter from a barbastelle held in the hand (David King).

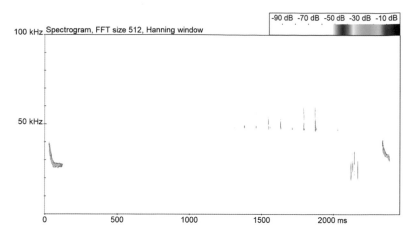

Figure 8.37.9   Social calls produced by a barbastelle repeatedly flying up to a perch within a large open barn (Jon Russ).

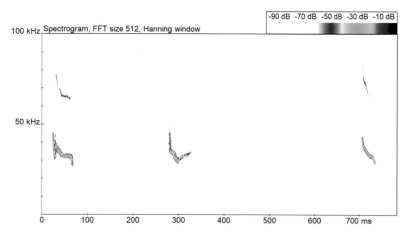

Figure 8.37.10   Social calls of a barbastelle flying within a large open barn (Jon Russ).

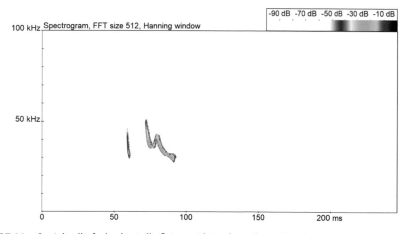

Figure 8.37.11   Social call of a barbastelle flying within a large barn (Jon Russ).

SPECIES WITH SIMILAR OR OVERLAPPING ECHOLOCATION CALLS

Plecotus spp. (see Chapter 7 species notes).

8.38 Brown long-eared bat *Plecotus auritus* (Linnaeus, 1758)

Stephanie Murphy and Jon Russ

© René Janssen

DISTRIBUTION

The brown long-eared bat in Europe extends westward to the west of Ireland (Shiel *et al.* 1991), as far south as central Spain (Fernandez 1989) and central Italy (Crucitti 1989), and as far north as Sweden, where the northern edge of the range appears to be 63°N (Ahlén and Gerell 1989). It was also thought to extend eastwards as far as Sakhalin and Japan (Corbet and Hill 1991), and it has been reported from Mongolia and northeast China (Zheng and Wang 1989). However, a study using a combination of morphological and molecular data revised the genus and concluded that the brown long-eared bat is restricted to Europe, as far east as the Ural and Caucasus mountains (Spitzenberger *et al.* 2006). The brown long-eared bat is common and widespread in Britain and is distributed throughout the country except in the far north and northwest of Scotland and offshore islands (Swift 1991).

EMERGENCE

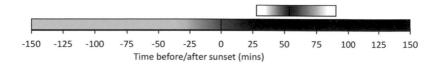

-150 -125 -100 -75 -50 -25 0 25 50 75 100 125 150

Time before/after sunset (mins)

FLIGHT AND FORAGING BEHAVIOUR

The wing shape of the brown long-eared bat allows for slow, fluttering, manoeuvrable flight in cluttered habitats, such as woodland. It frequently hovers (Entwistle *et al.* 1996) and can rise vertically for a few metres (Norberg 1976), and this enables the species to forage by gleaning (capturing prey that is crawling or at rest on solid surfaces rather than in flight). The large ears can pick up return echoes from short low-intensity calls (Waters and Jones 1995), and this species also uses passive listening to prey-generated sounds to locate prey (Anderson and Racey 1991). Laboratory experiments have also shown that brown long-eared bats use visual cues for detection, but exploit additional information, such as echolocation and passive listening, during the final pursuit (Eklöf and Jones 2003).

The wing and ear morphology, coupled with low-intensity echolocation, imply that brown long-eared bats are adapted to feeding in cluttered habitats (Norberg and Rayner 1987), and studies have shown that the species feeds predominately in woodland (Entwistle *et al.* 1996, Murphy *et al.* 2012). Brown long-eared bats emerge from their day roosts fairly late relative to many other vespertilionid species, and this late emergence is almost certainly connected with their habit of gleaning (Rydell *et al.* 1996, Swift 1998). Brown long-eared bats usually use landscape features, such as hedges or treelines, to fly between day roosts and foraging sites and tend to avoid flying in the open (Entwistle *et al.* 1996). This may reduce the risk of predation by nocturnal birds such as tawny owls (Speakman 1991, Lesinski *et al.* 2009a, 2009b), barn owls (Speakman 1991, Petrzelkova and Zukal 2003) and birds of prey flying at dusk (Speakman 1991, Fenton *et al.* 1994).

HABITAT

It has been recorded in a variety of habitat types but is usually associated with a range of woodland types including deciduous woodland (Entwistle *et al.* 1996, Murphy *et al.* 2012), coniferous woodland (Fuhrmann and Seitz 1992) through to birch scrub and gardens with mature trees (Swift and Racey 1983). It has also been recorded in orchards and parkland among meadows (Barataud 1990). Research in Germany and Sweden recorded frequent use of conifer forest (Fuhrmann and Seitz 1992, Ekman and DeJong 1996), while in northeast Scotland the species shows a significant preference for deciduous, broadleaved woodland, utilising only the edges of conifer plantations (Entwistle *et al.* 1996).

In western Europe the brown long-eared bat roosts almost exclusively in buildings such as attic spaces and open roof voids, behind cladding and in mortise holes. In eastern Europe it is more commonly found in trees. In the winter, underground sites are used but also tree holes.

ECHOLOCATION

Measured parameters	Mean (range) FM call (1st harmonic)
Inter-pulse interval (ms)	85.3 (25.0–189.0)
Call duration (ms)	3.1 (1.9–7.1)
Frequency of maximum energy (kHz)	34.4 (26.0–52.0)
Start frequency (kHz)	50.6 (31.5–64)
End frequency (kHz)	23.0 (16.0–38.0)

Heterodyne

Brown long-eared bat echolocation calls are characterised by their quietness and can only be detected in heterodyne if the bat is less than 5 m away (Swift 1998). The species is more commonly observed before it is heard. The call is a short FM sweep with a fast pulse rate that can be heard most clearly on a heterodyne detector set between 30 kHz and 40 kHz, the frequency of maximum energy (FmaxE) is about 33 kHz. It is not possible to separate the brown long-eared bat from the grey long-eared bat using a heterodyne bat detector. Ahlén (1981) identified a loud call, detectable at 40 m or more, that was longer in duration (7.1 ms) and descended in frequency from about 42 kHz to about 12 kHz, ending with a short (1 ms) CF portion at about 12 kHz, which was also the FmaxE, followed by a short downward FM sweep. Ahlén (1981) noted that this call type was emitted intermittently when brown long-eared bats were flying inside barns and mines and more regularly when they are flying in the open.

Time expansion/full spectrum

Brown long-eared bat echolocation calls are frequency-modulated signals often consisting of two harmonics (Figure 8.38.1). The first harmonic on average starts at around 55 kHz and ends at around 23 kHz, while the second harmonic typically starts at around 73 kHz and sweeps down to around 51 kHz. The FmaxE is typically produced in the first harmonic and is usually around 34 kHz. The duration of this call is on average 3.1 ms.

In cluttered environments, the inter-pulse interval decreases (Figure 8.38.2) and the FmaxE may be contained in the second harmonic. When flying in relatively uncluttered situations, call duration becomes very long and FmaxE drops to about 20 kHz. Occasionally while hunting in the open, or flying in barns, brown long-eared bats use loud low-frequency FM-CF sweeps which are regularly repeated. These are about 8 ms long and sweep down from 45 kHz to about 15 kHz (Figure 8.38.3), and occasionally as low as 12 kHz. Coles *et al.* (1989) demonstrated that this FmaxE of 12 kHz coincided with the centre of the most sensitive range of brown long-eared bats' hearing, and proposed that this call could be a long-distance communication call. However, the function of this call was not determined.

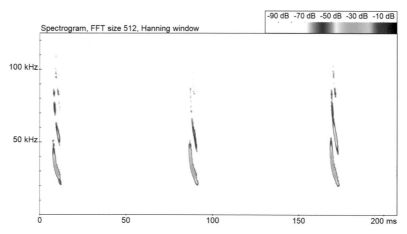

Figure 8.38.1   Echolocation calls of a brown long-eared bat foraging around a woodland edge (Stephanie Murphy).

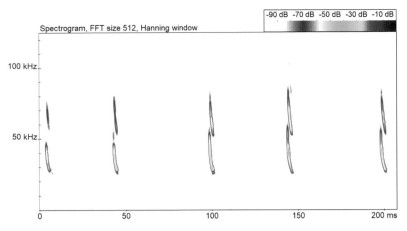

Figure 8.38.2 Echolocation calls of a brown long-eared bat flying within dense vegetation (Jon Russ).

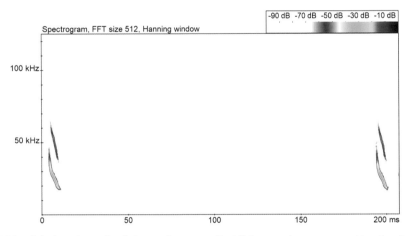

Figure 8.38.3 Echolocation calls of a brown long-eared bat flying over long grass next to a treeline (Jon Russ).

SOCIAL CALLS

Heterodyne
Loud social calls at irregular intervals at mating roost sites, in spring and autumn. These calls sound like rapid clicks, and FmaxE is around 21 kHz.

Time expansion/full spectrum
Several call types have been described, including a call first described by Ahlén (1981) (e.g. Figure 8.38.4). Murphy (2012) recorded brown long-eared bat social calls at 20 maternity roost sites in southern England and found that almost 97% of the calls recorded were similar in call structure to these. These calls, however, show significant variability in measured parameters, often commencing between 60 and 40 kHz and sweeping down to terminate between 12 and 13 kHz (e.g. Figures 8.38.4–8.38.6). The FmaxE may be between 16 and 17 kHz. They are usually between 8 and 12 ms in length with variable inter-pulse intervals. The calls can be recorded as single emitted calls or as part of a sequence of calls.

Less frequently recorded calls are characterised by an initial upward FM sweep and then sweeping down through a greater range of frequencies (Figures 8.38.7 and 8.38.8). This type of call has been recorded by Furmankiewicz (2004) in Poland during spring swarming activity and has been found to occur significantly more frequently at roost sites in September in southern England, suggesting that it may be related to mating activity (Murphy 2012).

V-shaped calls, recorded at swarming sites (Furmankiewicz 2004, Furmankiewicz *et al.* 2013), have now been confirmed as actually being the echolocation calls of Natterer's bats flying in clutter (J. Furmankiewicz, personal communication).

Distress calls, produced when an individual is under physical duress, consist of a series of FM components with FmaxE at about 18 kHz (Figure 8.38.9).

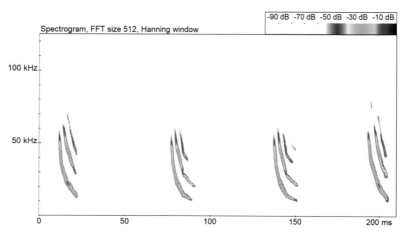

Figure 8.38.4   Social calls of a brown long-eared bat recorded outside a barn (Stephanie Murphy).

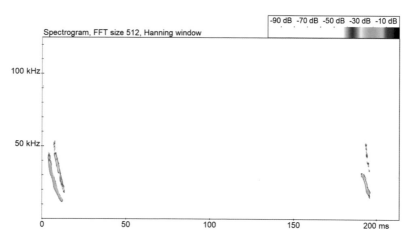

Figure 8.38.5   Social calls of a brown long-eared bat recorded at a maternity site (Stephanie Murphy).

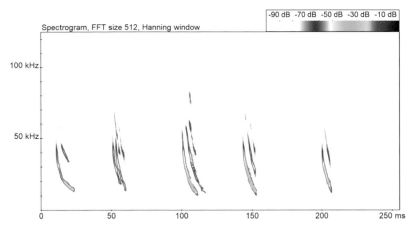

Figure 8.38.6   Social calls of a brown long-eared bat recorded within a small maternity roost in a barn (Jon Russ).

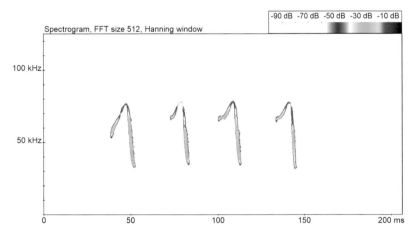

Figure 8.38.7   Social calls of a brown long-eared bat recorded within a church belfry (Jon Russ).

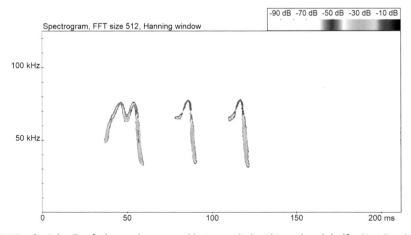

Figure 8.38.8   Social calls of a brown long-eared bat recorded within a church belfry (Jon Russ).

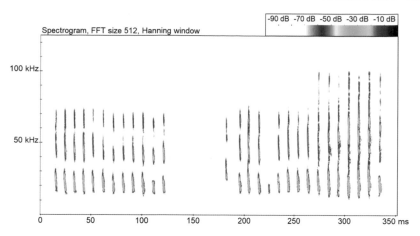

Figure 8.38.9 Distress calls emitted by a brown long-eared bat held in the hand (Jon Russ).

SPECIES WITH SIMILAR OR OVERLAPPING ECHOLOCATION CALLS

Plecotus spp., *Barbastella barbastellus* (see Chapter 7 species notes).

8.39 Alpine long-eared bat

Plecotus macrobullaris (Kuzjakin, 1965)

Jon Russ and Panagiotis Georgiakakis

© Panagiotis Georgiakakis

DISTRIBUTION

Generally limited to mountain areas, although at lower altitudes in Slovenia, Istria and Crete. Distributed from the Pyrenees, Corsica, the Alps from France, Switzerland, Liechtenstein, Italy and Austria to Slovenia. Also, in the Dinaric mountains from Croatia to Albania. In Greece, it is found mostly in the west mainland and on Crete.

EMERGENCE

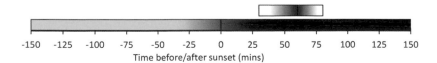

Time before/after sunset (mins)

FLIGHT AND FORAGING BEHAVIOUR

The Alpine long-eared bat has short, broad wings and is a slow flier but highly manoeu-vrable. Usually forages close to or within vegetation. Feeds primarily on moths which it captures by slow aerial hawking, rarely gleaning.

HABITAT

In the Pyrenees, appears to be an open-space forager, foraging over grassland. In northern Italy, transition habitats in rural areas appear to be frequently selected, with shrubland, wetlands and agricultural areas used opportunistically. However, in the Swiss Alps the Alpine long-eared bat selects mainly deciduous forest and intensive grassland. Hunting areas can be very large, exceeding 10 km^2, although most foraging occurs within 1 km of the roost.

Most known summer roosts are in attics, with churches being frequently used. Also in rock crevices, and males and nulliparous females make use of scree deposits as roosts. Have also been found in caves in the southeastern part of its range. Occasionally in tree holes.

ECHOLOCATION

Measured parameters (based on limited data)	Mean (range) FM call (1st harmonic)
Inter-pulse interval (ms)	87.3 (76.4–98.2)
Call duration (ms)	3.2 (1.4–25.2)
Frequency of maximum energy (kHz)	30.7 (28.5–32.9)
Start frequency (kHz)	44.2 (39.0–52.0)
End frequency (kHz)	20.5 (18.0–25.0)

Heterodyne
Quiet, short, narrowband FM sweeps. More commonly seen before it is heard. A very brief and dry 'tick'.

Time expansion/full spectrum
Uses low-intensity multi-harmonic downward frequency-modulated (FM) signals consisting mostly of the first and lower part of the second harmonic (Figure 8.39.1). The first harmonic begins at around 44 kHz and ends at around 22 kHz, with a duration of around 2 ms. In slightly cluttered environments the duration decreases and the call straightens (e.g. Figure 8.39.2). In clutter, the bandwidth of the calls increases and the inter-pulse interval decreases (Figure 8.39.3). In open habitats, such as when flying over a meadow, signals can be up to 7.3 ms and are more shallowly modulated, starting at about 42 kHz and ending around 15 kHz. They occasionally lack the second harmonic.

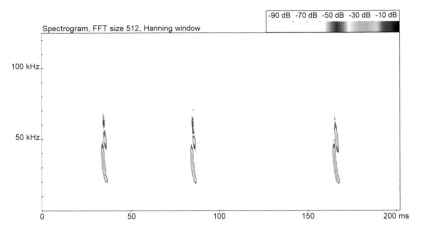

Figure 8.39.1   Typical echolocation calls of the Alpine long-eared bat (Panagiotis Georgiakakis).

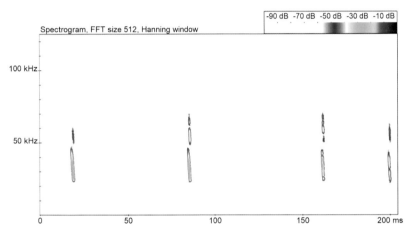

Figure 8.39.2   Echolocation calls of the Alpine long-eared bat upon emergence (Panagiotis Georgiakakis).

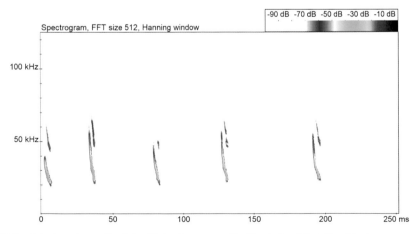

Figure 8.39.3   Echolocation calls of the Alpine long-eared bat in clutter (Christian Diez).

SOCIAL CALLS

Heterodyne
No information available.

Time expansion/full spectrum
No information available.

SPECIES WITH SIMILAR OR OVERLAPPING ECHOLOCATION CALLS

Plecotus spp., *Barbastella barbastellus* (see Chapter 7 species notes).

8.40 Sardinian long-eared bat

Plecotus sardus (Mucedda, Kiefer, Pidinchedda and Veith, 2002)

Mauro Mucedda, Gaetano Fichera, Ermanno Pidinchedda and Andreas Kiefer

© Mauro Mucedda

DISTRIBUTION

The Sardinian long-eared bat, discovered in 2002, is an endemic bat species known only from the island of Sardinia. The population of this species is very small, and it has a limited distribution, being found only in the central part of the island (Mucedda *et al.* 2002).

EMERGENCE

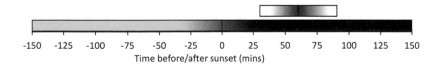

Time before/after sunset (mins)

FLIGHT AND FORAGING BEHAVIOUR

No information available.

HABITAT

The Sardinian long-eared bat occurs in wooded areas of the central part of Sardinia, mainly in karst areas, in both mountain and sea-coast localities.

It roosts in natural caves and also in attics and roofs of buildings (Mucedda *et al.* 2002, Bosso *et al.* 2016). The Sardinian long-eared bat has a very small, declining population and usually forms monospecific nursery colonies.

ECHOLOCATION

Measured parameters	Mean (range) FM call (1st harmonic)
Inter-pulse interval (ms)	67.6 (28.0–94.2)
Call duration (ms)	3.5 (2.8–4.1)
Frequency of maximum energy (kHz)	33.6 (28.1–40.2)
Start frequency (kHz)	47.5 (44.2–53.3)
End frequency (kHz)	21.9 (19.8–24.8)

Heterodyne
The Sardinian long-eared bat emits weak 'ticks' between 25 and 53 kHz that can only be detected within about 5 m.

Time expansion/full spectrum
The Sardinian long-eared bat emits frequency-modulated (FM) signals (sweeps), usually formed by two harmonics (Figure 8.40.1). The first harmonic starts at about 47 kHz and ends at about 20 kHz. The second harmonic starts at about 75 kHz and ends at about 45 kHz. The frequency containing maximum energy (FmaxE) of the first harmonic is about 33 kHz and the average pulse duration is about 3.5 ms. The FmaxE of the second harmonic is about 60 kHz and the average pulse duration is about 2.8 ms. When the bat is flying far away from the bat detector only the first harmonic can be detected. On release from the hand, it generally emits repeated pairs of signals, with two signals separated by a short interval followed by a longer interval.

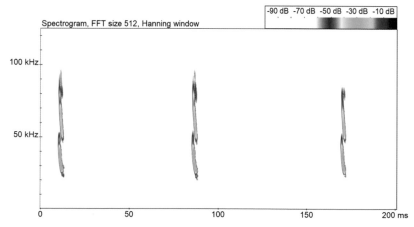

Figure 8.40.1   Echolocation calls of Sardinian long-eared bat recorded in open space after emerging from the roost (Mauro Mucedda).

SOCIAL CALLS

Heterodyne
Social calls are loud and can be heard between 10 and 50 kHz. Often emitted in the vicinity of roosts.

Time expansion/full spectrum
Social calls produced by the Sardinian long-eared bat consist of FM pulses, longer than the normal calls and ending at a lower frequency, emitted usually in groups of two or three (Figure 8.40.2).

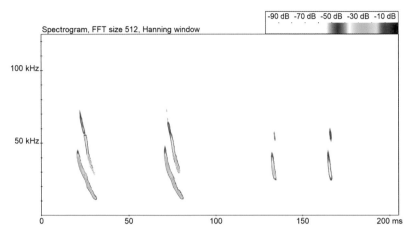

Figure 8.40.2 Social calls of a Sardinian long-eared bat emerging from the roost (Mauro Mucedda).

SPECIES WITH SIMILAR OR OVERLAPPING ECHOLOCATION CALLS

Plecotus spp., *Barbastella barbastellus* (see Chapter 7 species notes).

NOTES

It is currently not possible to distinguish the sounds of *Plecotus sardus* from those of the other *Plecotus* species.

8.41 Grey long-eared bat *Plecotus austriacus* (Fischer, 1829)

Orly Razgour, Erika Dahlberg and Jon Russ

© René Janssen

DISTRIBUTION

The grey long-eared bat is primarily restricted to Europe, but it is also found on the island of Madeira. The distribution extends from the Mediterranean in the south, where this species is relatively common and widespread, to England, Germany and Poland in the north, up to a latitude of 53°N, where it is rare (Juste *et al.* 2008).

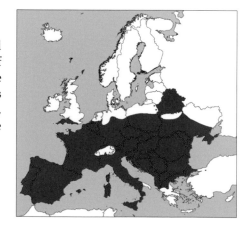

EMERGENCE

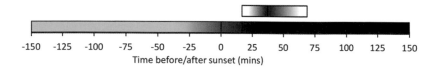

-150   -125   -100   -75   -50   -25   0   25   50   75   100   125   150
Time before/after sunset (mins)

FLIGHT AND FORAGING BEHAVIOUR

The grey long-eared bat has broad and short wings, with relatively low wing-loading values (7.9 N·m^{-2}) and average to low aspect ratio (6.1 A). Its wing morphology favours manoeuvrable slow flight, usually close to the vegetation, with limited long-distance dispersal abilities (Norberg and Rayner 1987). The grey long-eared bat is a sedentary species (Dietz *et al.* 2009). Most individuals remain near their natal area, and maximum recorded migration distance is 62 km (Hutterer *et al.* 2005). The grey long-eared bat can adjust its flight mode to the foraging habitat, flying fast and straight in open habitats but adopting a slower, fluttering flight when foraging in cluttered environments (Flückiger and Beck 1995, Razgour *et al.* 2011a). It generally flies at low altitudes. Based on wing morphology and diet composition (Razgour *et al.* 2011b), the foraging style includes a combination of gleaning and slow aerial hawking.

HABITAT

The grey long-eared bat is primarily an open-edge habitat forager, though it tends to forage low near the vegetation. In the UK, it primarily forages in semi-unimproved (unmanaged) lowland grasslands (including meadows and marshes), woody riparian vegetation and broadleaved woodlands (Razgour *et al.* 2011a). In central Europe it is found primarily in warm lowlands and appears to prefer unforested managed mosaic landscapes and steppe. In southern Europe, it is found in open landscapes as well as open woodlands and is not as closely associated with urban areas (Horáček *et al.* 2004). Nevertheless, unlike its cryptic congener the brown long-eared bat, the grey long-eared bat is not a woodland specialist and tends to use a variety of more open habitat types.

In the northern part of its range, the grey long-eared bat tends to live near human settlements and roost almost exclusively in anthropogenic structures, primarily in the roof spaces of buildings, including churches and barns. Roosts are often located at the edge of a village, surrounded by open grasslands, well-developed hedgerows and woodland patches (Razgour *et al.* 2013). In southern Europe, it roosts in a combination of caves, rock crevices and anthropogenic structures, often churches and old or abandoned buildings near forests. Across Europe, the grey long-eared bat hibernates in summer maternity roosts, cellars, attics, underground galleries, mines, quarries, caves and rock crevices (Stebbings 1970, Horáček 1975, Swift 1998, Dietz *et al.* 2009). Hibernation sites are usually located less than 30 km from summer roosts (Hutterer *et al.* 2005). There is no evidence of the use of swarming sites for mating in this species. Mating probably occurs in maternity colonies.

ECHOLOCATION

Measured parameters	Mean (range) FM call (1st harmonic)
Inter-pulse interval (ms)	105.0 (35.8–194.0)
Call duration (ms)	3.8 (1.4–7.0)
Frequency of maximum energy (kHz)	30.4 (26.3–50.5)
Start frequency (kHz)	43.4 (35.4–50.0)
End frequency (kHz)	21.6 (17.0–31.7)

Heterodyne
The grey long-eared bat uses low-intensity, short, multi-harmonic, frequency-modulated (FM) echolocation calls, characteristic of whispering gleaning bats. Because of their quietness, the grey long-eared bat calls can only be detected from short distances, generally less than 5 m. The bat is more commonly seen before it is heard. The calls sound like a very light purring. The frequency containing maximum energy (FmaxE) is, on average, about 33 kHz. It is not possible to separate brown long-eared bats from grey long-eared bats using a heterodyne detector, and in very cluttered conditions there is often confusion with the *Myotis* species.

Time expansion/full spectrum

Produces FM signals usually consisting of two harmonics (Figure 8.41.1). The first harmonic starts at around 45 kHz and ends at around 21 kHz, and the second harmonic starts at around 59 kHz and sweeps down to about 44 kHz. The FmaxE is usually in the first harmonic and is about 33 kHz. Call duration is about 3.8 ms. In very cluttered situations call duration and inter-pulse interval decrease and FmaxE may be contained in the second harmonic (Figures 8.41.2 and 8.41.3). In relatively uncluttered situations, such as when flying at least 4 m from a structure or the ground, the call duration becomes very long and the FmaxE drops to about 20 kHz (Figure 8.41.4). Occasionally in the open, the second harmonic is dropped altogether (Figure 8.41.5).

Generally, the echolocation calls of grey and brown long-eared bats differ in their start frequency, but the recording must be of good quality and the bat must be within a few metres of the detector. For brown long-eared bats the start frequency is generally above 48 kHz whereas for the grey long-eared bat the start frequency is generally below 48 kHz.

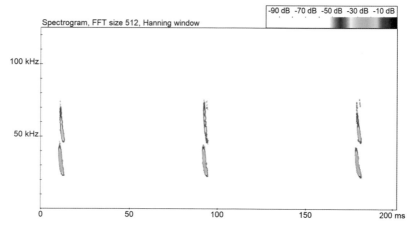

Figure 8.41.1 Typical echolocation calls of a grey long-eared bat flying outside a roost in England (Erika Dahlberg).

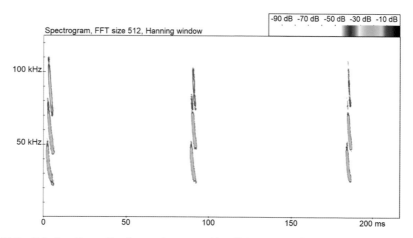

Figure 8.41.2 Echolocation calls of a grey long-eared bat flying in a small barn in England (Erika Dahlberg).

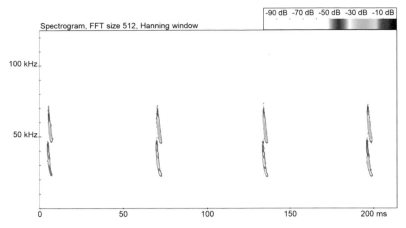

Figure 8.41.3   Typical echolocation calls of a grey long-eared bat flying outside a roost in England, with maximum energy in the second harmonic (Erika Dahlberg).

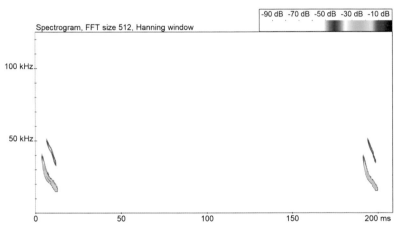

Figure 8.41.4   Echolocation calls of a grey long-eared bat flying in the open in Portugal (Plecotus – Estudos Ambientais, Unip).

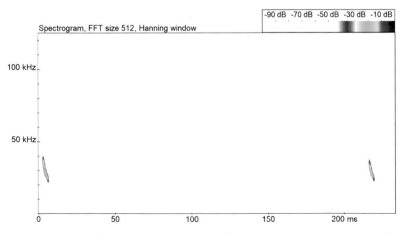

Figure 8.41.5   Echolocation calls of a grey long-eared bat flying in an open area next to woodland in Portugal (Plecotus – Estudos Ambientais, Unip).

SOCIAL CALLS

Heterodyne
Occasionally loud social calls may be heard, particularly around roost sites. The loudest frequency of these calls is about 17 kHz. A variety of types are emitted but it is not possible to differentiate these using a heterodyne detector.

Time expansion/full spectrum
Before emergence, individuals emit low-frequency FM calls in rapid succession (e.g. Figure 8.41.6). Similar calls are also occasionally emitted during dawn swarming (e.g. Figure 8.41.7).

The most common type of social call, which is probably associated with autumnal (and possibly spring) mating, is a call which sweeps down in frequency from about 41 kHz to 11 kHz. The FmaxE is around 17 kHz (Figure 8.41.8).

Distress calls, consisting of a series of rapidly repeated FM sweeps, are often produced when bats are restrained, such as when caught in a trap, held in the hand or caught by a predator (Figure 8.41.9).

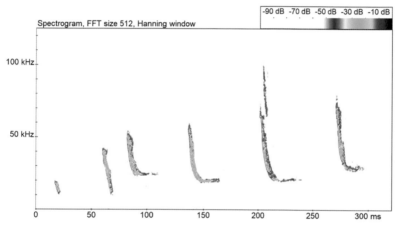

Figure 8.41.6   Variable social calls of a grey long-eared bat emitted from the roost before evening emergence (Erika Dahlberg).

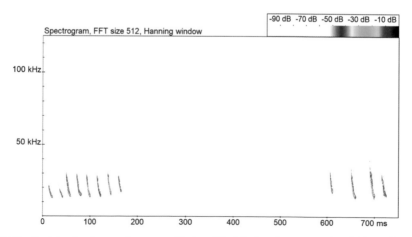

Figure 8.41.7   Social calls of a grey long-eared bat emitted during dawn swarming (Erika Dahlberg).

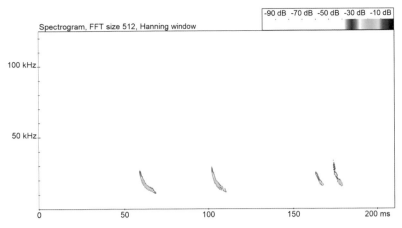

Figure 8.41.8   Social calls of a grey long-eared bat emitted during dawn near the roost (Erika Dahlberg).

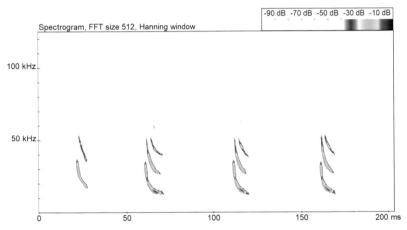

Figure 8.41.9   Social calls of a grey long-eared bat recorded from an individual flying outside the entrance to a roost in an attic space (Erika Dahlberg).

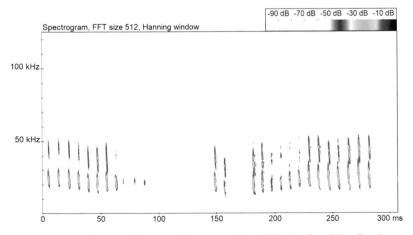

Figure 8.41.10   Distress calls emitted by a grey long-eared bat held in the hand (Jon Russ).

SPECIES WITH SIMILAR OR OVERLAPPING ECHOLOCATION CALLS

Plecotus spp., *Barbastella barbastellus* (see Chapter 7 species notes).

NOTES

There are slight regional differences in the measured parameters of echolocation calls of *Plecotus austriacus* (duration, FmaxE, start frequency, end frequency): UK 2.1 ± 0.7, 30.8 ± 2.8, 44 ± 2.9, 21.8 ± 1.5; Italy 3.8 ± 1.4, 32.6 ± 8.7, 41.4 ± 2.1, 23.6 ± 2.9 (Russo and Jones 2002); Switzerland 5.8 ± 1.4, 27.6 ± 2.5, 45.3 ± 3.3, 18 ± 2.3 (Obrist *et al.* 2004).

We thank *Plecotus* – Estudos Ambientais, Unip for providing recording files from Portugal.

8.42 Mediterranean long-eared bat

Plecotus kolombatovici (Đulić, 1980)

Dina Rnjak and Daniela Hamidović

© Boris Krstinić

DISTRIBUTION

Distributed from Croatia through the Adriatic islands and a narrow coastal strip along the Adriatic Sea, an area near the Greek coast and on some islands in the Aegean Sea. Also found in Crete, Rhodes and Cyprus.

EMERGENCE

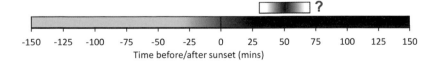

Time before/after sunset (mins)

FLIGHT AND FORAGING BEHAVIOUR

The Mediterranean long-eared bat forages partly in dense vegetation but also flies along vegetation edges and in open habitats. It is suggested that it forages by aerial hawking, since diet analysis so far does not indicate that it also hunts by gleaning (Pavlinić 2008). Bats forage 5–8 m above bushes and forest clearings, thus indicating a similar foraging behaviour to the grey long-eared bat (Flückiger and Beck 1995), as well as flying at a height of 2–3 m in open habitats (Pavlinić 2008). During radio-tracking of 10 adult females in Istria (Croatia) an average distance of 1.5 km was recorded between roosting sites and foraging areas, while the maximum distance was 4.1 km (Pavlinić 2008).

HABITAT

It seems that the Mediterranean long-eared bat almost exclusively inhabits Mediterranean coastal karst areas, usually at lower altitudes, and is often found on islands. It forages in open broadleaved deciduous and Mediterranean evergreen oak woodland, over dense scrub vegetation, thermo-Mediterranean shrubland and grasslands and cultivated habitats (Kiefer and von Helversen 2004, Tvrtković *et al.* 2005, Pavlinić 2008). It also often visits inland surface water such as ponds and wells for drinking, but it is rarely recorded in coniferous woodland and urban areas, especially in comparison to the Alpine long-eared bat, which can more often be found foraging near settlements (Pavlinić 2008).

At higher elevations the Alpine long-eared bat primarily forages in montane and alpine habitats, such as high mountain meadows and rocky areas (Alberdi *et al.* 2012). In comparison, the brown long-eared bat is considered a woodland species and tends to avoid flying in open spaces (Entwistle *et al.* 1996, Kiefer and von Helversen 2004) and sometimes uses hedgerows (Murphy *et al.* 2012), while grey long-eared bats prefer open spaces in river valleys, human settlements and managed agricultural areas (Dietz *et al.* 2009).

Individual bats roost in rock crevices, wall cracks of abandoned buildings and bridges (Dietz *et al.* 2009). Bats are also found in caves, even above 1,000 m above sea level during the swarming and hibernation period (Benda *et al.* 2009). Nursery roosts are recorded in church towers (Tvrtković *et al.* 2005), railway tunnels (Kiefer and von Helversen 2004) and monasteries (D. Hamidović, unpublished data).

ECHOLOCATION

Measured parameters	Mean (range) FM call (1st harmonic)
Inter-pulse interval (ms)	100.6 (29.9–325.0)
Call duration (ms)	2.3 (1.3–4.9)
Frequency of maximum energy (kHz)	33.0 (27.5–41.5)
Start frequency (kHz)	44.2 (34.2–54.9)
End frequency (kHz)	28.3 (23.2–33.6)

Heterodyne
Quiet, short, narrowband FM sweeps, and therefore no variation in sonority can be heard. More commonly seen before it is heard. A very brief and dry 'tick'.

Time expansion/full spectrum

Uses low-intensity FM calls and very often produces harmonics (Figures 8.42.1–8.42.3). It has been reported that the two harmonics usually do not overlap (final frequency of the second harmonic lies above the initial frequency of the first harmonic) (Dietz *et al.* 2009). However, recordings from Lokrum (Croatia) suggest that harmonics overlap most of the time, although more often the initial frequency of the second harmonic starts after the initial frequency of the first harmonic. Sometimes the second harmonic is louder than the first harmonic. Signals in clutter are very brief, under 2 ms in duration, while in open space signals can be up to 4.9 ms. They often sweep down from 40–50 kHz to 25–30 kHz, with the frequency containing maximum energy (FmaxE) usually at 30–35 kHz.

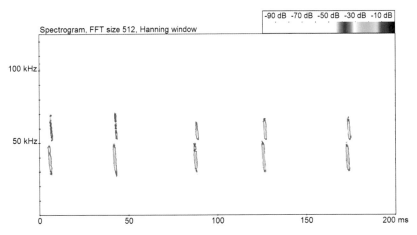

Figure 8.42.1 Echolocation calls of a Mediterranean long-eared bat flying around a tree near the water surface (Daniela Hamidović).

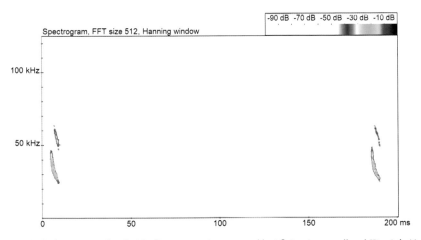

Figure 8.42.2 Echolocation calls of a Mediterranean long-eared bat flying in woodland (Daniela Hamidović).

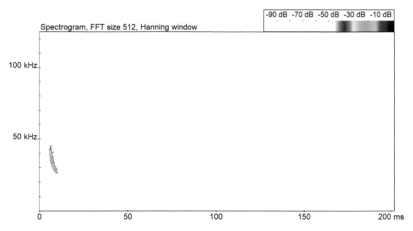

Figure 8.42.3   Echolocation call of a Mediterranean long-eared bat flying high above scrub vegetation and pasture (inter-pulse interval 300 ms) (Daniela Hamidović).

SOCIAL CALLS

Heterodyne
Social calls are rarely heard except around roost sites and consist of loud repeated 'chonks' between 17 kHz and 40 kHz.

Time expansion/full spectrum
Social calls of the Mediterranean long-eared bat have not yet been fully described. Several social calls have been recorded near a roost on Lokrum island, Croatia, and these consist of 2–3 FM calls, sometimes with a sigmoidal shape, somewhat interrupted or with low intensity toward the end of a call (e.g. Figure 8.42.4). Based on five measured calls, they are 8–12 ms long, and sweep down from 41 to 17 kHz with FmaxE around 30 kHz.

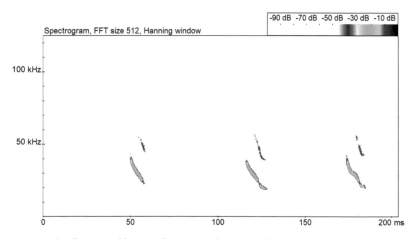

Figure 8.42.4   Social calls emitted by a Mediterranean long-eared bat outside the roost (Daniela Hamidović).

SPECIES WITH SIMILAR OR OVERLAPPING ECHOLOCATION CALLS

Plecotus spp., *Barbastella barbastellus* (see Chapter 7 species notes).

8.43 Schreiber's bent-winged bat

Miniopterus schreibersii (Kuhl, 1817)

Danilo Russo and Jon Russ

© James Shipman

DISTRIBUTION

The species has a broad geographical range in Europe, occurring from the Iberian peninsula east to Russia, while the latitudinal limits are represented by France and Slovakia (north), and Crete (south).

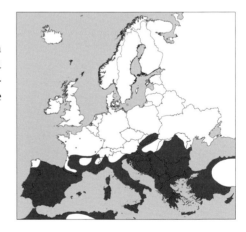

EMERGENCE

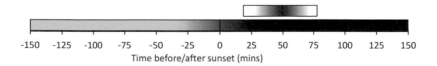

Time before/after sunset (mins)

FLIGHT AND FORAGING BEHAVIOUR

A fast-flying bat foraging by aerial hawking, Schreiber's bent-winged bat may reach flight speeds averaging over 13.9 m·s⁻¹. Generally, flies over straight routes at 5–10 m above the ground or above the forest canopy, but may also fly near forest edges or hedgerows, especially on windy nights (Lugon and Roué 1999, Vincent *et al.* 2011). Bats emerging from a hill or mountain cave typically emerge downhill following the contours of the hill or a dry river bed at high speed into the valley.

HABITAT

Typical foraging habitats are broadleaved forest, orchards, riparian vegetation and hedgerows, where moths (the species' staple food) may be found. May also hunt near street-lamps. Rivers are also used as commuting landmarks (Serra-Cobo *et al.* 2000).

The species typically roosts in underground sites year-round, where it forms large colonies (up to several thousand individuals).

ECHOLOCATION

Measured parameters	Mean (range) FM-qCF call and qCF call
Inter-pulse interval (ms)	85.0 (40.0–209.3)
Call duration (ms)	5.8 (2.0–13.8)
Frequency of maximum energy (kHz)	54.2 (49.4–58.5)
Start frequency (kHz)	85.2 (59.3–113.5)
End frequency (kHz)	51 (47.5–55.7)

Heterodyne
Fast repetition rate, loud 'slaps' peaking at around 54 kHz. As the detector is tuned up from 55 kHz, 'clicks' are heard. Best heard frequencies may range between *c.*50 kHz (in open space) and over 60 kHz (in clutter). Sequences recall those of a common or soprano pipistrelle, and confusion with such species may arise.

Time expansion/full spectrum
Echolocation calls have often an FM-qCF structure and sweep down from *c.*80 kHz to 51 kHz with a frequency of maximum energy (FmaxE) approaching 54 kHz, similar to those of soprano pipistrelle (Figure 8.43.1). End frequencies are usually 50–52 kHz, rarely above 53 kHz. Mean call duration is *c.*6 ms and inter-pulse interval around 80–90 ms. When flying in open space, spectrograms become very narrowband and both FmaxE and end frequency approach 47–49 kHz (Figure 8.43.2). Conversely, in cluttered situations such as in a forest, when approaching prey or from hand release, both pulse rate and frequency bandwidth increase, the FM component largely dominates over the qCF portion, and FmaxE will increase up to more than 60 kHz (Figure 8.43.3). It is common for such short-duration calls to start with a hook (Figure 8.43.4). Compared to pipistrelle species, call duration is quite often more than 8 ms and the pulse repetition is very irregular. Calls are occasionally kinked, particularly during a transition from cluttered to more open habitats (Figure 8.43.5).

In feeding buzzes, the approach phase may be less pronounced than in pipistrelles (e.g. Figures 8.43.6–8.43.8). Social calls broadcast by this species also resemble feeding buzzes, so confusion may arise between these call types.

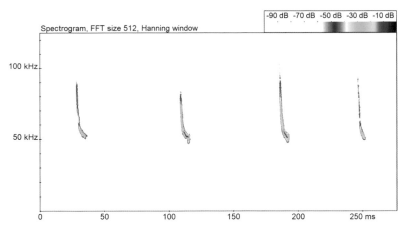

Figure 8.43.1 Typical echolocation calls of Schreiber's bent-winged bat (Danilo Russo).

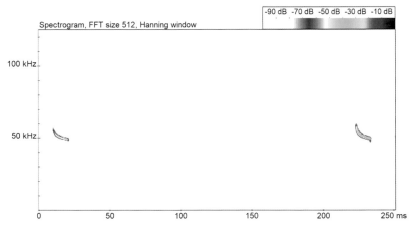

Figure 8.43.2 Echolocation calls of Schreiber's bent-winged bat flying in the open (Danilo Russo).

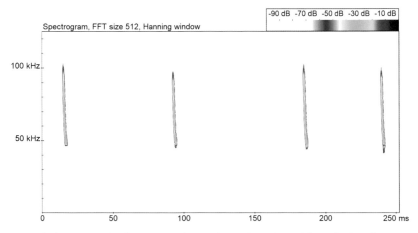

Figure 8.43.3 Echolocation calls of Schreiber's bent-winged bat released from the hand (Jon Russ).

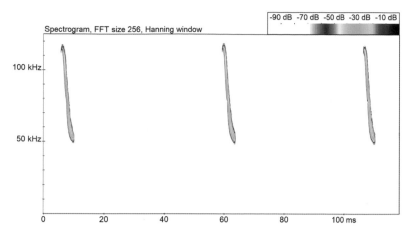

Figure 8.43.4   Echolocation calls of Schreiber's bent-winged bat in a cave, showing hooked start to the call (Jon Russ).

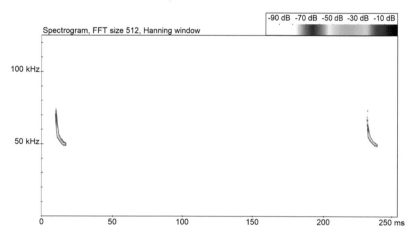

Figure 8.43.5   Echolocation calls of Schreiber's bent-winged bat emerging from a cave (Jon Russ).

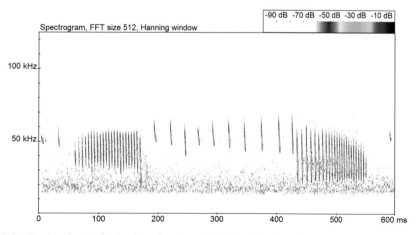

Figure 8.43.6   Feeding buzz of Schreiber's bent-winged bat (Paulo Barros).

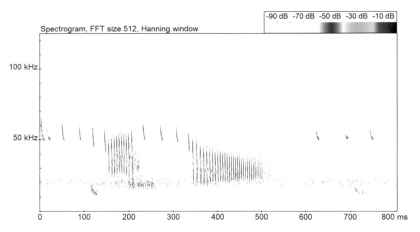

Figure 8.43.7   Feeding buzz of Schreiber's bent-winged bat (Paulo Barros).

SOCIAL CALLS

Heterodyne
Buzz-like social calls. When heard in heterodyne mode they may be mistaken for feeding buzzes.

Time expansion/full spectrum
Schreiber's bent-winged bat broadcasts social calls that are very similar to feeding buzzes and often confused with them (Figure 8.43.8) (Russo and Papadatou 2014). They show a variable number (3–24) of components, with an average of around 8–9. Mean maximum and minimum frequencies are respectively 49.5 kHz (range 38.0–63.3 kHz) and 21.7 kHz (range 19.0–26.0 kHz). These social calls are probably employed in agonistic contexts, as they have frequently been recorded when bats are chasing each other. They are very similar to feeding buzzes (see above), although occasionally they include variable components (e.g. Figures 8.43.9 and 8.43.10).

Social calls are occasionally recorded in the vicinity of cave roosts, where they may consist of undulating trills (e.g. Figure 8.43.11) and short FM sweeps (e.g. Figure 8.43.12).

Distress calls, produced while a bat is under physical duress, are similar in structure to those produced by many of the other vespertilionids (e.g. Figure 8.43.13).

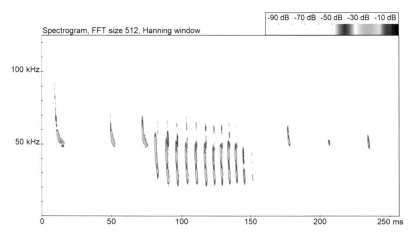

Figure 8.43.8   A social call produced by Schreiber's bent-winged bat during chasing behaviour (Danilo Russo).

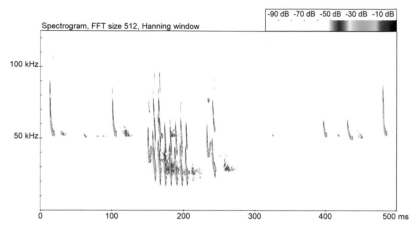

Figure 8.43.9   Social calls produced by Schreiber's bent-winged bat (Paulo Barros).

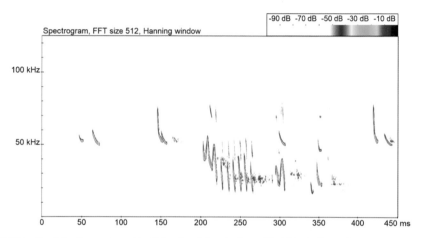

Figure 8.43.10   Social calls produced by Schreiber's bent-winged bat (Paulo Barros).

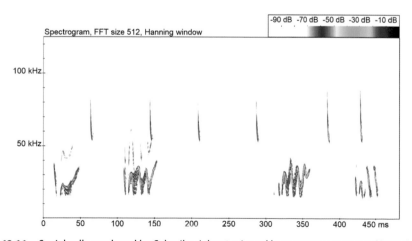

Figure 8.43.11   Social calls produced by Schreiber's bent-winged bat at a cave entrance (Arjan Boonman).

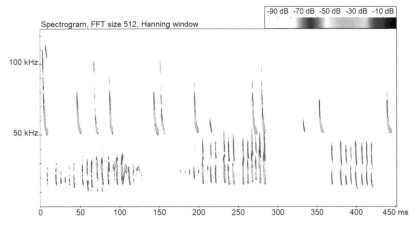

Figure 8.43.12 Social calls produced by Schreiber's bent-winged bat at a cave entrance (Arjan Boonman).

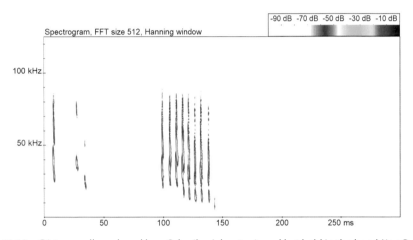

Figure 8.43.13 Distress call produced by a Schreiber's bent-winged bat held in the hand (Jon Russ).

SPECIES WITH SIMILAR OR OVERLAPPING ECHOLOCATION CALLS

Pipistrellus pygmaeus, Pipistrellus hanaki (see Chapter 7 species notes).

NOTES

Confusion is possible with similarly calling pipistrelles (*Pipistrellus pygmaeus* and *P. pipistrellus*). The ambiguity can be resolved if the bat is observed in flight, when its unmistakable long and narrow wing profile and fast flight will clearly distinguish it from the pipistrelles.

8.44 European free-tailed bat
Tadarida teniotis (Rafinesque, 1814)

Leonardo Ancillotto, Francisco Amorim and Ricardo Pérez-Rodríguez

© James Shipman

DISTRIBUTION

The European free-tailed bat has a typically Mediterranean distribution, being frequent and widespread from the entire Iberian peninsula to the west, to the Balkans (Croatia, Bulgaria) and Turkey to the east, ranging through southern regions of France and Switzerland, Italy and Greece, including islands and archipelagos such as the Balearics, Sardinia, Sicily, Crete and Cyprus. Occasionally, the species has been recorded in southern Germany and the UK, presumably vagrants.

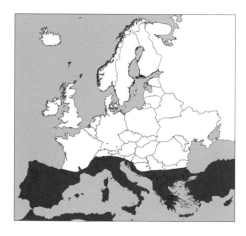

EMERGENCE

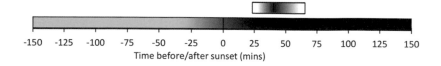

Time before/after sunset (mins)

-150 -125 -100 -75 -50 -25 0 25 50 75 100 125 150

FLIGHT AND FORAGING BEHAVIOUR

This large bat flies fast and high, usually 100–300 m above the ground; it mainly adopts straight trajectories, with shallow and frequent wingbeats, being poor at manoeuvring. Rarely observed directly but easily heard during commuting and foraging; prey captured by aerial hawking.

HABITAT

Flying high above the ground, this species has a weak relationship with land cover; its fast, aerial-hawking foraging style though makes open areas such as fields, pastures and water bodies favoured foraging habitats for this species (Marques *et al.* 2004). Coastal and mountain cliffs are also preferred areas, as well as urban habitats, even in big cities (Russo and Ancillotto 2015). Occasionally, this bat uses artificial lights as foraging areas. In the Mediterranean, it is not unusual to hear it flying along the coastline close to rocky cliffs, as well as over the sea (Ancillotto *et al.* 2014).

Roosts are found in buildings and rock crevices, usually more than 20 m from the ground and 2.5–4 cm in width, preferentially in vertically oriented fissures (Dietz *et al.* 2009, Balmori 2017a, 2017b).

ECHOLOCATION

Measured parameters	Mean (range) FM-qCF call and qCF call
Inter-pulse interval (ms)	635.2 (175.5–1110.0)
Call duration (ms)	16.5 (8.5–36.1)
Frequency of maximum energy (kHz)	14.1 (9.5–18.1)
Start frequency (kHz)	17.0 (13.2–32.3)
End frequency (kHz)	11.9 (9.0–15.2)

Heterodyne

Echolocation and social calls emitted by the European free-tailed bat fall within the hearing range of most people, being identified as loud, high-pitched 'chirps' and 'tweets' coming high from above or from roosts. Heterodyne detecting is thus of no use for this species. Repetition rate is similar to that of noctules but generally slower.

Time expansion/full spectrum

Typical echolocation calls have a narrowband FM-qCF structure with the frequency of maximum energy (FmaxE) around 14 kHz (Figure 8.44.1), usually with no harmonics (but see Zbinden and Zingg 1986). As for other FM-qCF echolocators, in open habitats, when the bat is flying very high above the ground and far from obstacles, such as over water bodies or the sea, the FM portion may be extremely limited and most signals consist of a strong qCF final portion, with inter-pulse intervals in the highest part of the range (Figure 8.44.2). This is the most common commuting call. More strongly FM calls with shorter inter-pulse intervals may be found in the vicinity of roosts, or when animals fly closer to obstacles (e.g. walls of buildings, cliff faces) (Figures 8.44.3 and 8.44.4). Due to the frequency ranges, this species cannot be confused with any other FM-qCF echolocating bat.

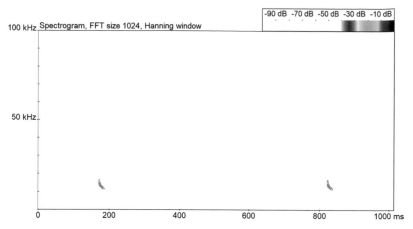

Figure 8.44.1 Typical echolocation calls of the European free-tailed bat (Leonardo Ancillotto).

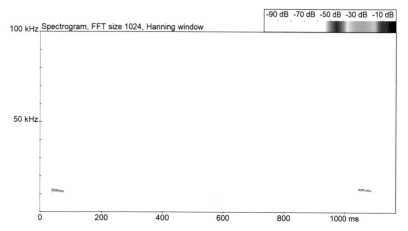

Figure 8.44.2 Echolocation calls of the European free-tailed bat recorded commuting over the Tyrrhenian Sea (Leonardo Ancillotto).

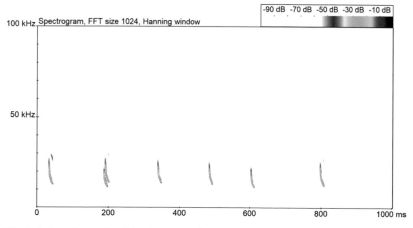

Figure 8.44.3 Echolocation calls of the European free-tailed bat flying in semi-clutter near to a cliff face (Ricardo Pérez-Rodríguez).

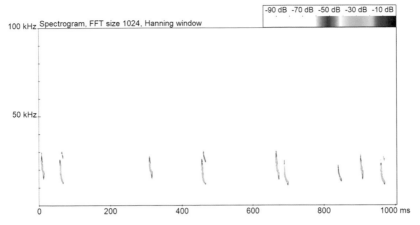

Figure 8.44.4   Variation in echolocation call structure between several individuals of the European free-tailed bat flying in semi-clutter (Ricardo Pérez-Rodríguez).

SOCIAL CALLS

Heterodyne
Social calls can usually be heard without the aid of a bat detector. However, in heterodyne, the calls usually comprise a rapid rasping sound, repeated frequently.

Time expansion/full spectrum
Social calls emitted in flight during foraging are typical 'buzzes' composed of short sequences (3–8 elements) of steep FM-qCF calls, with higher ending frequency and shorter inter-pulse intervals than echolocation signals (Figures 8.44.5 and 8.44.6). V-shaped social calls emitted from bats in flight are also common and can vary in frequency (e.g. Figures 8.44.7 and 8.44.8)

Contact calls are emitted by juveniles and isolated adults both in flight and from roosts, and are typically made of sequences of 1–5 U- or W-shaped modulated calls (Figures 8.44.9 and 8.44.10).

Social calls from roosts are usually single sweeps (long, downward FM-qCF calls), but they can also be combined in long sequences of modulated shorter calls ('trills') (e.g. Figures 8.44.11 and 8.44.12), and these calls can be highly variable. In any case, frequencies of these calls are lower than echolocation calls.

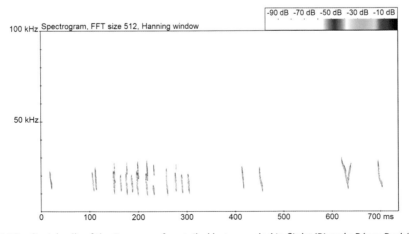

Figure 8.44.5   Social calls of the European free-tailed bat recorded in flight (Ricardo Pérez-Rodríguez).

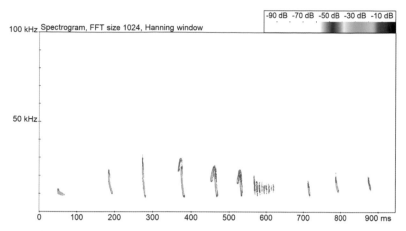

Figure 8.44.6   A further example of social calls of the European free-tailed bat recorded in flight (Yves Bas).

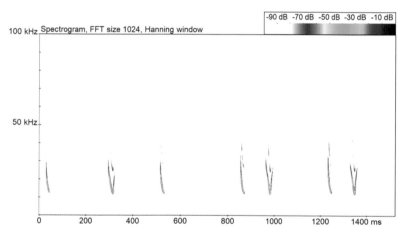

Figure 8.44.7   V-shaped social calls produced by a European free-tailed bat recorded in flight (Marc Van De Sijpe).

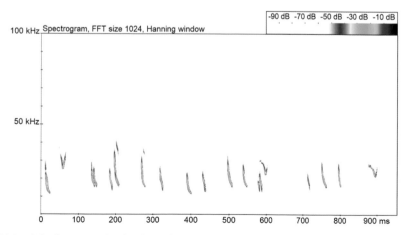

Figure 8.44.8   A further example of V-shaped social calls produced by a European free-tailed bat recorded in flight (Ricardo Pérez-Rodríguez).

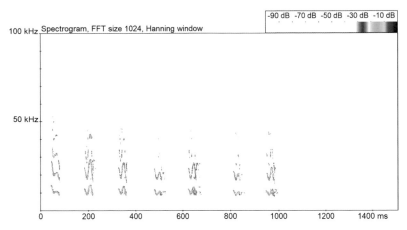

Figure 8.44.9   Contact calls emitted by an isolated adult European free-tailed bat within a roost (Leonardo Ancillotto).

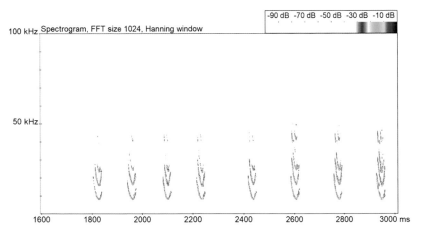

Figure 8.44.10   Further contact calls emitted by an isolated adult European free-tailed bat within a roost (continuation of Figure 8.44.9) (Leonardo Ancillotto).

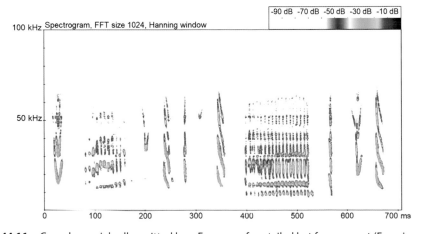

Figure 8.44.11   Complex social calls emitted by a European free-tailed bat from a roost (Francisco Amorin).

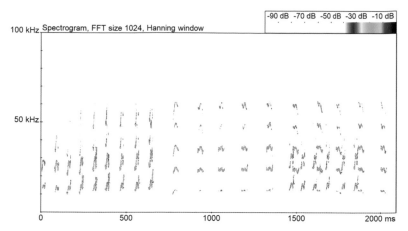

Figure 8.44.12   A further example of complex social calls of a European free-tailed bat emitted from a roost (Francisco Amorin).

SPECIES WITH SIMILAR OR OVERLAPPING ECHOLOCATION CALLS

Nyctalus lasiopterus (see Chapter 7 species notes).

NOTES

Notably, some authors report the highest variability in echolocation call parameters in this species, with significant variation among individuals, sites and geographical regions (Bayefsky-Anand *et al.* 2008).

References and Further Reading

Aellen, V. 1983. Migrations des chauves-souris en Suisse [Migrations of bats in Switzerland]. *Bonner zoologische Beiträge* 34: 3–27.

Ahlén, I. 1981. Identification of Scandinavian bats by their sounds. Department of Wildlife Ecology, SLU, Uppsala, Sweden, report 6: 1–56.

Ahlén, I. and Baagøe, H.J. 1999. Use of ultrasound detectors for bat studies in Europe: experiences from field identification, surveys, and monitoring. *Acta Chiropterologica* 1: 137–150.

Ahlén, I. and Gerell, R. 1989. Distribution and status of bats in Sweden. In Hanàk, V., Horáček, I. and Gaisler, J. (eds), *European Bat Research 1987*. Charles University Press, Praha, pp. 319–325.

Aihartza, J.R., Almenar, D, Salsamendi, E., Goiti, U. and Garin, I. 2008. Fishing behaviour in the longfingered bat *Myotis capaccinii* (Bonaparte, 1837): an experimental approach. *Acta Chiropterologica* 10: 287–301.

Aizpurua, O., Aihartza, J., Alberdi, A., Baagøe, H.J. and Garin, I. 2014. Fine-tuned echolocation and capture-flight of *Myotis capaccinii* when facing different-sized insect and fish prey. *Journal of Experimental Biology* 217: 3318–3325.

Alberdi, A. and Aizpurua, O. 2018. *Plecotus macrobullaris* (Chiroptera: Vespertilionidae). *Mammalian Species* 50(958): 26–33.

Alberdi, A., Garin, I., Aizpurua, O. and Aihartza, J. 2012. The foraging ecology of the mountain long-eared bat *Plecotus macrobullaris* revealed with DNA mini-barcodes. *PLoS ONE* 7(4): e35692.

Alcaldé, J., Benda, P. and Juste, J. 2016. *Rhinolophus mehelyi*. The IUCN Red List of Threatened Species 2016: e.T19519A21974380.

Almenar, D., Aihartza, J.R., Goiti, U., Salsamendi, E. and Garin, I. 2006. Habitat selection and spatial use by the trawling bat *Myotis capaccinii* (Bonaparte, 1837). *Acta Chiropterologica* 8: 157–167.

Almenar, D., Aihartza, J.R., Goiti, U., Salsamendi, E. and Garin, I. 2008. Diet and prey selection in the trawling long-fingered bat. *Journal of Zoology* 274: 340–348.

Altringham, J.D. and Fenton, M.B. 2003. Sensory ecology and communication in the Chiroptera. In Kunz, T.H. and Fenton, M.B. (eds), *Bat Ecology*. University of Chicago Press, Chicago, IL, pp. 90–127.

Ancillotto, L. and Russo, D. 2014. Selective aggressiveness in European free-tailed bat (*Tadarida teniotis*): influence of familiarity, age, and sex. *Naturwissenschaften* 101: 221–228.

Ancillotto, L., Rydell, J., Nardone, V. and Russo, D. 2014. Coastal cliffs on islands as foraging habitat for bats. *Acta Chiropterologica* 16: 103–108.

Ancillotto, L., Santini, L., Ranc, N., Maiorano, L. and Russo. D. 2016. Extraordinary range expansion in a common bat: the potential roles of climate change and urbanisation. *Naturwissenschaften* 103(3–4): 15.

Anderson, M.E. and Racey, P.A. 1991. Feeding behaviour of captive brown long-eared bats, *Plecotus auritus. Animal Behaviour* 42: 489–493.

Andreassen, T., Surlykke, A. and Hallam, J. 2014. Semi-automatic long-term acoustic surveying: a case study with bats. *Ecological Informatics* 21: 13–24.

Andrews, M.M. and Andrews, P.T. 2003. Ultrasound social calls made by greater horseshoe bats (*Rhinolophus ferrumequinum*) in a nursery roost. *Acta Chiropterologica* 5: 221–234.

Andrews, M.M. and Andrews, P.T. 2016. Greater horseshoe bat (*Rhinolophus ferrumequinum*) ultrasound calls outside a nursery roost indicate social interaction not light sampling. *Mammal Communications* 2: 1–8.

Andrews, M.M., Andrews, P.T., Wills, D.F. and Bevis, S.M. 2006. Ultrasound social calls of greater horseshoe bats (*Rhinolophus ferrumequinum*) in a hibernaculum. *Acta Chiropterologica* 8: 197–212.

Andrews, M.M., McOwat, T.P., Andrews, P.T. and Haycock, R.J. 2011. The development of the ultrasound social calls of (*Rhinolophus ferrumequinum*) from infant ultrasound calls. *Bioacoustics* 20: 297–316.

Andrews, M.M., Hodnett, A.M. and Andrews, P.T. 2017. Social activity of lesser horseshoe bats (*Rhinolophus hipposideros*) at nursery roosts and a hibernaculum in north Wales, U.K. *Acta Chiropterologica* 19: 161–174.

Andrews, P.T. 1995. Rhinolophid acoustic orientation. *Myotis* 32–33: 88–90.

Arlettaz, R. 1996. Feeding behaviour and foraging strategy of free-living mouse-eared bats, *Myotis myotis* and *Myotis blythii*. *Animal Behaviour* 51: 1–11.

Arlettaz, R. 1999. Habitat selection as a major resource partitioning mechanism between the two sympatric sibling bat species *Myotis myotis* and *Myotis blythii*. *Journal of Animal Ecology* 68: 460–471.

Arlettaz, R., Perrin, N. and Hausser, J. 1997a. Trophic resource partitioning and competition between the two sibling bat species *Myotis myotis* and *Myotis blythii*. *Journal of Animal Ecology* 66: 897–911.

Arlettaz, R., Ruedi, M., Ibáñez, C., Palmeirim, J. and Hausser, J. 1997b. A new perspective on the zoogeography of the sibling mouse-eared bat species *Myotis myotis* and *Myotis blythii*: morphological, genetical and ecological evidence. *Journal of Zoology* 242: 45–62.

Arlettaz, R., Jones, G. and Racey, P.A. 2001. Effect of acoustic clutter on prey detection by bats. *Nature* 414: 742–745.

Arnold, B.D. and Wilkinson, G.S. 2011. Individual specific contact calls of pallid bats (*Antrozous pallidus*) attract conspecifics at roosting sites. *Behavioral Ecology and Sociobiology* 65: 1581–1593.

Arthur, L. and Lemaire, M. 2009. *Les Chauves-Souris de France, Belgique, Luxembourg et Suisse*. Biotope, Paris.

Artyushin, I.V., Kruskop, S.V., Lebedev, V.S. and Bannikova, A.A. 2018. Molecular phylogeny of serotines (Mammalia, Chiroptera, *Eptesicus*): evolutionary and taxonomical aspects of the *E. serotinus* species group. *Biology Bulletin* 45(5): 469–477.

Ashrafi, S., Beck, A., Rutishauser, M., Arletta, R. and Bontadina, F. 2011. Trophic niche partitioning of cryptic species of long-eared bats in Switzerland: implications for conservation. *European Journal of Wildlife Research* 57: 843–849.

Audet, D. 1990. Foraging behaviour and habitat use by a gleaning bat, *Myotis myotis* (Chiroptera, Vespertilionidae). *Journal of Mammalogy* 71: 420–427.

August, P.V. and Anderson, J.G.T. 1987. Mammal sounds and motivation-structural rules: a test of the hypothesis. *Journal of Mammalogy* 68: 1–9.

Aulagnier, S. and Thevenot, M. 1986. Catalogue des mammiferes sauvages du Maroc. *Travaux de l'intitut Scientifique, Série Zoologie* 41: 1–163.

Azam, C., Le Viol, I., Bas, Y. *et al.* 2018. Evidence for distance and illuminance thresholds in the effects of artificial lighting on bat activity. *Landscape and Urban Planning* 175: 123–135.

Baagøe, H. 2001. Eptesicus serotinus – Breitflügelfledermaus. In Krapp, F. (ed.), *Handbuch der Säugetiere Europas, Fledertiere (Chiroptera) I*, 4th edn. Aula-Verlag, Wiebesheim, pp. 519–559.

Balcombe, J.P. 1990. Vocal recognition of pups by mother Mexican free-tailed bat, *Tadarida brasiliensis mexicana*. *Animal Behaviour* 39: 960–966.

Balcombe, J.P. and McCracken, G.F. 1992. Vocal recognition in Mexican free-tailed bats: do pups recognize mothers? *Animal Behaviour* 43: 79–87.

Balmori, A. 2017a. Murciélago rabudo – *Tadarida teniotis* (Rafinesque, 1814). In Salvador, A. and Barja, I. (eds), *Enciclopedia Virtual de los Vertebrados Españoles*. Museo Nacional de Ciencias Naturales, Madrid.

Balmori, A. 2017b. Advances on the group composition, mating system, roosting and flight behaviour of the European free-tailed bat (*Tadarida teniotis*). *Mammalia* 82: 460–468.

Barataud, M. 1990. Elements sur le comportement alimentaire des oreillards brun et gris, *Plecotus auritus* (Linnaeus, 1758) et *Plecotus austriacus* (Fischer, 1829). *Le Rhinolophe* 7: 3–10.

Barataud, M. 2015. Acoustic ecology of European bats. Species identification, study of their habitats and foraging behaviour. Biotope-Muséum National d'Histoire Naturelle (Inventaires et biodiversité series), 352 pp.

Barlow, K.E. and Jones, G. 1997a. Function of pipistrelle social calls: field data and a playback experiment. *Animal Behaviour* 53: 991–999.

Barlow, K.E. and Jones, G. 1997b. Differences in songflight calls and social calls between two phonic types of the vespertilionid bat *Pipistrellus pipistrellus*. *Journal of the Zoological Society of London* 241: 315–324.

Baron, B. and Vella, A. 2010. A preliminary analysis of the population genetics of *Myotis punicus* in the Maltese islands. *Hystrix* 21: 65–72.

Bartoničková, L., Antonín, R. and Bartonička, T. 2016. Mating and courtship behaviour of two sibling bat species (*Pipistrellus pipistrellus*, *P. pygmaeus*) in the vicinity of a hibernaculum. *Acta Chiropterologica* 8: 467–475.

Bartsch, J. 2012. Das Echoortungsverhalten der Nymphenfledermaus *Myotis alcathoe*. Institut für Neurobiologie. Eberhard Karls Universität, Tübingen.

Bayefsky-Anand, S., Skowronski, M.D., Fenton, M.B., Korine, C. and Holderied, M.W. 2008. Variations in the echolocation calls of the European free-tailed bat. *Journal of Zoology* 275: 115–123.

Behr, O. and von Helversen, O. 2004. Bat serenades: complex courtship songs of the sac-winged bat (*Saccopteryx bilineata*). *Behavioral Ecology and Sociobiology* 56: 106–115.

Behr, O., von Helversen, O., Heckel, G. *et al.* 2006. Territorial songs indicate male quality in the sac-winged bat *Saccopteryx bilineata* (Chiroptera, Emballonuridae). *Behavioral Ecology* 17: 810–817.

Behr, O., Knörnschild, M. and von Helversen, O. 2009. Territorial counter-singing in male sac-winged bats (*Saccopteryx bilineata*): low-frequency songs trigger a stronger response. *Behavioral Ecology and Sociobiology* 63: 433–442.

Benda, P. and Tsytsulina, K.A. 2000. Taxonomic revision of *Myotis mystacinus* group (Mammalia: Chiroptera) in the western Palaearctic. *Acta Societatis Zoologicae Bohemicae* 64: 331–398.

Benda, P. and Uhrin, M. 2017. First records of bats from four Dodecanese islands, Greece (Chiroptera). *Lynx*, n.s. 48: 15–38.

Benda, P., Andreas, M., Kock, D. *et al.* 2006. Bats (Mammalia: Chiroptera) of the Eastern Mediterranean. Part 4. Bat fauna of Syria: distribution, systematics, ecology. *Acta Societatis Zoologicae Bohemicae* 70: 1–329.

Benda, P., Hanak, V., Horacek, I. *et al.* 2007. Bats (Mammalia: Chiroptera) of the Eastern Mediterranean. Part 5. Bat fauna of Cyprus: review of records with confirmation of six species new for the island and description of a new subspecies. *Acta Societatis Zoologicae Bohemicae* 71: 71–130.

Benda, P., Georgiakakis, P., Dietz, C. *et al.* 2009. Bats (Mammalia: Chiroptera) of the Eastern Mediterranean and Middle East. Part 7. The bat fauna of Crete, Greece. *Acta Societatis Zoologicae Bohemicae* 72: 105–190.

Benda, P., Gazaryan, S. and Vallo, P. 2016. On the distribution and taxonomy of bats of the *Myotis mystacinus* morphogroup from the Caucasus region (Chiroptera: Vespertilionidae). *Turkish Journal of Zoology* 40: 842–863.

Benda, P., Satterfield, L., Gücel, S. *et al.* 2018. Distribution of bats in Northern Cyprus (Chiroptera). *Lynx*, n.s. 49: 91–138.

Beuneux, G. 2004. Morphometrics and ecology of *Myotis* cf. punicus (Chiroptera, Vespertilionidae) in Corsica. *Mammalia* 68: 269–273.

Bohn, K.M., Moss, C.F. and Wilkinson, G.S. 2006. Correlated evolution between hearing sensitivity and social calls in bats. *Biology Letters* 2: 561–564.

Bohn, K.M., Wilkinson, G.S. and Moss, C.F. 2007. Discrimination of infant isolation calls by female greater spear-nosed bats, *Phyllostomus hastatus*. *Animal Behaviour* 73: 423–432.

Bohn, K.M., Schmidt-French, B., Ma, S.T. and Pollak, G.D. 2008. Syllable acoustics, temporal patterns, and call composition vary with behavioral context in Mexican free-tailed bats. *Journal of the Acoustical Society of America* 124: 1838–1848.

Bohn, K.M., Smarsh, G.C. and Smotherman, M. 2013. Social context evokes rapid changes in bat song syntax. *Animal Behaviour* 85: 1485–1491.

Boonman, A., Bar-On, Y., Cvikel, N. and Yovel, Y. 2013. It's not black or white: on the range of vision and echolocation in echolocating bats. *Frontiers in Physiology* 4: 248.

Boonman, A., Bumrungsri, S. and Yovel, Y. 2014. Nonecholocating fruit bats produce biosonar clicks with their wings. *Current Biology* 24: 2962–2967.

Boonman, A., Fenton, B. and Yovel, Y. 2019. The benefits of insect-swarm hunting to echolocating bats, and its influence on the evolution of bat echolocation signals. *PLoS Computational Biology* 15(12): e1006873.

Bosso, L., Mucedda, M., Fichera, G., Kiefer, A. and Russo, D. 2016. A gap analysis for threatened bat populations on Sardinia. *Hystrix* 27: 212–214.

Boughman, J.W. 1997. Greater spear-nosed bats give group-distinctive calls. *Behavioral Ecology and Sociobiology* 40: 61–70.

Boughman, J.W. 1998. Vocal learning by greater spear-nosed bats. *Proceedings of the Royal Society B* 265: 227–233.

Boughman, J.W. and Wilkinson, G.S. 1998. Greater spear-nosed bats discriminate groupmates by vocalizations. *Animal Behaviour* 55: 1717–1732.

Bradbury, J.W. and Vehrencamp, S.L. 2011. *Principles of Animal Communication*, 2nd edition. Sinauer Associates, Sunderland, MA.

Braun de Torrez, E.C., Wallrichs, M.A., Ober, H.K. and McCleery, R.A. 2017. Mobile acoustic transects miss rare bat species: implications of survey method and spatio-temporal sampling for monitoring bats. *PeerJ* 5: e3940.

Brenowitz, E.A. 1986. Environmental influences on acoustic and electric animal communication. *Brain, Behavior, and Evolution* 28: 32–42.

Brinkløv, S., Fenton, M.B. and Ratcliffe, J.M. 2013. Echolocation in Oilbirds and swiftlets. *Frontiers in Physiology* 4: 123.

Britton, A.R.C., Jones, G., Rayner, J.M.V., Boonman, A.M. and Verboom, B. 1997. Flight performance, echolocation and foraging behaviour in pond bats, *Myotis dasycneme* (Chiroptera: Vespertilionidae). *Journal of Zoology* 241: 503–522.

Brosset, A. 1955. Observations sur la biologie des chiroptères du Maroc oriental. *Bulletin de la Société des Sciences Naturelles et Physiques du Maroc* 35: 295–306.

Brown, P.E., Brown, T.W. and Grinnell, A.D. 1983. Echolocation, development, and vocal communication in the lesser bulldog bat, *Noctilio albiventris*. *Behavioral Ecology and Sociobiology* 13: 287–298.

Bruckner, A. 2016. Recording at water bodies increases the efficiency of a survey of temperate bats with stationary, automated detectors. *Mammalia* 80: 645–653.

Bruderer, B. and Popa-Lisseanu, A. 2005. Radar data on wing-beat frequencies and flight speeds of two bat species. *Acta Chiropterologica* 7: 73–82.

Carrier, B. 2009. Identification, description et cartographie des habitats de chasse du Murin du Maghreb (*Myotis punicus*, Felten 1977). Master Conservation Restauration Ecosystèmes, Univ. Metz, 69 pp.

Carter, G.G. and Wilkinson, G.S. 2013. Food sharing in vampire bats: reciprocal help predicts donations more than relatedness or harassment. *Proceedings of the Royal Society B* 280: 20122573.

Carter, G.G. and Wilkinson, G.S. 2016. Common vampire bat contact calls attract past food-sharing partners. Animal Behaviour 11: 45–51.

Carter, G.G., Skowronski, M.D., Faure, P.A. and Fenton, B. 2008. Antiphonal calling allows individual discrimination in white-winged vampire bats. *Animal Behaviour* 76: 1343–1355.

Carter, G.G., Logsdon, R., Arnold, B.D., Menchaca, A. and Medellin, R.A. 2012. Adult vampire bats produce contact calls when isolated: acoustic variation by species, population, colony, and individual. *PLoS ONE* 7(6): e38791.

Carter, G.G., Schoeppler, D., Manthey, M., Knörnschild, M. and Denzinger, A. 2015. Distress calls of a fast-flying bat (*Molossus molossus*) provoke inspection flights but not cooperative mobbing. *PLoS ONE* 10(9): e0136146.

Carter, G.G., Wilkinson, G.S. and Page, R.A. 2017. Food-sharing vampire bats are more nepotistic under conditions of perceived risk. *Behavioral Ecology* 28: 565–569.

Castella, V., Ruedi, M., Excoffier, L. *et al.* 2000. Is the Gibraltar Strait a barrier to gene flow for the bat *Myotis myotis* (Chiroptera: Vespertilionidae)? *Molecular Ecology* 9: 1761–1772.

Catchpole, C.K. and Slater, P.J.B. 2008. *Bird Song: Biological Themes and Variations*, 2nd edition. Cambridge University Press, Cambridge.

Catto, C., Hutson, A., Racey, P. and Stephenson, P. 1996. Foraging behaviour and habitat use of the serotine bat (*Eptesicus serotinus*) in southern England. *Journal of Zoology* 238: 623–633.

Centeno-Cuadros, A., Razgour, R., García-Mudarra, J. *et al.* 2019. Comparative phylogeography and asymmetric hybridization between cryptic bat species. *Journal of Zoological Systematics and Evolutionary Research* 57: 1004–1018.

Chaverri, G. and Gillam, E.H. 2016. Acoustic communication and group cohesion in Spix's disc-winged bats. In Ortega, J. (ed.), *Sociality in Bats*. Springer, Switzerland, pp. 161–178.

Chaverri, G., Gillam, E.H. and Vonhof, M.J. 2010. Social calls used by a leaf-roosting bat to signal location. *Biology Letters* 6: 441–444.

Chen, Y., Liu, Q., Su, Q. *et al.* 2016. 'Compromise' in echolocation calls between different colonies of the intermediate leaf-nosed bat (*Hipposideros larvatus*). *PLoS ONE* 11(3): e0151382.

Chenaval, N. 2010. Identification, description et cartographie des habitats de chasse du Murin du Maghreb (*Myotis punicus*, Felten 1977) en période de reproduction, cavité d'Oletta, Corse (2B). Master Expertise Faune, Flore, Inventaires et Indicateurs de Biodiversité, MNHN – Université Pierre et Marie Curie, Paris.

Clement, M.J. and Kanwal, J.S. 2012. Simple syllabic calls accompany discrete behavior patterns in captive *Pteronotus parnelli*: an illustration of the motivation-structure hypothesis. *Scientific World Journal* 2012: 1–15.

Clement, M.J., Murray, K.L., Solick, D.I. and Gruver, J.C. 2014. The effect of call libraries and acoustic filters on the identification of bat echolocation. *Ecology and Evolution* 4: 3482–3493.

Coles, R.B., Guppy, A., Anderson, M.E. and Schlegel, P. 1989. Frequency sensitivity and directional hearing in the gleaning bat, *Plecotus auritus* (Linnaeus 1758). *Journal of Comparative Physiology A* 165: 269–280.

Collins, J. and Jones, G. 2009. Differences in bat activity in relation to bat detector height: implications for bat surveys at proposed windfarm sites. *Acta Chiropterologica* 11: 343–350.

Çoraman, E., Furman, A., Karataş, A. and Bilgin, R. 2013. Phylogeographic analysis of Anatolian bats highlights the importance of the region for preserving the Chiropteran mitochondrial genetic diversity in the Western Palaearctic. *Conservation Genetics* 14: 1205–1216.

Corben, C. 2010. Feeding buzzes. *Australasian Bat Society Newsletter* 35: 40–44.

Corbet, G.B. and Hill, F.E. 1991. *A World list of Mammalian Species*, 3rd edition. British Museum (Natural History), London.

Coroiu, I., Juste, J. and Paunović, M. 2016. *Myotis myotis*. The IUCN Red List of Threatened Species 2016: e.T14133A22051759.

Crucitti, P. 1989. Distribution, diversity and abundance of cave bats in Latium (central Italy). In Hanàk, V., Horáček, I. and Gaisler, J. (eds), *European Bat Research 1987*. Charles University Press, Praha.

Dalhoumi, R., Aissa, P. and Aulagnier, S. 2016. Bat species richness and activity in Bou Hedma National Park (central Tunisia). *Turkish Journal of Zoology* 40: 864–875.

Dalhoumi, R., Morellet, N., Aissa, P. and Aulagnier, S. 2017. Seasonal activity pattern and habitat use by the Isabelline serotine bat (*Eptesicus isabellinus*) in an arid environment of Tunisia. *Acta Chiropterologica* 19: 141–153.

Dechmann, D.K.N., Heucke, S.L., Giuggioli, L. *et al.* 2009. Experimental evidence for group hunting via eavesdropping in echolocating bats. *Proceedings of the Royal Society B* 276: 2721–2728.

Dechmann, D.K.N., Kranstauber, B., Gibbs, D. and Wikelski, M. 2010. Group hunting: a reason for sociality in molossid bats? *PLoS ONE* 5(2): e9012.

Dekker, J.J.A., Regelink, J.R., Jansen, E.A., Brinkmann, R. and Limpens, H.J.G.A. 2013. Habitat use by female Geoffroy's bats (*Myotis emarginatus*) at its two northernmost maternity roosts and the implications for their conservation. *Lutra* 56: 111–120.

Denzinger, A. and Schnitzler, H.-U. 2013. Bat guilds, a concept to classify the highly diverse foraging and echolocation behaviors of microchiropteran bats. *Frontiers in Physiology* 4: 1–16.

Dietrich, S., Dzameitat, D.P., Kiefer, A., Schnitzler, H.U. and Denzinger, A. 2006. Echolocation signals of the plecotine bat, *Plecotus macrobullaris* Kuzjakin, 1965. *Acta Chiropterologica* 8: 465–475.

Dietz, C. and Kiefer, A. 2014. *Die Fledermäuse Europas: Kennen, Bestimmen, Schützen*. Kosmos-Verlag, Stuttgart.

Dietz, C. and Kiefer, A. 2016. *Bats of Britain and Europe*. Bloomsbury, London.

Dietz, C., von Helversen, O. and Nill, D. 2007. *Handbuch der Fledermäuse Europas und Nordwestafrikas*. Kosmos-Verlag, Stuttgart.

Dietz, C., von Helversen, O. and Nill, D. 2009. *Bats of Britain, Europe and Northwest Africa*. A. & C. Black, London.

Dijkgraaf, S. 1960. Spallanzani's unpublished experiments on the sensory basis of object perception in bats. *Isis* 51: 9–20.

Dinger, G. 1991. Winternachweise von Breitflügelfledermausen (*Eptesicus serotinus*) in Kirchen. *Nyctalus* N.F. 1: 521–530.

Disca, T., Allegrini, B. and Prié, V. 2014. Caractéristiques acoustiques des cris d'écholocation de 16 espèces de chiroptères (Mammalia, Chiroptera) du Maroc. *Vespère* 3 (Février 2014).

Dondini, G. and Vergari, S. 2000. Carnivory in the greater noctule bat (*Nyctalus lasiopterus*) in Italy. *Journal of Zoology* 251: 233–236.

Dondini, G., Tomassini, A., Inguscio, S. and Rossic, E. 2014. Rediscovery of Mehely's horseshoe bat (*Rhinolophus mehelyi*) in peninsular Italy. *Hystrix* 25: 59–60.

Dundarova, H., Diez, C., Gazaryan, S, Çoraman, E. and Mayer, F. 2017. The curious phylogenetic pattern of *Myotis mystacinus* and *Myotis davidii* in the Balkan peninsula. doi: 10.13140/RG.2.2.26952.62725.

Eckenweber, M. and Knörnschild, M. 2013. Social influences on territorial signaling in greater sac-winged bats. *Behavioral Ecology and Sociobiology* 67: 639–648.

Eckenweber, M. and Knörnschild, M. 2016. Responsiveness to conspecific distress calls is influenced by day-roost proximity in bats (Saccopteryx bilineata). *Royal Society Open Science* 3: 160151.

Egert-Berg, K., Hurme, E.R. Greif, S. *et al.* 2018. Resource ephemerality drives social foraging in bats. *Current Biology* 28: 3667–3673.

Eklöf, J. and Jones, G. 2003. Use of vision in prey detection by brown long-eared bats, *Plecotus auritus*. *Animal Behaviour* 66: 949–953.

Ekman, M. and DeJong, J. 1996. Local patterns of distribution and resource utilization of four bat species (*Myotis brandti, Eptesicus nilssonii, Plecotus auritus* and *Pipistrellus pipistrellus*) in patchy and continuous environments. *Journal of Zoology* 238: 571–580.

Endler, J.A. 1992. Signals, signal conditions, and the direction of evolution. *American Naturalist* 139: S125–S153.

Engler, S., Rose, A. and Knörnschild, M. 2017. Isolation call ontogeny in bat pups (*Glossophaga soricina*). *Behaviour* 154: 267–286.

Entwistle, A.C., Racey, P.A. and Speakman, J.R. 1996. Habitat exploitation by a gleaning bat, *Plecotus auritus*. *Philosophical Transactions of the Royal Society of London B* 351: 921–931.

Esser, K.-H. and Schmidt, U. 1989. Mother–infant communication in the lesser spear-nosed bat *Phyllostomus discolor* (Chiroptera, Phyllostomidae): evidence for acoustic learning. *Ethology* 82: 156–168.

Estók, P. and Siemers, B.M. 2009. Calls of a bird-eater: the echolocation behaviour of the enigmatic greater noctule, *Nyctalus lasiopterus*. *Acta Chiropterologica* 11: 405–414.

Estók, P., Gombkötő, P. and Cserkész, T. 2007. Roosting behaviour of Greater Noctule *Nyctalus lasiopterus* Schreber, 1780 (Chiroptera, Vespertilionidae) in Hungary as revealed by radio-tracking. *Mammalia* 71: 86–88.

Fanis, E. de and Jones, G. 1995. Postnatal growth, mother–infant interactions and development of vocalizations in the vespertilionid bat, *Plecotus auritus*. *Journal of Zoology* 235: 85–97.

Fanis, E. de and Jones, G. 1996. Allomaternal care and recognition between mothers and young in pipistrelle bats (*Pipistrellus pipistrellus*). *Journal of Zoology* 240: 781–787.

Farina, F., Gori, E., Lazzari, R. *et al.* 1999. Study on a mixed nursery of *Myotis capaccinii* and *Myotis daubentonii* on the Como Lake (Lombardy). *Atti Io Concegno Italiano sui Chirotteri*: 197–210.

Felten, H. 1971. Eine neue Art der Fledermaus-Gattung *Eptesicus* aus Kleinasien (Chiroptera: Vespertilionidae). *Senckenbergiana Biologica* 52(6): 371–376.

Fenton, M.B. and Bell, G.B. 1981. Recognition of species of insectivorous bats by their echolocation calls. *Journal of Mammalogy* 62: 233–243.

Fenton, M.B. and Simmons, N. 2015. *Bats: A World of Science and Mystery*. University of Chicago Press, Chicago, IL.

Fenton, M.B., Rautenbach, I.L., Smith, S.E. *et al.* 1994. Raptors and bats: threats and opportunities. *Animal Behaviour* 48: 9–18.

Fenton, M.B., Portfors, C.V., Rautenbach, I.L. and Waterman, M. 1998. Compromises: sound frequencies used in echolocation by aerial-feeding bats. *Canadian Journal of Zoology* 76: 1174–1182.

Fernandez, A.A., Fasel, N., Knörnschild, M. and Richner, H. 2014. When bats are boxing: aggressive behaviour and communication in male Seba's short-tailed fruit bat. *Animal Behaviour* 98: 149–156.

Fernandez, R. 1989. Patterns of distribution of bats in the Iberian peninsula. In Hanàk, V., Horáček, I. and Gaisler, J. (eds), *European Bat Research 1987*. Charles University Press, Praha.

Flaquer, C., Puig-Montserrat, X., Burgas, A. and Russo, D. 2008. Habitat selection by Geoffroy's bats (*Myotis emarginatus*) in a rural Mediterranean landscape: implications for conservation. *Acta Chiropterologica* 10: 61–67.

Flückiger, P. and Beck, A. 1995. Observations on the habitat use for hunting by *Plecotus austriacus* (Fischer 1829). *Myotis* 32–33: 121–122.

Froidevaux, J.S.P., Zellweger, F., Bollmann, K. and Obrist, M.K. 2014. Optimizing passive acoustic sampling of bats in forests. *Ecology and Evolution* 4: 4690–4700.

Froidevaux, J.S.P., Fialas, P.C. and Jones, G. 2018. Catching insects while recording bats: impacts of light trapping on acoustic sampling. *Remote Sensing in Ecology and Conservation* 4: 240–247.

Fuhrmann, M. and Seitz, A. 1992. Nocturnal activity of the brown long-eared bat (*Plecotus auritus* L. 1758): data from radiotracking in the Lenneburg forest near Mainz (Germany). In Priede, I.G. and Swift, S.M. (eds), *Wildlife Telemetry: Remote Monitoring and Tracking of Animals*. Ellis Horwood, New York, pp. 538–548.

Fujita, K. and Boeckx, C. 2016. The biolinguistic program: a new beginning. In Fujita, K. and Boeckx, C. (eds), *Advances in Biolinguistics: The Human Language Faculty and Its Biological Basis*. Routledge, Abingdon, pp. 1–5.

Fulmer, A.G. and Knörnschild, M. 2012. Intracolonial social distance, signaling modality, and association choice in the greater sac-winged bat (*Saccopteryx bilineata*). *Journal of Ethology* 30: 117–124.

Furmankiewicz, J. 2003. The vocal activity of *Pipistrellus nathusii* (Vespertilionidae) in SW Poland. *Acta Chiropterologica* 5: 97–105.

Furmankiewicz, J. 2004. Social calls and vocal activity of the brown long-eared bat *Plecotus auritus* in SW Poland. *Le Rhinolophe* 17: 101–120.

Furmankiewicz, J., Ruczynski, G.J.I. and Radoslaw, U. 2011. Social calls provide tree-dwelling bats with information about location of conspecifics at roosts. *Ethology* 117: 480–489.

Furmankiewicz, J., Duma, K., Manias, K. and Borowiec, M., 2013. Reproductive status and vocalisation in swarming bats indicate a mating function of swarming and an extended mating period in Plecotus auritus. *Acta Chiropterologica* 15: 371–385.

Gelfand, D.L. and McCracken, G.F. 1986. Individual variation in the isolation calls of Mexican free-tailed bat pups (*Tadarida brasiliensis mexicana*). *Animal Behaviour* 34: 1078–1086.

Georgiakakis, P. and Russo, D. 2012. The distinctive structure of social calls by Hanák's dwarf bat *Pipistrellus hanaki*. *Acta Chiropterologica* 14: 167–174.

Georgiakakis, P., Poursandis, D., Kantzaridou, M., Kontogeorgos, G. and Russo, D. 2018. The importance of forest conservation for the survival of the range-restricted *Pipistrellus hanaki*, an endemic bat from Crete and Cyrenaica. Mammalian Biology 93: 109–117.

Gerell, R. and Lundberg, K. 1985. Social organization in the bat *Pipistrellus pipistrellus*. *Behavioral Ecology and Sociobiology* 16: 177–184.

Gerell-Lundberg, K. and Gerrell, R. 1994. The mating behaviour of pipistrelle and the Nathusius' pipistrelle (Chiroptera): a comparison. *Folia Zoologica* 43: 315–324.

Gillam, E.H. and Chaverri, G. 2012. Strong individual signatures and weaker group signatures in contact calls of Spix's disc-winged bat, *Thyroptera tricolor*. *Animal Behaviour* 83: 269–276.

Gillam, E.H., Chaverri, G., Montero, K. and Sagot, M. 2013. Social calls produced within and near the roost in two species of tent-making bats, *Dermanura watsoni* and *Ectophylla alba*. *PLoS ONE* 8(4): e61731.

Gillooly, J.F. and Ophir, A.G. 2010. The energetic basis of acoustic communication. *Proceedings of the Royal Society B* 277: 1325–1331.

Gjerde, L. 2004. Methods in surveying advertisement calling *Vespertilio murinus* L., 1758, and notes on its fall distribution in Europe. *Le Rhinolophe* 17: 127–132.

Goiti, U., Garin, I., Almenar, D., Salsamendi, E. and Aihartza, J. 2008. Foraging by Mediterranean horseshoe bats (*Rhinolophus euryale*) in relation to prey distribution and edge habitat. *Journal of Mammalogy* 89: 493–502.

Goiti, U., Aihartza, J., Guiu, M. *et al.* 2011. Geoffroy's bat, *Myotis emarginatus*, preys preferentially on spiders in multistratified dense habitats: a study of foraging bats in the Mediterranean. *Folia Zoologica* 60: 17–24.

Gottfried, I. 2009. Use of underground hibernacula by the barbastelle (*Barbastella barbastellus*) outside the hibernation season. *Acta Chiropterologica* 11: 363–373.

Götze, S., Denzinger, A. and Schnitzler, H.-U. 2020. High frequency social calls indicate food source defense in foraging common pipistrelle bats. *Scientific Reports* 10(1): 5764.

Gould, E. 1971. Studies of maternal–infant communication and development of vocalizations in the bats *Myotis* and *Eptesicus*. *Communications in Behavioral Biology* 5: 263–313.

Gould, E. 1975. Experimental studies of ontogeny of ultrasonic vocalizations in bats. *Developmental Psychobiology* 8: 333–346.

Gray, L.J. 1993. Response of insectivorous birds to emerging aquatic insects in riparian habitats of a tallgrass prairie stream. *American Midland Naturalist* 129: 288–300.

Greenaway, F. and Hill, D.A. 2004. Woodland management advice for Bechstein's bat and barbastelle bat. English Nature Research Reports 658.

Greif, S. and Yovel, Y. 2019. Using on-board sound recordings to infer behaviour of free-moving wild animals. *Journal of Experimental Biology* 222: jeb184689.

Griffin. D.R. 1944. Echolocation by blind men, bats, and radar. *Science* 100: 589–590.

Griffin, D.R. 1958. *Listening in the Dark: The Acoustic Orientation of Bats and Men.* Yale University Press, New Haven, CT.

Griffin, D.R. 1971. The importance of atmospheric attenuation for the echolocation of bats. *Animal Behaviour* 19: 55–61.

Griffin, D.R. and Novick, A. 1955. Acoustic orientation of neotropical bats. *Journal of Experimental Zoology* 130: 251–299.

Griffin, D.R. and Thompson, D. 1982. Echolocation by cave swiftlets. *Behavioral Ecology and Sociobiology* 10: 119–123.

Guillen, A. 1999. *Myotis capaccinii* (Bonaparte, 1837). In Mitchell-Jones, A.J., Amori, G., Bogdanowicz, W. *et al.* (eds), *The Atlas of European Mammals.* T. & A.D. Poyser, London: 106–107.

Gustin, M.K. and McCracken, G.F. 1987. Scent recognition between females and pups in the bat *Tadarida brasiliensis mexicana. Animal Behaviour* 35: 13–19.

Güttinger, R., Lustenberger, J., Beck, A. and Weber, U. 1998. Traditionally cultivated wetland meadows as foraging habitats of the grass-gleaning lesser mouse-eared bat (*Myotis blythii*). *Myotis* 36: 41–49.

Hafner, J., Dietz, C., Schnitzler, H.-U. and Denzinger, A. 2015. Das Echoortungsverhalten der Nymphenfledermaus *Myotis alcathoe.* In *Tagungsband Verbreitung und Ökologie der Nymphenfledermaus.* Bayerisches Landesamt für Umwelt, Augsburg, pp. 27–34.

Hamidović, D. 2005. Echolocation and wing morphology in the long-fingered bat, *Myotis capaccinii* (Bonaparte, 1837) (Mammalia: Chiroptera). MSc thesis, Zagreb.

Hanák, V., Benda, P., Ruedi, E., Horacek, I. and Sofianidou, T. 2001. Bats (Mammalia: Chiroptera) of the Eastern Mediterranean. Part 2. New records and review of distribution of bats in Greece. *Acta Societatis Zoologicae Bohemicae* 65: 279–346.

Haplea, S., Covey, E. and Casseday, J.H. 1994. Frequency tuning and response latencies at three levels in the brainstem of the echolocating bat, *Eptesicus fuscus. Journal of Comparative Physiology A* 174: 671–683.

Harbusch, C. 2003. Aspects of the ecology of serotine bat (*Eptesicus serotinus*) in contrasting landscapes in southwest Germany and Luxembourg. PhD thesis, University of Aberdeen.

Hartridge, H. 1920. The avoidance of objects by bats in their flight. *Journal of Physiology* 54: 54–57.

Hayes, J.P. 1997. Temporal variation in activity of bats and the design of echolocation-monitoring studies. *Journal of Mammalogy* 78: 514–524.

Heller, K.-G. and von Helversen, O. 1989. Resource partitioning of sonar frequency bands in rhinolophoid bats. *Oecologia* 80: 178–186.

Hill, D.A. and Greenaway, F. 2006. Putting Bechstein's bat on the map. Final report to Mammals Trust UK. London.

Hof, D. and Podos, J. 2013. Escalation of aggressive vocal signals: a sequential playback study. *Proceedings of the Royal Society B* 280: 20131553.

Holderied, M.W., Korine, C., Fenton, M.B. *et al.* 2005. Echolocation call intensity in the aerial hawking bat *E. bottae* (Vespertilionidae) studied using stereo videogrammetry. *Journal of Experimental Biology* 208: 1321–1327.

Horáček, I. 1975. Notes on the ecology of bats of the genus *Plecotus* Geoffroy, 1818 (Mammalia: Chiroptera). *Vestnik Ceskoslovenske Spolecnosti Zoologicke* 34: 195–210.

Horáček, I., Bogdanowicz, W. and Dulid, B. 2004. *Plecotus austriacus* (Fischer, 1829) – Graues Langohr. In Krapp, F. (ed.), *Handbuch der Säugetiere Europas, Fledertiere (Chiroptera) II,* 4th edn. Aula-Verlag, Wiebesheim, pp. 1001–1049.

Horta, P., Raposeira, H., Santos, H. *et al.* 2015. Bats' echolocation call characteristics of cryptic Iberian *Eptesicus* species. *European Journal of Wildlife Research* 61: 813–818.

Huang, X., Metzner, W., Zhang, K. *et al.* 2018. Acoustic similarity elicits responses to heterospecific distress calls in bats (Mammalia: Chiroptera). *Animal Behaviour* 146: 143–154.

Hulva, P., Benda, P., Hanak, V., Evin, A. and Horacek, I. 2007. New mitochondrial lineages within the *Pipistrellus pipistrellus* complex from Mediterranean Europe. *Folia Zoologica* 56: 378–388.

Hutson, A.M. 2018. Review of species to be listed on the Annex to the Agreement. 8th Session of the Meeting of the Parties. Inf.EUROBATS.MoP8.9.

Hutterer, R., Ivanova, T., Meyer-Cords, C. and Rodrigues, L. 2005. *Bat Migration in Europe: A Review of Banding Data and Literature*. Naturschutz und Biologische Vielfault Heft 28. Federal Agency for Nature Conservation, Bonn.

Ibáñez, C., Juste, J., García-Mudarra, J.L. and Agirre-Mendi, P.T. 2001. Bat predation on nocturnally migrating birds. *PNAS* 98: 9700–9702.

Ibáñez, C., Guillén, A. and Bogdanowicz, W. 2004. *Nyctalus lasiopterus* (Schreber, 1780) – Riesenabendsegler. In Krapp, F. (ed.), *Handbuch der Säugetiere Europas, Fledertiere (Chiroptera) II*, 4th edn. Aula-Verlag, Wiebesheim, pp. 695–716.

Ibáñez, C., García-Mudarra, J.L., Ruedi, M., Stadelmann, B. and Juste, J. 2006. The Iberian contribution to cryptic diversity in European bats. *Acta Chiropterologica* 8: 277–297.

Irwin, N.R. and Speakman, J.R. 2003. Azorean bats *Nyctalus azoreum* cluster as they emerge from roosts, despite the lack of avian predators. *Acta Chiropterologica* 5: 185–192.

Jackson, J.K. and Fisher, S.G. 1986. Secondary production, emergence, and export of aquatic insects of a Sonoran Desert stream. *Ecology* 67: 629–638.

Jahelková, H. 2011. Unusual social calls of Nathusius' pipistrelle (Vespertilionidae, Chiroptera) recorded outside the mating season. *Journal of Vertebrate Zoology* 60: 25–30.

Jahelková, H. and Horáček, I. 2011. Mating system of a migratory bat, Nathusius' pipistrelle (*Pipistrellus nathusii*): different male strategies. *Acta Chiropterologica* 13: 123–137.

Jahelková, H., Horáček, I. and Bartonicka, T. 2008. The advertisement song of *Pipistrellus nathusii* (Chiroptera, Vespertilionidae): a complex message containing acoustic signatures of individuals. *Acta Chiropterologica* 10: 103–126.

Janssen, S. and Schmidt, S. 2009. Evidence for a perception of prosodic cues in bat communication: contact call classification by *Megaderma lyra*. *Journal of Comparative Physiology A* 195: 663–672.

Jensen, M. and Miller, L. 1999. Echolocation signals of the bat *Eptesicus serotinus* recorded using a vertical microphone array: effect of flight altitude on searching signals. *Behavioral Ecology and Sociobiology* 47: 60–69.

Jensen, M.E., Miller, L.A. and Rydell, J. 2001. Detection of prey in a cluttered environment by the northern bat *Eptesicus nilssonii*. *Journal of Experimental Biology* 204: 199–208.

Jesus, J., Teixeira, S., Teixeira, D., Freitas, T. and Russo, D. 2009. Vertebrados terrestres autóctones dos Arquipélagos da Madeira e Selvagens. Répteis e mamíferos. Biodiversidade Madeirense: Avaliação e Conservação. Direcção Regional do Ambiente, Funchal.

Jesus, J., Teixeira, S., Freitas, T., Teixeira, D. and Brehm, A. 2013. Genetic identity of *Pipistrellus maderensis* from the Madeira archipelago: a first assessment, and implications for conservation. *Hystrix* 24: 177–180.

Jetz, W., Steffen, J. and Linsenmair, K.E. 2003. Effects of light and prey availability on nocturnal, lunar and seasonal activity of tropical nightjars. *Oikos* 103: 627–639.

Jin, L., Yang, S., Kimball, R.T. *et al.* 2015. Do pups recognize maternal calls in pomona leaf-nosed bats, *Hipposideros pomona*? *Animal Behaviour* 100: 200–207.

Jones, G. 1995. Flight performance, echolocation and foraging behaviour in noctule bats *Nyctalus noctula*. *Journal of Zoology* 237: 303–312.

Jones, G. and Rayner, J.M.V. 1988. Flight performance, foraging tactics and echolocation in free-living Daubenton's bats *Myotis daubentonii* (Chiroptera: Vespertilionidae). *Journal of Zoology* 215: 113–132.

Jones, G. and Rayner, J.M.V. 1989. Foraging behaviour and echolocation of wild horseshoe bats, *Rhinolophus ferrumequinum* and *Rhinolophus hipposideros* (Chiroptera, Rhinolophidae) *Behavioral Ecology and Sociobiology* 25: 183–191.

Jones, G. and Rydell, J. 1994. Foraging strategy and predation risk as factors influencing emergence time in echolocating bats. *Philosophical Transactions of the Royal Society of London B* 346: 445–455.

Jones, G. and Siemers, B.M. 2011. The communicative potential of bat echolocation pulses. *Journal of Comparative Physiology A* 197: 447–457.

Jones, G. and Teeling, E.C. 2006. The evolution of echolocation in bats. *Trends in Ecology and Evolution* 21: 149–156.

Jones, G., Hughes, P.M. and Rayner, J.M.V. 1991. The development of vocalizations in *Pipistrellus pipistrellus* (Chiroptera: Vespertilionidae) during postnatal growth and the maintenance of individual vocal signatures. *Journal of Zoology* 225: 71–84.

Jones, G., Gordon, G. and Nightingale, J. 1992. Sex and age differences in the echolocation calls of the lesser horseshoe bat, *Rhinolophus hipposideros, Mammalia* 56: 189–193.

Jones, G., Vaughan, N. and Parsons, S. 2000. Acoustic identification of bats directly sampled and time-expanded recordings of vocalizations. *Acta Chiropterologica* 2: 155–170.

Jones, K.E., Russ, J.A., Bashta, A.-T. *et al.* 2013. Indicator bats program: a system for the global acoustic monitoring of bats. In Collen, B., Pettorelli, N.J.E., Baillie, M. and Durant, S.M. (eds), *Biodiversity Monitoring and Conservation.* Wiley-Blackwell, Oxford, pp. 213–248.

Jones, P.L., Page, R.A., Hartbauer, M. and Siemers, B.M. 2011. Behavioral evidence for eavesdropping on prey song in two Palearctic sibling bat species. *Behavioral Ecology and Sociobiology* 65: 333–340.

Juste, J. and Paunović, M. 2016a. *Myotis blythii. The IUCN Red List of Threatened Species* 2016: e.T14124A22053297.

Juste, J. and Paunović, M. 2016b. *Myotis punicus. The IUCN Red List of Threatened Species* 2016: e.T44864A22073410.

Juste, J., Karatas, A., Palmeirim, J. *et al.* 2008. *Plecotus austriacus. The IUCN Red List of Threatened Species,* version 2010.4.

Juste, J., Benda, P., García-Mudarra, J.L. and Ibáñez, C. 2013. Phylogeny and systematics of Old World serotine bats (genus *Eptesicus,* Vespertilionidae, Chiroptera): an integrative approach. *Zoologica Scripta* 42: 441–457.

Juste, J., Ruedi, M., Puechmaille, S.J., Salicini, I. and Ibáñez, C. 2019. Two new cryptic bat species within the *Myotis nattereri* species complex (Vespertilionidae, Chiroptera) from the Western Palaearctic. *Acta Chiropterologica* 20: 285–300.

Kalko, E.K.V. 1990. Field study on the echolocation and hunting behaviour of the long-fingered bat, *Myotis capaccinii. Bat Research News* 31(3): 42–43.

Kalko, E.K.V. and Schnitzler, H.-U. 1989. The echolocation and hunting behaviour of Daubenton's bat, *Myotis daubentonii* (Chiroptera, Vespertilionidae). *Behavioral Ecology and Sociobiology* 24: 225–238.

Kapteyn, K. and Lina, P.H.C. 1994. First record of a nursery roost of Nathusius' pipistrelle, *Pipistrellus nathusii,* in the Netherlands. *Lutra* 37: 106–109.

Kerth, G. 2008. Causes and consequences of sociality in bats. *Bioscience* 58: 737–746.

Kiefer, A. and von Helversen, O. 2004. *Plecotus kolombatovici* – Balkanlangohr. In Krapp, F. (ed.), *Handbuch der Säugetiere Europas, Fledertiere (Chiroptera) II,* 4th edn. Aula-Verlag, Wiebesheim, pp. 1059–1066.

King, R.S. and Wrubleski, D.A. 1998. Spatial and diel availability of flying insects as potential duckling food in prairie wetlands. *Wetlands* 18: 100–114.

Knörnschild, M. 2014. Vocal production learning in bats. *Current Opinion in Neurobiology* 28: 80–85.

Knörnschild, M. and Tschapka, M. 2012. Predator mobbing behaviour in the Greater Spear-Nosed Bat, *Phyllostomus hastatus. Chiroptera Neotropical* 18: 1132–1135.

Knörnschild, M. and von Helversen, O. 2008. Nonmutual vocal mother–pup recognition in the greater sac-winged bat. *Animal Behaviour* 76: 1001–1009.

Knörnschild, M., Behr, O. and von Helversen, O. 2006. Babbling behaviour in the sac-winged bat (*Saccopteryx bilineata*). *Naturwissenschaften* 93: 451–454.

Knörnschild, M., von Helversen, O. and Mayer, F. 2007. Twin siblings sound alike: isolation call variation in the noctule bat, *Nyctalus noctula. Animal Behaviour* 74: 1055–1063.

Knörnschild, M., Feidel, M. and Kalko, E.K.V. 2013. Mother–offspring recognition in the bat *Carollia perspicillata. Animal Behaviour* 86: 941–948.

Knörnschild, M., Bluml, S., Steidl, P., Eckenweber, M. and Nagy, M. 2017. Bat songs as acoustic beacons: male territorial songs attract dispersing females. *Scientific Reports* 7: 13918.

Kondo, N. and Watanabe, S. 2009. Contact calls: information and social function. *Japanese Psychological Research* 51: 197–208.

Konstantinov, A.I. 1973. Development of echolocation in bats in postnatal ontogenesis. *Periodicum Biologorum.* 75: 13–19.

Konstantinov, A.I., Dzenaevich, O.S. and Ivaruian, E.G. 1990. The development of lesser horseshoe bats (*Rhinolophus hipposideros*) and the formation of the location signal in ontogeny. *Nernaia Sistema* 29: 127–135.

Krapp, F. (ed.) 2001. *Handbuch der Säugetiere Europas, Fledertiere I*, 4th edn. Aula-Verlag, Wiebesheim.

Krull, D., Schumm, A., Metzner, W. and Neuweiler, G. 1991. Foraging areas and foraging behaviour in the notch eared bat, *Myotis emarginatus* (Vespertilionidae). *Behavioral Ecology and Sociobiology* 28: 247–253.

Kubista, C.E. and Bruckner, A. 2017. Within-site variability of field recordings from stationary, passively working detectors. *Acta Chiropterologica* 19: 189–197.

Kuc, R. 1994. Sensorimotor model of bat echolocation and prey capture. *Journal of the Acoustical Society of America* 96: 1965–1978.

Kuhl, W., Schodder, G.R. and Schröder, F.-K. 1954. Condenser transmitters and microphones with solid dielectric for airborne ultrasonics. *Acustica* 4: 519–532.

Law, B.S., Reinhold, L. and Pennay, M. 2002. Geographic variation in the echolocation calls of *Vespadelus* spp. (Vespertilionidae) from New South Wales and Queensland, Australia. *Acta Chiropterologica* 4: 201–215.

Lawrence, B.D. and Simmons, J.A. 1982. Measurements of atmospheric attenuation at ultrasonic frequencies and the significance for echolocation by bats. *Journal of the Acoustical Society of America* 71: 585–590.

Leippert, D. 1994. Social behaviour on the wing in the false vampire, *Megaderma lyra*. *Ethology* 98: 111–127.

Leippert, D., Goymann, W., Hofer, H., Marimuthu, G. and Balasingh, J. 2000. Roost-mate communication in adult Indian false vampire bats (*Megaderma lyra*): an indication of individual in temporal and spectral pattern. *Animal Cognition* 3: 99–106.

Lemen, C., Freeman, P.W., White, J.A. and Andersen, B.R. 2015. The problem of low agreement among automated identification programs for acoustical surveys of bats. *Western North American Naturalist* 75: 218–225.

Leonardo, M. and Medeiros, F.M. 2011. Preliminary data about the breeding cycle and diurnal activity of the Azorean bat (*Nyctalus azoreum*). *Açoreana* 7: 139–148.

Lesinski, G., Gryz, J. and Kowalski, M. 2009a. Bat predation by tawny owls *Strix aluco* in differently human-transformed habitats. *Italian Journal of Zoology* 76: 415–421.

Lesinski, G., Ignaczak, M. and Manias, J. 2009b. Opportunistic predation on bats by the tawny owl *Strix aluco*. *Animal Biology* 59: 283–288.

Lewis, S. and Llewellyn-Jones, L. 2018. *The Culture of Animals in Antiquity*. Routledge, New York.

Limpens, H.G.J.A. and Feenstra, M. 1997. Franjestaart *Myotis nattereri* (Kuhl, 1817). In Limpens, H.G.J.A., Mostert, K. and Bongers, F. (eds), *Atlas van de Nederlandse Vleermuizen*. Uitgeverij KNNV, Utrecht, pp. 91–100.

Limpens, H.J.G.A. and Roschen, A. 1995. *Bestimmung der mitteleuropäischen Fledermausarten anhand ihrer Rufe: Lern- und Übungskassette mit Begleitheft*. BAG Fledermausschutz im Naturschutzbund Deutschland/NABU Projectgruppe Fledermauserfassung Niedersachsen, Bremervörde, Germany.

Lin, A., Jiang, T., Kanwal, J.S. *et al*. 2015. Geographical variation in echolocation vocalizations of the Himalayan leaf-nosed bat: contribution of morphological variation and cultural drift. *Oikos* 124: 364–371.

Lin, H.-J., Kanwal, J.S., Jiang, T.-L., Liu, Y. and Feng, J. 2015. Social and vocal behavior in adult greater tube-nosed bats (*Murina leucogaster*). *Zoology (Jena)* 118: 192–202.

Liu, Y., Feng, J., Jiang, Y.L., Wu, L. and Sun, K.P. 2007. Vocalization development of greater horseshoe bat, *Rhinolophus ferrumequinum* (Rhinolophidae, Chiroptera). *Folia Zoologica* 56: 126–136.

López-Baucells, A., Torrent, L., Rocha, R. *et al*. 2019. Stronger together: combining automated classifiers with manual post-validation optimizes the workload vs reliability trade-off of species identification in bat acoustic surveys. *Ecological Informatics* 49: 45–53.

Lučan, R.K. and Šálek, M. 2013. Observation of successful mobbing of an insectivorous bat, *Taphozous nudiventris* (Emballonuridae), on an avian predator, *Tyto alba* (Tytonidae). *Mammalia* 77: 235–236.

Lugon, A. and Roué, S.Y. 1999. *Miniopterus schreibersii* forages close to vegetation, results from faecal analysis from two eastern French maternity colonies. *Bat Research News* 40: 126–127.

Lundberg, K. and Gerell, R. 1986. Territorial advertisement and mate attraction in the bat *Pipistrellus pipistrellus*. *Ethology* 71: 115–124.

Luo, B., Huang, X., Li, Y. *et al.* 2017. Social call divergence in bats: a comparative analysis. *Behavioral Ecology* 28: 533–540.

Lynch, R.J., Bunn, S.E. and Catterall, C.P. 2002. Adult aquatic insects: potential contributors to riparian food webs in Australia's wet–dry tropics. *Austral Ecology* 27: 515–526.

Ma, J., Kobayasi, K., Zhang, S. and Metzner, W. 2006. Vocal communication in adult greater horseshoe bat, *Rhinolophus ferrumequinum*. *Journal of Comparative Physiology A* 192: 535–550.

Macias, S., Mora, E.C., Coro, F. and Kössl, M. 2006. Threshold minima and maxima in the behavioral audiograms of the bats *Artibeus jamaicensis* and *Eptesicus fuscus* are not produced by cochlear mechanics. *Hearing Research* 212: 245–250.

Maluleke, T., Jacobs, D.S. and Winker, H. 2017. Environmental correlates of geographic divergence in a phenotypic trait: a case study using bat echolocation. *Ecology and Evolution* 7: 7347–7361.

Manley, G.A. 2012. *Peripheral Hearing Mechanisms in Reptiles and Birds*. Springer, Berlin.

Marques, J.T., Rainho, A., Carapuço, M., Oliveira, P. and Palmeirim, J.M. 2004. Foraging behaviour and habitat use by the European free-tailed bat *Tadarida teniotis*. *Acta Chiropterologica* 6: 99–110.

Matsumara, S. 1979. Mother–infant communication in a horseshoe bat (*Rhinolophus ferrumequinum nippon*): Development of vocalization. *Journal of Mammalogy* 60: 76–84.

Matsuo, I., Kunugiyama, K. and Yano, M. 2004. An echolocation model for range discrimination of multiple closely spaced objects: transformation of spectrogram into the reflected intensity distribution. *Journal of the Acoustical Society of America* 115: 920–928.

Mayberry, H.W. and Faure, P.A. 2015. Morphological, olfactory, and vocal development in big brown bats. *Biology Open* 4: 22–34.

Mayer, F., Dietz, C. and Kiefer, A. 2007. Molecular species identification boosts bat diversity. *Frontiers in Zoology* 4: 4.

McCracken, G. and Bradbury, J. 2000. Bat mating systems. In Crichton, E. and Krutzsch, P. (eds), *Reproductive Biology of Bats*. Academic Press, San Diego, CA, pp. 321–362.

McWilliam, A. 1987. Territoriality and pair behaviour of the African false vampire bat, *Cardioderma cor* (Chiroptera: Megadermatidae), in coastal Kenya. *Journal of Zoology* 213: 243–252.

Mehdizadeh, R., Eghbali, H. and Sharifi, M. 2018. Postnatal growth and vocalization development in the long-fingered bat, *Myotis capaccinii* (Chiroptera, Vespertilionidae). *Zoological Studies* 57: e37.

Middleton, N. 2020. *Is That a Bat? A Guide to Non-Bat Sounds Encountered During Bat Surveys*. Pelagic Publishing, Exeter.

Middleton, N., Froud, A. and French, K. 2014. *Social Calls of the Bats of Britain and Ireland*. Pelagic Publishing, Exeter.

Miller, L.A. and Degn, H.J. 1981. The acoustic behaviour of four Vespertilionid bats studied in the field. *Journal of Comparative Physiology* 142: 67–74.

Mitchell-Jones, A.J., Amori, G., Bogdanowicz, W. *et al.* (eds). 1999. *The Atlas of European Mammals*. T. & A.D. Poyser, London.

Moehres, F.P. 1952. Die Ultraschall-Orientierung der Fledermäuse. *Naturwissenschaften* 39: 273–279.

Moehres, F.P. and Kulzer, E. 1955. Untersuchungen fiber die Ultraschallorientierung von vier afrikanischen Fledermausfamilien. *Verhandlungen der deutschen Zoologischen Gesellschaft* 49: 59–65.

Moermans, T. 2000. Kolonieplaatsselectie en dieet van de Ingekorven vlermuis, *Myotis emarginatus* in Vlaanderen, MSc thesis, University of Antwerp, Belgium.

Monroy, J.A., Carter, M.E., Miller, K.E. and Covey, E. 2011. Development of echolocation and communication vocalizations in the big brown bat, *Eptesicus fuscus*. *Journal of Comparative Physiology A* 197: 459–467.

Morton, E.S. 1975. Ecological sources of selection on avian sounds. *American Naturalist* 109: 17–85.

Morton, E.S. 1977. On the occurrence and significance of motivation-structural rules in some bird and mammal sounds. *American Naturalist* 111: 855–869.

Mucedda, M., Murittu, G., Oppes, A. and Pidinchedda, E. 1995. Osservazioni sui Chirotteri troglofili della Sardegna. *Bollettino della Società Sarda di Scienze Naturali* 30: 97–129.

Mucedda, M., Kiefer, A., Pidinchedda, E. and Veith, M. 2002. A new species of long-eared bat (Chiroptera, Vespertilionidae) from Sardinia (Italy). *Acta Chiropterologica* 4: 121–135.

Mucedda, M., Pidinchedda, E. and Bertelli, M.L. 2009. Status del Rinolofo di Mehely (*Rhinolophus mehelyi*) (Chiroptera, Rhinolophidae) in Italia. *Atti del 2° Convegno Italiano sui Chirotteri, Serra San Quirico (AN)*, 21–23 novembre 2008: 89–98.

Murphy, S.E. 2012. Function of social calls in brown long-eared bats *Plecotus auritus*. DPhil thesis, University of Sussex.

Murphy, S.E., Greenaway, F. and Hill, D.A. 2012. Patterns of habitat use by female brown long-eared bats presage negative impacts of woodland conservation management. *Journal of Zoology* 288: 177–183.

Murray, K.L., Britzke, E.R. and Robbins, L.W. 2001. Variation in search-phase calls of bats. *Journal of Mammalogy* 82: 728–737.

Mutumi, G.L., Jacobs, D.S. and Winker, H. 2016. Sensory drive mediated by climatic gradients partially explains divergence in acoustic signals in two horseshoe bat species, *Rhinilophus swinnyi* and *Rhinolophys simulator*. *PLoS ONE* 11(1): e0148053.

Nardone, V., Ancillotto, L. and Russo, D. 2017. A flexible communicator: social call repertoire of Savi's pipistrelle, *Hypsugo savii*. *Hystrix* 28: 68–72.

Necker, R. 2000. The avian ear and hearing. In Whttow, G.C. (ed.), *Sturkie's Avian Physiology*, 5th edition. Academic Press, New York, pp. 21–38.

Neuweiler, G. 1989. Foraging ecology and audition in echolocating bats. *Trends in Ecology and Evolution* 4: 6–12.

Neuweiler, G. 2000. Echolocation. In *The Biology of Bats*. Oxford University Press, Oxford, pp. 151–156.

Newson, S.E. 2017. How should static detectors be deployed to produce robust national population trends for British bat species? British Trust for Ornithology.

Norberg, U.M. 1976. Aerodynamics of hovering flight in Long-eared bat *Plecotus auritus*. *Journal of Experimental Biology* 65: 459–470.

Norberg, U.M. and Rayner, J.M.V. 1987. Ecological morphology and flight in bats (Mammalia; Chiroptera): wing adaptations, flight performance, foraging strategy and echolocation. *Philosophical Transactions of the Royal Society of London B* 316: 335–427.

Novick, A. 1958. Orientation in paleotropical bats. I. Microchiroptera. *Journal of Experimental Zoology* 138: 81–153.

Novick, A. 1963. Orientation in neotropical bats. II. Phyllostomatidae and Desmodontidae. *Journal of Mammalogy* 44: 44–56.

Noyes, A. and Pierce, G.W. 1937. Apparatus for acoustic research in the supersonic frequency range. *Journal of the Acoustical Society of America* 9: 205–211.

Obrist, M.K. and Giavi, S. 2016. Bioakustisches Monitoring von Fledermäusen – Methoden, Aufwand und Grenzen. *Natur und Landschaft Inside* 4: 17–21.

Obrist, M.K., Boesch, R. and Flückiger, P.F. 2004. Variability in echolocation call design of 26 Swiss bat species: consequences, limits and options for automated field identification with a synergetic pattern recognition approach. *Mammalia* 68: 307–322.

O'Farrell, M.J., Corben, C. and Gannon, W.L. 2000. Geographic variation in the echolocation calls of the hoary bat (*Lasiurus cinereus*). *Acta Chiropterologica* 2: 185–196.

O'Mara, M.T., Wikelski, M., Kranstauber, B. and Dechmann, D.K.N. 2019. Common noctules exploit low levels of the aerosphere. *Royal Society Open Science* 6: 181942.

Omer, D.B., Maimon, S.R., Las, L. and Ulanovsky, N. 2018. Social place-cells in the bat hippocampus. *Science* 359: 218–224.

Page, R.A. and Ryan, M.J. 2006. Social transmission of novel foraging behavior in bats: frog calls and their referents. *Current Biology* 16: 1201–1205.

Panyutina, A.A., Kuznetsov, A.N., Volodin, I.A., Abramov, A.V. and Soldatova, I.B. 2017. A blind climber: the first evidence of ultrasonic echolocation in arboreal mammals. *Integrative Zoology* 12: 172–184.

Papadatou, E., Butlin, R.K. and Altringham, J.D. 2008a. Seasonal roosting habits and population structure of the long-fingered bat *Myotis capaccinii* in Greece. *Journal of Mammalogy* 89: 503–512.

Papadatou, E., Butlin, R.K. and Altringham, J.D. 2008b. Identification of bat species in Greece from their echolocation calls. *Acta Chiropterologica* 10: 127–143.

Parijs, S.M.V. and Corkeron, P.J. 2002. Ontogeny of vocalisations in infant black flying foxes, *Pteropus alecto*. *Behaviour* 139: 1111–1124.

Paunović, M. 2016. *Myotis capaccinii. The IUCN Red List of Threatened Species* 2016: e.T14126A22054131.

Pavlinić, I. 2008. Ekologija gorskog dugouhog i Kolombatovićevog dugouhoh šišmiša. Doktorska disertacija. Biološki odsjek Prirodoslovno-matematičkog fakulteta Sveučilišta u Zagrebu, Zagreb.

Pérez, J. and Ibáñez, C. 1991. Preliminary results on activity rhythms and space use obtained by radio-tracking a colony of *Eptesicus serotinus. Myotis* 29: 61–66.

Peterson, H., Finger, N., Bastian, A. and Jacobs, D. 2019. The behaviour and vocalisations of captive Geoffroy's horseshoe bats, *Rhinolophus clivosus* (Chiroptera: Rhinolophidae). *Acta Chiropterologica* 20: 439–453.

Petrzelkova, K.J. and Zukal, J. 2003. Does a live barn owl (*Tyto alba*) affect emergence behaviour of serotine bats (*Eptesicus serotinus*)? *Acta Chiropterologica* 5: 177–184.

Pfalzer, G. 2017. Inter- and intra-specific variability of social calls from native bat species. An English translation of the results from the thesis of Dr Guido Pfalzer.

Pfalzer, G. and Kusch, J. 2003. Structure and variability of bat social calls: implications for specificity and individual recognition. *Journal of Zoology* 261: 21–33.

Pfeiffer, B., Hammer, M., Marckmann, U. *et al.* 2015. Die Verbreitung der Nymphenfledermaus *Myotis alcathoe* in Bayern. In *Tagungsband Verbreitung und Ökologie der Nymphenfledermaus*. Bayerisches Landesamt für Umwelt, Augsburg, pp. 98–114.

Pinheiro, A.D., Wu, M. and Jen, P.H.S. 1991. Encoding repetition rate and duration in the inferior colliculus of the big brown bat, *Eptesicus fuscus. Journal of Comparative Physiology A* 169: 69–85.

Piraccini, R. 2016. *Plecotus sardus. The IUCN Red List of Threatened Species* 2016: e.T136503A518549.

Poon, P.W.F., Sun, X., Kamada, T. and Jen, P.H.S. 1990. Frequency and space representation in the inferior colliculus of the FM bat, *Eptesicus fuscus. Experimental Brain Research* 79: 83–91.

Popa-Lisseanu, A.G., Delgado, A., Forero, M. *et al.* 2007. Bats' conquest of a formidable foraging niche: the myriads of nocturnally migrating songbirds. *PLoS ONE* 2: e205.

Popa-Lisseanu, A.G., Bontadina, F. and Ibáñez, C. 2009. Giant noctule bats face conflicting constraints between roosting and foraging in a fragmented and heterogeneous landscape. *Journal of Zoology* 278: 126–133.

Poupart, T. 2011. Le Murin du Maghreb (*Myotis punicus*, Felten 1977) en Corse: identification et description des terrains de chasse – colonie d'Aghione (2B): 47 pp.

Prat, Y., Taub, M. and Yovel, Y. 2015. Vocal learning in a social mammal demonstrated by isolation and playback experiments in bats. *Science Advances* 1(2): e1500019.

Prat, Y., Taub, M. and Yovel, Y. 2016. Everyday bat vocalizations contain information about emitter, addressee, context, and behaviour. *Scientific Reports* 6: 39419.

Prat, Y., Azoulay, L., Dor, R. and Yovel, Y. 2017. Crowd vocal learning induces vocal dialects in bats: Playback of conspecifics shapes fundamental frequency usage by pups. *PLoS Biology* 15(10): e2002556.

Preatoni, D.G., Spada, M., Wauters, L.A. and Martinoli, A. 2011. Habitat use in the female Alpine long-eared bat (*Plecotus macrobullaris*): does breeding make the difference? *Acta Chiropterologica* 13: 355–364.

Puechmaille, S.J., Allegrini, B., Boston, E.S.M. *et al.* 2012. Genetic analyses reveal further cryptic lineages within the *Myotis nattereri* species complex. *Mammalian Biology – Zeitschrift für Säugetierkunde* 77: 224–228.

Pye, J.D. 1968. Animal sonar in air. *Ultrasonics* 6: 32–38.

Pye, J.D. 1980. Adaptiveness of echolocation signals in bats: flexibility in behaviour and in evolution. *Trends in Neurosciences* 3(10): 232–235.

Rainho, A. and Palmeirim, J.M. 1999. Foraging behaviour and habitat selection in *Myotis myotis*. In *VIIIth European Bat Research Symposium*. Chiropterological Information Center, Krakow, p. 53.

Rainho, A. and Palmeirim, J.M. 2013. Prioritizing conservation areas around multispecies bat colonies using spatial modeling. *Animal Conservation* 16: 438–448.

Rainho, A., Marques, J.T. and Palmeirim, J., 2002. Os morcegos dos arquipélagos dos Açores e da Madeira: um contributo para a sua conservação. Centro de Biologia Ambiental, Instituto de Conservação de Natureza, Lisboa.

Rainho, A., Augusto, A.M. and Palmeirim, J.M. 2010. Influence of vegetation clutter on the capacity of ground foraging bats to capture prey. *Journal of Applied Ecology* 47: 850–858.

Ramakers, J.J.C., Dechmann, D.K.N., Page, R.A. and O'Mara, M.T. 2016. Frugivorous bats prefer information from novel social partners. *Animal Behaviour* 116: 83–87.

Ramos Pereira, M.J., Rebelo, H., Rainho, A. and Palmeirim, J.M. 2002. Prey selection by *Myotis myotis* (Vespertilionidae) in a Mediterranean region. *Acta Chiropterologica* 4: 183–193.

Ransome, R.D. 1971. The effect of ambient temperature on the arousal frequency of the hibernating greater horseshoe bat, *Rhinolophus ferrumequinum*, relating to site selection and the hibernation state. *Journal of Zoology* 164: 357–371.

Ransome, R.D. 2008. Bats, Chiroptera, Greater horseshoe bat, *Rhinolophus ferrumequinum*. In Harris, S. and Yaldon, D.W. (eds), *Mammals of the British Isles, Handbook*, 4th edition. Mammal Society, London.

Rasmuson, T.M. and Barclay, R.M.R. 1992. Individual variation in the isolation calls of newborn big brown bats (*Eptesicus fuscus*): is variation genetic? *Canadian Journal of Zoology* 70: 698–702.

Razgour, O., Hanmer, J. and Jones, G. 2011a. Using multi-scale modelling to predict habitat suitability for species of conservation concern: the grey long-eared bat as a case study. *Biological Conservation* 144: 2922–2930.

Razgour, O., Clare, E.L. and Zeale, M.R.K. 2011b. High-throughput sequencing offers insight into mechanisms of resource partitioning in cryptic bat species. *Ecology and Evolution* 1: 556–570.

Razgour, O., Whitby, D., Dahlberg, E. *et al.* 2013. *Conserving Grey Long-Eared Bats (Plecotus austriacus) in our Landscape: a Conservation Management Plan*. University of Bristol and Bat Conservation Trust. Available at www.bats.org.uk (accessed September 2020).

Redgwell, R.D., Szewczak, J.M., Jones, G. and Parsons, S. 2009. Classification of echolocation calls from 14 species of bat by support vector machines and ensembles of neural networks. *Algorithms* 2: 907–924.

Reyes, G. 2013. The surprising social calls of a 'solitary' bat. *BCI BATS magazine* 31.

Richardson, S.M., Lintott, P.R., Hosken, D.J. and Mathews, F. 2019. An evidence-based approach to specifying survey effort in ecological assessments of bat activity. *Biological Conservation* 231: 98–102.

Robinson, M. and Stebbings, R. 1997. Homerange and habitat use by the serotine bat, *Eptesicus serotinus*, in England. *Journal of Zoology* 243: 117–136.

Rodrigues, L., Zahn, A., Rainho, A., Palmeirim, J.M. 2003. Contrasting the roosting behaviour and phenology of an insectivorous bat (*Myotis myotis*) in its southern and northern distribution ranges. *Mammalia* 67: 321–335.

Rudolph, B.U., Liegl, A. and von Helversen, O. 2009. Habitat selection and activity patterns in the greater mouse-eared bat *Myotis myotis*. *Acta Chiropterologica* 11: 351–361.

Russ, J. 2012. *British Bat Calls: A Guide to Species Identification*. Pelagic Publishing, Exeter.

Russ, J.M. and Racey, P.A. 2007. Species-specificity and individual variation in the song of male Nathusius' pipistrelles (*Pipistrellus nathusii*). *Behavioral Ecology and Sociobiology* 61: 669–677.

Russ, J.M., Racey, P.A. and Jones, G. 1998. Intraspecific responses to distress calls of the pipistrelle bat, *Pipistrellus pipistrellus*. *Animal Behaviour* 55: 705–713.

Russ, J.M., Briffa, M. and Montgomery, W.I. 2003. Seasonal patterns in activity and habitat use by bats (*Pipistrellus* spp. and *Nyctalus leisleri*) in Northern Ireland determined using a driven transect. *Journal of Zoology* 259: 289–299.

Russ, J.M., Jones, G., Mackie, I. and Racey, P.A. 2004. Interspecific responses to distress calls in bats (Chiroptera: Vespertilionidae): a function for convergence in call design? *Animal Behaviour* 67: 1005–1014.

Russo, D. and Ancillotto, L. 2015. Sensitivity of bats to urbanization: a review. *Mammalian Biology – Zeitschrift für Säugetierkunde* 80: 205–212.

Russo, D. and Jones, G. 1999. The social calls of Kuhl's pipistrelles *Pipistrellus kuhlii* (Kuhl, 1819): structure and variation. *Journal of Zoology* 249: 476–481.

Russo, D. and Jones, G. 2002. Identification of twenty-two bat species (Mammalia: Chiroptera) from Italy by analysis of time-expanded recordings of echolocation calls. *Journal of Zoology* 258: 91–103.

Russo, D. and Jones, G. 2003. Use of foraging habitats by bats in a Mediterranean area determined by acoustic surveys: conservation implications. *Ecography* 26: 197–209.

Russo, D. and Papadatou, E. 2014. Acoustic identification of free-flying Schreiber's bat *Miniopterus schreibersii* by social calls. *Hystrix* 25: 119–120.

Russo, D. and Voigt, C. 2016. The use of automated identification of bat echolocation calls in acoustic monitoring: a cautionary note for a sound analysis. *Ecological Indicators* 66: 598–602.

Russo, D., Jones, G. and Mucedda, M. 2001. Influence of age, sex and body size on echolocation calls of Mediterranean and Mehely's horseshoe bats, *Rhinolophus euryale* and *R. mehelyi* (Chiroptera: Rhinolophidae). *Mammalia* 65: 429–436.

Russo, D., Jones, G. and Migliozzi, A. 2002. Habitat selection by the Mediterranean horseshoe bat, *Rhinolophus euryale* (Chiroptera: Rhinolophidae), in a rural area of southern Italy and implications for conservation. *Biological Conservation* 107: 71–81.

Russo, D., Almenar, D., Aihartza, J. *et al.* 2005. Habitat selection in sympatric *Rhinolophus mehelyii* and *R. euryale* (Mammalia: Chiroptera). *Journal of Zoology* 266: 327–332.

Russo, D., Jones, G. and Arlettaz, R. 2007a. Echolocation and passive listening by foraging mouse-eared bats *Myotis myotis* and *M. blythii*. *Journal of Experimental Biology* 210: 166–176.

Russo, D., Mucedda, M., Bello, M. *et al.* 2007b. Divergent echolocation call frequencies in insular rhinolophids (Chiroptera): a case of character displacement? *Journal of Biogeography* 34: 2129–2138.

Russo, D., Teixeira, S., Cistrone, L. *et al.* 2009. Social calls are subject to stabilizing selection in insular bats. *Journal of Biogeography* 36: 2212–2221.

Rydell, J. 1986. Feeding territoriality in female northern bats *Eptesicus nilssonii*. *Ethology* 72: 329–337.

Rydell, J. 1990. Occurrence of bats in northernmost Sweden 65°N and their feeding ecology in summer. *Journal of Zoology* 227: 517–529.

Rydell, J. and Baagøe, H.J. 1994. *Vespertilio murinus*. *Mammalian Species* 467: 1–6.

Rydell, J., Entwistle, A. and Racey, P.A. 1996. Timing of foraging flights of three species of bats in relation to insect activity and predation risk. *Oikos* 76: 243–252.

Rydell, J., Nyman, S., Eklöf, J., Jones, G. and Russo, D. 2017. Testing the performances of automated identification of bat echolocation calls: a request for prudence. *Ecological Indicators* 78: 416–420.

Salicini, I., Ibáñez, C. and Juste, J. 2011. Multilocus phylogeny and species delimitation within the Natterer's bat species complex in the Western Palearctic. *Molecular Phylogenetics and Evolution* 61: 888–898.

Salsamendi, E., Aihartza, J., Goiti, U., Almenar, D. and Garin, I. 2005. Echolocation calls and morphology in the Mehelyi's (*Rhinolophus mehelyi*) and Mediterranean (*R. euryale*) horseshoe bats: implications for resource partitioning. *Hystrix* 16: 149–158.

Santos, H., Juste, J., Ibáñez, C. *et al.* 2014. Influences of ecology and biogeography on shaping the distributions of cryptic species: three bat tales in Iberia. *Biological Journal of the Linnean Society* 112: 150–162.

Schaub, A. and Schnitzler, H.U. 2007. Echolocation behaviour of the bat *Vespertilio murinus* reveals the border between the habitat types 'edge' and 'open space'. *Behavioral Ecology and Sociobiology* 61: 513.

Scherrer, J.A. and Wilkinson, G.S. 1993. Evening bat isolation calls provide evidence for heritable signatures. *Animal Behaviour* 46: 847–860.

Schmieder, D.A., Zsebők, S. and Siemers, B.M. 2014. The tail plays a major role in the differing manoeuvrability of two sibling species of mouse-eared bats (*Myotis myotis* and *Myotis blythii*). *Canadian Journal of Zoology* 92: 965–977.

Schnitzler, H.-U. and Kalko, E.K.V. 2001. Echolocation by insect-eating bats. *BioScience* 51: 557–569.

Schnitzler, H.-U., Kalko, E., Miller, L. and Surlykke, A. 1987. The echolocation and hunting behaviour of the bat, *Pipistrellus kuhlii*. *Journal of Comparative Physiology A* 161: 267–274.

Schofield, H.W. 1999. Lesser Horseshoe Bat, *Rhinolophus hipposideros*. In Mitchell-Jones, A.J., Amori, G., Bogdanowicz, W. *et al.* (eds), *The Atlas of European Mammals*. T. & A.D. Poyser, London, pp. 96–97.

Schofield, H.W. 2008. *The Lesser Horseshoe Bat Conservation Handbook*. Vincent Wildlife Trust, Ledbury.

Schofield, H.W. and McAney, C.M. 2008. Lesser horseshoe bat. In Harris, S. and Yaldon, D.W. (eds), *Mammals of the British Isles, Handbook*, 4th edition. Mammal Society, London, pp. 306–310.

Schöner, C.R., Schöner, M.G. and Kerth, G. 2010. Similar is not the same: social calls of conspecifics are more effective in attracting wild bats to day roosts than those of other bat species. *Behavioral Ecology and Sociobiology* 64: 2053–2063.

Schuller, G. 1977. Echo delay and overlap with emitted orientation sounds and Doppler-shift compensation in the bat, *Rhinolophus ferrumequinum*. *Journal of Comparative Physiology* 114: 103–114.

Schumm, A., Krull, D. and Neuweiler, G. 1991. Echolocation in the notch-eared bat, *Myotis emarginatus*. *Behavioral Ecology and Sociobiology* 28: 255–261.

Sedláček, J. and Kolomaznik, P. 2015. Observation of the mobbing behaviour of insectivorous bats (Chiroptera: Vespertilionidae) towards *Falco peregrinus* (Aves: Falconiformes). *Lynx* 46: 145–147.

Seibert, A.-M., Koblitz, J.C., Denzinger, A. and Schnitzler, H.-U. 2013. Scanning behavior in echolocating common pipistrelle bats (*Pipistrellus pipistrellus*). *PLoS ONE* 8(4): e60752.

Serra-Cobo, J., López-Roig, M., Marquès-Lopez, T. and Lahuerta, E. 2000. Rivers as possible landsmarks in the orientation flight of *Miniopterus schreibersii*. *Acta Theriologica* 45: 347–352.

Shiel, C.B. and Fairley, J.S. 1998. Activity of Leisler's bat *Nyctalus leisleri* Kuhl in the field in south-east County Wexford, as revealed by a bat detector. *Biology and Environment: Proceedings of the Royal Irish Academy* 98B: 105–112.

Shiel, C.B. and Fairley, J.S. 1999. Evening emergence of two nursery colonies of Leisler's bat (*Nyctalus leisleri*) in Ireland. *Journal of Zoology* 247: 439–447.

Shiel, C.B., McAney, C.M. and Fairley, J.S. 1991. Analysis of the diet of Natterers bat, *Myotis nattereri* and the common long-eared bat *Plecotus auritus* in the west of Ireland. *Journal of Zoology* 223: 299–305.

Shiel, C.B., Duverge, P.L., Smiddy, P. and Fairley, J.S. 1998. Analysis of the diet of Leisler's bat (*Nyctalus leisleri*) in Ireland with some comparative analyses from England and Germany. *Journal of Zoology* 246: 417–425.

Shiel, C.B., Shiel, R.E. and Fairley, J.S. 1999. Seasonal changes in the foraging behaviour of Leisler's bats (*Nyctalus leisleri*) in Ireland as revealed by radio-telemetry. *Journal of Zoology* 249: 347–358.

Siemers, B.M. and Güttinger, R. 2006. Prey conspicuousness can explain apparent prey selectivity. *Current Biology* 16: R157–R159.

Siemers, B.M. and Ivanova, T. 2004. Ground gleaning in horseshoe bats: comparative evidence from *Rhinolophus blasii*, *R. euryale* and *R. mehelyi*. *Behavioral Ecology and Sociobiology* 56: 464–471.

Siemers, B.M. and Schnitzler, H.-U. 2000. Natterer's bat (*Myotis nattereri* Kuhl 1818) hawks for prey close to vegetation using echolocation signals of very broad bandwidth. *Behavioral Ecology and Sociobiology* 47: 400–412.

Siemers, B.M. and Schnitzler, H.-U. 2004. Echolocation signals reflect niche differentiation in five sympatric congeneric bat species. *Nature* 429: 657–661.

Siemers, B.M. and Swift, S.M. 2006. Differences in sensory ecology contribute to resource partitioning in the bat *Myotis bechsteinii* and *Myotis nattereri* (Chiroptera: Vespertilionidae). *Behavioral Ecology and Sociobiology* 59: 373–380.

Siemers, B.M., Baur, E. and Schnitzler, H.-U. 2005a. Acoustic mirror effect increases prey detection distance in trawling bats. *Naturwissenschaften* 92: 272–276.

Siemers, B.M, Beedholm, K., Dietz, C., Dietz, I. and Ivanova, T. 2005b. Is species identity, sex, age or individual quality conveyed by echolocation call frequency in European horseshoe bats? *Acta Chiropterologica* 7: 259–274.

Siemers, B.M., Greif, S., Borissov, I., Voigt-Heucke, S. and Voigt, C. 2011. Divergent trophic levels in two cryptic sibling bat species. *Oecologia* 166: 69–78.

Simmons, J.A., Fenton, M.B. and O'Farrell, M.J. 1979. Echolocation and pursuit of prey by bats. *Science* 203: 16–21.

Simmons, N.B. and Geisler, J.H. 1998. Phylogenetic relationships of *Icaronycteris*, *Archaeonycteris*, *Hassianycteris*, and *Palaeochiropteryx* to extant bat lineages, with comments on the evolution of echolocation and foraging strategies in Microchiroptera. *Bulletin of the AMNH* 235.

Skalak, S.L., Sherwin, R.E. and Brigham, R.M. 2012. Sampling period, size and duration influence measures of bat species richness from acoustic surveys. *Methods in Ecology and Evolution* 3: 490–502.

Skiba, R. 1996. Nachweis einer Zwergfledermaus, *Pipistrellus pipistrellus* (Schreber, 1774), auf der Azorinsel Flores (Portugal). *Myotis*, 34: 81–84.

Skiba, R. 2003. *Europäische Fledermäuse*. Westarp Wissenschaften, Hohenwarsleben.

Skowronski, M.D. and Harris, J.G. 2006. Acoustic detection and classification of Microchiroptera using machine learning: lessons learned from automatic speech recognition. *Journal of the Acoustical Society of America* 119: 1817–1833.

Sleeman, D.P. 1988. Recent records of Leisler's bat from the south coast. *Irish Naturalists' Journal* 22: 416.

Smarsh, G.C. and Smotherman, M. 2015. Singing away from home: songs are used on foraging territories in the heart-nosed bat, *Cardioderma cor*. *Proceedings of Meetings on Acoustics* 25: 010002.

Smarsh, G.C. and Smotherman, M. 2017. Behavioral response to conspecific songs on foraging territories of the heart-nosed bat. *Behavioral Ecology and Sociobiology* 71: 142.

Smotherman, M., Knörnschild, M., Smarsh, G. and Bohn, K. 2016. The origins and diversity of bat songs. *Journal of Comparative Physiology A* 202: 535–554.

Sommer, R. and Sommer, S. 1997. Ergebninsse zur Kotanalyse bei Teichfledermausen, *Myotis dasycneme* (Biose, 1825). *Myotis* 35: 103–107.

Spada, M. 2009. Environment, biodiversity and rare species: analysis of factors affecting bat conservation. PhD thesis, Università degli Studi dell'Insubria.

Speakman, J.R. 1991. The impact of predation by birds on bat populations in the British Isles. *Mammal Review* 21: 123–142.

Speakman, J.R. and Racey, P.A. 1991. No cost of echolocation for bats in flight. *Nature* 350: 421–423.

Spitzenberger, F. 1994. The genus *Eptesicus* (Mammalia, Chiroptera) in southern Anatolia. *Folia Zoologica* 43: 437–454.

Spitzenberger, F. and von Helversen, O. 2001. *Myotis capaccinii* (Bonaparte, 1837) – Langfussfledermaus. In Krapp, F. (ed.), *Handbuch der Säugetiere Europas, Fledertiere (Chiroptera) I*, 4th edition. Aula-Verlag, Wiebesheim, pp. 281–302.

Spitzenberger, F., Strelkov, P.P., Winkler, H. and Haring, E. 2006. A preliminary revision of the genus *Plecotus* (Chiroptera, Vespertilionidae) based on genetic and morphological results. *Zoologica Scripta* 35: 187–230.

Springall, B.T., Li, H. and Kalcounis-Rueppell, M.C. 2019. The in-flight social calls of insectivorous bats: species specific behaviours and contexts of social call production. *Frontiers in Ecology and Evolution* 7: 441.

Stahlschmidt, P. and Brühl, C. 2012. Bats as bioindicators: the need of a standardized method for acoustic bat activity surveys. *Methods in Ecology and Evolution* 3: 503–508.

Stebbings, R.E. 1970. A comparative study of *Plecotus auritus* and *Plecotus austriacus* (Chiroptera, Vespertilionidae) inhabiting one roost. *Bijdragen tot de Dierkunde* 40: 91–94.

Steck, C. and Brinkmann, R. 2006. The trophic niche of the Geoffroy's bat (*Myotis emarginatus*) in southwestern Germany. *Acta Chiropterologica* 8: 445–450.

Steck, C. and Brinkmann, R. 2015. *Wimperfledermaus, Bechsteinfledermaus und Mopsfledermaus: Einblicke in die Lebensweise gefährdeter Arten in Baden-Württemberg*. Haupt-Verlag, Bern.

Stone, E.L. and Jones, G. 2009. Street lighting disturbs commuting bats. *Current Biology* 19: 1123–1127.

Strelkov, P.P. 1969. Migratory and stationary bats (Chiroptera) of the European part of the Soviet Union. *Acta Zoologica Cracoviensia* 14: 393–440.

Sun, C., Jiang, T., Kanwal, J. *et al.* 2018. Great Himalayan leaf-nosed bats modify vocalizations to communicate threat escalation during agonistic interactions. *Behavioural Processes* 157: 180–187.

Suthers, R.A., Thomas, S.P. and Suthers, B.J. 1972. Respiration, wingbeat and ultrasonic pulse emission in an echolocating bat. *Journal of Experimental Biology* 56: 37–48.

Swift, S.M. 1991. Genus *Plecotus*. In Corbet, G.B. and Harris, S. (eds), *Handbook of British Mammals*. Blackwell, Oxford, pp. 131–138.

Swift, S.M. 1998. *Long-Eared Bats*. Cambridge University Press, Cambridge.

Swift, S.M. and Racey, P.A. 1983. Resource partitioning in two species of vespertilionid bats (Chiroptera) occupying the same roost. *Journal of Zoology* 200: 249–259.

Swift, S.M. and Racey, P.A. 2002. Gleaning as a foraging strategy in Natterer's bat *Myotis nattereri*. *Behavioral Ecology and Sociobiology* 52: 408–416.

Taylor, P. and Jackson, H.D. 2003. A review of foraging and feeding behaviour, and associated anatomical adaptations in Afrotropical nightjars. *Ostrich* 74: 187–204.

Teixeira, S. 2005. Os morcegos (Mammalia: Chiroptera) do Arquipélago da Madeira: identificação morfológica e acústica. Um contributo para a sua conservação. Tese de Licenciatura, Universidade da Madeira, Funchal.

Teixeira, S. 2008. Chordata – Chiroptera (Vespertilionidae, Molossidae). In Borges, P.A.V., Abreu, C., Aguiar, A.M.F. *et al.* (eds), *A List of the Terrestrial Fungi, Flora and Fauna of Madeira and Selvagens Archipelagos*. Direcção Regional do Ambiente da Madeira and Universidade dos Açores, Funchal and Angra do Heroísmo, pp. 366–368.

Teixeira, S. 2014. Agreement on the conservation of populations of European bats: report on implementation of the agreement in Autonomous Region of Madeira. In *Agreement on The Conservation of Populations of European Bats. Report on Implementation of the Agreement in Portugal* 2014/7th MoP. ICNF. Ministério do Agricultura, Mar, Ambiente e Ordenamento do Território.

Teixeira, S. and Jesus, J. 2009. Echolocation calls of bats from Madeira Island: Acoustic characterization and implications for surveys. *Acta Chiropterologica* 11: 183–190.

Thaler, L., Reich, G.M., Zhang, X. *et al.* 2017. Mouth-clicks used by blind expert human echolocators: signal description and model-based signal synthesis. *PLoS Computational Biology* 13(8): e1005670.

Thomas, J.A. and Jalili, M. 2004. Review of echolocation in insectivores and rodents. In Thomas, J.A., Moss, C.F. and Vater, M. (eds), *Echolocation in Bats and Dolphins*. University of Chicago Press, Chicago, IL, pp. 547–564.

Thomson, C.E., Fenton, M.B. and Barclay, R.M.R. 1985. The role of infant isolation calls in mother–infant reunions in the little brown bat, *Myotis lucifugus* (Chiroptera: Vespertilionidae). *Canadian Journal of Zoology* 63: 1982–1988.

Tibbetts, E.A. and Dale, J. 2007. Individual recognition: it is good to be different. *Trends in Ecology and Evolution* 22: 529–537.

Toth, C.A. and Parsons, S. 2013. Is lek breeding rare in bats? *Journal of Zoology* 291: 3–11.

Toth, C.A. and Parsons, S. 2018. The high-output singing displays of a lekking bat encode information on body size and individual identity. *Behavioral Ecology and Sociobiology* 72: 102.

Toth, C.A., Dennis, T.E., Pattemore, D.E. and Parsons, S. 2015. Females as mobile resources: communal roosts promote the adoption of lek breeding in a temperate bat. *Behavioral Ecology* 26: 1156–1163.

Trujillo, D. 1991. *Murciélagos de las Islas Canarias*. Ministerio de Agricultura, Pesca y Alimentación.

Trujillo, D. and Gonzalez, C. 2011. *Pipistrellus maderensis* (Dobson, 1878) (Chiroptera: Vespertilionidae), una nueva adicion a la fauna de las islas Azores (oceano Atlantico). *Vieraea* 39: 215–218.

Turner, D.A., Shaughnessy, A. and Gould, E. 1972. Individual recognition between mother and infant bats (*Myotis*). In Galler, S.R., Schmidt-Koenig, K., Jacobs, G.J. and Belleville, R.E. (eds), *Animal Orientation and Navigation*. SP-262. National Aeronautics and Space Administration, Washington DC, pp. 365–371.

Tvrtković, N., Pavlinić, I. and Haring, E. 2005. Four species of long-eared bats (*Plecotus*, Geoffroy, 1818; Mammalia, Vespertilionidae) in Croatia: field identification and distribution. *Folia Zoologica* 54: 75–88.

UNEP/EUROBATS. 2016. Conservation of key underground sites: the database. www.eurobats.org/activities/intersessional_working_groups/underground_sites (accessed September 2020).

Van De Sijpe, M. 2011. Differentiating the echolocation calls of Daubenton's bats, pond bats and long-fingered bats in natural flight conditions. *Lutra* 54: 17–38.

van der Elst, D., Steckel, J., Boen, A., Peremans, H. and Holderied, M.W. 2016. Place recognition using batlike sonar. *eLife* 5: e14188.

Vaughan, N., Jones, G. and Harris, S. 1997. Identification of British bat species by multivariate analysis of echolocation call parameters. *Bioacoustics* 7: 189–207.

Vaughan, T.A. 1976. Nocturnal behavior of the African false vampire bat (*Cardioderma cor*). *Journal of Mammalogy* 57: 227–248.

Vaughan, T.A. and O'Shea, T.J. 1976. Roosting ecology of the pallid bat, *Antrozous pallidus*. *Journal of Mammalogy* 57: 19–42.

Vaughan, T.A. and Vaughan, R.P. 1986. Seasonality and the behavior of the African yellow-winged bat. *Journal of Mammalogy* 67: 91–102.

Vernes, S.C. 2017. What bats have to say about speech and language. *Psychonomic Bulletin and Review* 24: 111–117.

Vincent, S., Némoz, M. and Aulagnier, S. 2011. Activity and foraging habitats of *Miniopterus schreibersii* (Chiroptera, Miniopteridae) in Southern France: implications for its conservation. *Hystrix* 22: 57–72.

Voigt, C.C. and Lewanzik, D. 2012. No cost of echolocation for flying bats' revisited. *Journal of Comparative Physiology B* 182: 831–840.

Voigt-Heucke, S.L., Zimmer, S. and Kipper, S. 2016. Does interspecific eavesdropping promote aerial aggregations in European pipistrelle bats during autumn? *Ethology* 122: 745–757.

von Helversen, O. 1998. *Eptesicus bottae* (Mammalia, Chiroptera) auf der Insel Rhodos. *Bonner Zoologische Beiträge* 48: 113–121.

von Helversen, O. and von Helversen, D. 1994. The 'advertisement song' of the lesser noctule bat (*Nyctalus leisleri*). *Folia Zoologica* 43: 331–338.

Walter, M.H. and Schnitzler, H.-U. 2019. Spectral call features provide information about the aggression level of greater mouse-eared bats (*Myotis myotis*) during agonistic interactions. *Bioacoustics* 28: 1–25.

Walters, C.L., Freeman, R., Collen, A. *et al.* 2012. A continental-scale tool for acoustic identification of European bats. *Journal of Applied Ecology* 49: 1064–1074.

Waters, D.A. and Jones, G. 1995. Echolocation call structure and intensity in five species of insectivorous bats. *Journal of Experimental Biology* 198: 475–489.

Weid, R. 1994. Social calls of male noctule bats (*Nyctalus noctula*). *Bonner Zoologische Beiträge* 45: 33–38.

Weid, R. and von Helversen, O. 1987. Ortungsrufe Europäischer Fledermäuse beim Jagdflug im Freiland. *Myotis* 25: 5–27.

Wellig, S.D., Nusslé, S., Miltner, D. *et al.* 2018. Mitigating the negative impacts of tall wind turbines on bats: vertical activity profiles and relationships to wind speed. *PLoS ONE* 13(3): e0192493.

Wickler, W. and Uhrig, D. 1969. Verhalten und ökologische Nische der Gelbflügelfredermaus, *Lavia frons* (Geoffroy) (Chiroptera, Megadermatidae). *Zeitschrift für Tierphysiologie* 26: 726–736.

Wiley, R.H. and Richards, D.G. 1978. Physical constraints on acoustic communication in the atmosphere: implications for the evolution of animal vocalizations. *Behavioral Ecology and Sociobiology* 3: 69–94.

Wilkins, M.R., Seddon, N. and Safran, R.J. 2013. Evolutionary divergence in acoustic signals: causes and consequences. *Trends in Ecology and Evolution* 28: 156–166.

Wimmer, B. and Kugelschafter, K. 2017. *Akustische Erfassung von Fledermäusen in unterirdischen Quartieren*. GRIN Verlag.

Winkelmann, J.R., Bonaccorso, F.J., Goedeke, E.E. and Ballock, L.J. 2003. Home range and territoriality in the least blossom bat, *Macroglossus minimus*, in Papua New Guinea. *Journal of Mammalogy* 84: 561–570.

Wong, J.G. and Waters, D.A. 2001. The synchronisation of signal emission with wingbeat during the approach phase in soprano pipistrelles (*Pipistrellus pygmaeus*). *Journal of Experimental Biology* 204: 575–583.

Wright, G.S., Chiu, C., Xian, W., Wilkinson, G.S. and Moss, C.F. 2013. Social calls of flying big brown bats (*Eptesicus fuscus*). *Frontiers in Physiology* 4: 214.

Wright, G.S., Chiu, C., Xian, W., Wilkinson, G.S. and Moss, C.F. 2014. Social calls predict foraging success in big brown bats. *Current Biology* 24: 885–889.

Zagmajster, M. 2003. Display song of parti-coloured bat *Vespertilio murinus* Linnaeus, 1758 (Chiroptera, Mammalia) in southern Slovenia and preliminary study of its variability. *Natura Sloveniae* 5: 27–41.

Zahn, A. and Dippel, B. 1997. Male roosting habits and mating behaviour of *Myotis myotis*. *Journal of Zoology* 243: 659–674.

Zahn, A., Rottenwallner, A. and Güttinger, R. 2006. Population density of the greater mouse-eared bat (*Myotis myotis*), local diet composition and availability of foraging habitats. *Journal of Zoology* 269: 486–493.

Zahn, A., Bauer, S., Kriner, E. and Holzhaider, J. 2010. Foraging habitats of *Myotis emarginatus* in central Europe. *European Journal of Wildlife Research* 56: 395–400.

Zbinden, K. and Zingg, P.E. 1986. Search and hunting signals of echolocating European free-tailed bats, *Tadarida teniotis*, in southern Switzerland. *Mammalia* 50: 9–26.

Zheng, G. and Wang, S. 1989. On the bat fauna and bat conservation in China. In Hanàk, V., Horáček, I. and Gaisler, J. (eds), *European Bat Research 1987*. Charles University Press, Praha.

Zhigalin, A.V. and Moskvitina, N.S. 2017. Fecundity of the parti-coloured bat *Vespertilio murinus* L., 1758 (Chiroptera, Vespertilionidae) in urban and suburban environments. *International Journal of Environmental Studies* 74: 884–890.

Zhou, Q., Zheng, J., Onishi, S., Crommie, M.F. and Zettl, A.K. 2015. Graphene electrostatic microphone and ultrasonic radio. *PNAS* 112: 8942–8946.

Index

References to photos, plans and tables are given in *italics*. References to the full species accounts in Part 8 are given in **bold**.